CHOMSKY NOTEBOOK

Columbia Themes in Philosophy

COLUMBIA THEMES IN PHILOSOPHY

Series Editor: Akeel Bilgrami, Johnsonian Professor of Philosophy, Columbia University

Columbia Themes in Philosophy is a new series with a broad and accommodating thematic reach as well as an ecumenical approach to the outdated disjunction between analytical and European philosophy. It is committed to an examination of key themes in new and startling ways and to the exploration of new topics in philosophy.

Edward Said, *Humanism and Democratic Criticism*

Michael Dummett, *Truth and the Past*

John Searle, *Freedom and Neurobiology: Reflections on Free Will, Language, and Political Power*

Daniel Herwitz and Michael Kelly, eds., *Action, Art, History: Engagements with Arthur C. Danto*

CHOMSKY NOTEBOOK

Edited by JEAN BRICMONT + JULIE FRANCK

COLUMBIA UNIVERSITY PRESS NEW YORK

Columbia University Press
Publishers Since 1893
New York Chichester, West Sussex

Originally published in French as *Chomsky*, copyright © Éditions de l'Herne 2007
Chapters 3, 11, 13, 14, 15, and 17 translated by William McCuaig.

Library of Congress Cataloging-in-Publication Data
Chomsky. English
Chomsky notebook / edited by Jean Bricmont and Julie Franck.
p. cm. — (Columbia themes in philosophy)
Originally published in French as Chomsky.
Includes bibliographical references.
ISBN 978-0-231-14474-2 (cloth : alk. paper) — ISBN 978-0-231-14475-9 (pbk. : alk. paper) —
ISBN 978-0-231-51778-2 (e-book)
1. Chomsky, Noam — Philosophy. 2. Chomsky, Noam — Political and social views. 3. Linguistics.
4. World politics — 1989– I. Chomsky, Noam. II. Bricmont, J. (Jean) III. Franck, Julie. IV. Title.
P85.C47C4513 2010
410'.92—dc22

2009023929

∞
Columbia University Press books are printed on permanent and durable acid-free paper.
This book is printed on paper with recycled content.
Printed in the United States of America

C 10 9 8 7 6 5 4 3 2 1
P 10 9 8 7 6 5 4 3 2

References to Internet Web sites (URLs) were accurate at the time of writing. Neither the author nor Columbia University Press is responsible for URLs that may have expired or changed since the manuscript was prepared.

CONTENTS

PART IV. COGNITIVE SCIENCE AND THE PHILOSOPHY OF MIND

PART V. CHOMSKY AND THE INTELLIGENTSIA

PART VI. POLITICS: THEORY AND PRACTICE

CHOMSKY NOTEBOOK

PART I

CHOMSKY

I

THE MYSTERIES OF NATURE

HOW DEEPLY HIDDEN?

NOAM CHOMSKY

The title for these remarks is drawn from Hume's observations about the man he called "the greatest and rarest genius that ever arose for the ornament and instruction of the species": Isaac Newton. In Hume's judgment, Newton's greatest achievement was that while he "seemed to draw the veil from some of the mysteries of nature, he shewed at the same time the imperfections of the mechanical philosophy; and thereby restored [Nature's] ultimate secrets to that obscurity, in which they ever did and ever will remain." On different grounds, others reached similar conclusions. Locke, for example, had observed that motion has effects "which we can in no way conceive motion able to produce"—as Newton had in fact demonstrated shortly before. Since we remain in "incurable ignorance of what we desire to know" about matter and its effects, Locke concluded, no "science of bodies [is] within our reach," and we can only appeal to "the arbitrary determination of that All-wise Agent who has made them to be, and to operate as they do, in a way wholly above our weak understandings to conceive."[1]

I think it is worth attending to such conclusions, the reasons for them, their aftermath, and what that history suggests about current concerns and inquiries in philosophy of mind.

The mechanical philosophy that Newton undermined is based on our common-sense understanding of the nature and interactions of objects, in large part genetically determined and, it appears, reflexively yielding such perceived properties as persistence of objects through time and space (and as a corollary their cohesion and continuity)[2] and causality through contact, a fundamental feature of intuitive physics: "body, as far as we can conceive, being able only to strike and affect body, and motion, according to the utmost reach of our ideas, being able to produce nothing but motion," as Locke plausibly characterized the common-sense understanding of the world—the limits of our "ideas," in his sense. The theoretical counterpart was the materialist conception of the world that animated the seventeenth-century scientific revolution, the conception of the world as a machine, which was simply a far grander version of the automata that stimulated the imagination of thinkers of the time (much in the way programmed computers do today): the remarkable clocks, the artifacts constructed by master artisans like Jacques de Vaucanson that imitated animal behavior and internal functions like digestion, the hydraulically activated machines that played instruments and pronounced words when triggered by visitors walking through the royal gardens. The mechanical philosophy aimed to dispense with forms flitting through the air, sympathies and antipathies, and other occult ideas and to keep to what is firmly grounded in common-sense understanding and intelligible to it. As is well known, Descartes claimed to have explained the phenomena of the material world in mechanistic terms while also demonstrating that the mechanical philosophy is not all encompassing, not reaching to the domain of mind—again pretty much in accord with the common-sense dualistic interpretation of oneself and the world around us.

I. B. Cohen observes that "there is testimony aplenty in Newton's *Principia* and *Opticks* to his general adherence to the Cartesian mechanical philosophy."[3] The word "general" is important. Newton was much influenced by the neo-Platonic and alchemical traditions and also by the disturbing consequences of his own inquiries. For such reasons, he sometimes modified the more strict Cartesian dichotomy of matter and spirit, including in the latter category "the natural agencies responsible for the 'violent' motions of chemical and electrical action and even, perhaps, for accelerated motion in general," as Ernan McMullin shows in a careful analysis of the evolution of Newton's struggle with the paradoxes and conundrums he sought to resolve. In Newton's own words, "spirit" may be the cause of all movement in nature, including the "power of moving our body by our thoughts" and "the same power in other living creatures, [though] how this is done and by what laws we do not know. We cannot say that all nature is not alive."[4]

Going a step beyond, Locke added that we cannot say that nature does not think. In the formulation that has come down through history as "Locke's suggestion," he writes that "Whether Matter may not be made by God to think is more than man can know. For I see no contradiction in it, that the first Eternal thinking Being, or Omnipotent Spirit, should, if he pleased, give to certain systems of created senseless matter, put together as he thinks fit, some degrees of sense, perception, and thought." Furthermore, just as God had added inconceivable effects to motion, it is "not much more remote from our comprehension to conceive that GOD can, if he pleases, superadd to matter a faculty of thinking, than that he should superadd to it another substance with a faculty of thinking." There is no warrant, then, for postulating a second substance whose essence is thought. And elsewhere, it "involves no contradiction [that God should] give to some parcels of matter, disposed as he thinks fit, a power of thinking and moving [which] might properly be called spirits, in contradistinction to unthinking matter," a view that he finds "repugnant to the *idea* of senseless matter" but that we cannot reject, given our incurable ignorance and the limits of our ideas (cognitive capacities). Having no intelligible concept of "matter" (body, etc.), we cannot dismiss the possibility of living or thinking matter, particularly after Newton undermined common-sense understanding.[5]

Locke's suggestion was taken up through the eighteenth century, culminating in the important work of Joseph Priestley, to which we return. Hume, in the *Treatise*, reached the conclusion that "motion may be, and actually is, the cause of thought and perception," rejecting familiar arguments about absolute difference in kind and divisibility on the general grounds that "we are never sensible of any connexion betwixt causes and effects, and that 'tis only by our experience of their constant conjunction, we can arrive at any knowledge of this relation." In one or another form, it came to be recognized that since "thought, which is produced in the brain, cannot exist if this organ is wanting," and since there is no longer a reason to question the thesis of thinking matter, "it is necessary to consider the brain as a special organ designed especially to produce [thought], as the stomach and the intestines are designed to operate the digestion, the liver to filter bile," and so on through the bodily organs. Just as foods enter the stomach and leave it with

> new qualities, [so] impressions arrive at the brain, through the nerves; they are then isolated and without coherence. The organ enters into action; it acts on them, and soon it sends them back changed into ideas, which the language of physiognomy and gesture, or the signs of speech and writing, manifest outwardly. We conclude then, with the same certainty, that the brain digests, as it were, the impressions, i.e., that organically it makes the secretion of thought.

As Darwin put the matter succinctly, "Why is thought, being a secretion of the brain, more wonderful than gravity, a property of matter?"[6]

Qualifications aside, Newton did generally adhere to the mechanical philosophy, but he also showed its "imperfections"—in fact, he demolished it—though to the end of his life he sought to find some way to account for the mystical principle of action at a distance that he was compelled to invoke to account for the most elementary phenomena of nature. Perhaps, he thought, there might be "a most subtle spirit which pervades and lies hid in all gross bodies," which will somehow yield a physical account of attraction and cohesion and offer some hope of rescuing an intelligible picture of the world.[7]

We should not lightly ignore the concerns of "the greatest and rarest genius that ever arose for the ornament and instruction of the species," or those of Galileo and Descartes or Locke and Hume. Or of Newton's most respected scientific contemporaries, who "unequivocally blamed [Newton] for leading science back into erroneous ways which it seemed to have definitely abandoned," as E. J. Dijksterhuis writes in his classic study of the mechanistic world picture and its collapse as a substantive doctrine. Christiaan Huygens described Newton's principle of attraction as an "absurdity." Leibniz argued that Newton was reintroducing occult ideas similar to the sympathies and antipathies of the much-ridiculed Scholastic science and was offering no *physical* explanations for phenomena of the material world.[8]

Newton largely agreed with his scientific contemporaries. He wrote that the notion of action at a distance is "inconceivable." It is "so great an Absurdity, that I believe no Man who has in philosophical matters a competent Faculty of thinking, can ever fall into it." By invoking it, we concede that we do not understand the phenomena of the material world. As McMullin observes, "By 'understand' Newton still meant what his critics meant: understand in mechanical terms of contact action."[9]

To take a contemporary analogue, the absurd notion of action at a distance is as inconceivable as the idea that "mental states are states of the brain," a proposal "we do not really understand [because] we are still unable to form a conception of *how* consciousness arises in matter, even if we are certain that it does."[10] Similarly, Newton was unable to form a conception of how the simplest phenomena of nature could arise in matter—and they didn't, given his conception of matter, the natural-theoretical version of common-sense understanding. Locke and others agreed, and Hume carried that failure of conceivability a long step beyond by concluding that Newton had restored these ultimate secrets of nature "to that obscurity, in which they ever did and ever will remain"—a stand that we may interpret, naturalistically, as a speculation about the limits of human cognitive capacities. In the light of history, there seems to be little reason to be concerned about the inconceivability of relating mind to brain, or about conceivability altogether, at least in inquiry into the nature of the world. Nor is there any reason for qualms about an "explanatory gap" between *the physical* and consciousness, beyond the unification concerns that arise throughout efforts to understand the world. And unless *the physical* is given some new post-Newtonian sense, there is even less reason for qualms about an "explanatory gap" than in cases where there is some clear sense to the assumed reduction base. The most extreme of such concerns, and perhaps the most significant for the subsequent development of the sciences, is the explanatory gap that Newton unearthed and left unresolved, possibly a permanent mystery for humans, as Hume conjectured.[11]

Science, of course, did not end with the collapse of the notion of body (material, physical, etc.). Rather, it was reconstituted in a radically new way, with questions of conceivability and intelligibility dismissed as demonstrating nothing except about human cognitive capacities, though that conclusion has taken a long time to become firmly established. Later stages of science introduced more "absurdities." The legitimacy of the steps is determined by criteria of depth of explanation and empirical support, not conceivability and intelligibility of the world that is depicted.

Thomas Kuhn suggests that "It does not, I think, misrepresent Newton's intentions as a scientist to maintain that he wished to write a *Principles of Philosophy* like Descartes [that is, true science] but that his inability to explain gravity forced him to restrict his subject to the *Mathematical Principles of Natural Philosophy*, [which] did not even pretend to explain why the universe runs as it does," leaving the question in obscurity. For such reasons, "It was forty years before Newtonian physics firmly supplanted Cartesian physics, even in British universities," and some of the ablest physicists of the eighteenth century continued to seek a mechanical-corpuscular explanation of gravity—that is, what they took to be a *physical* explanation—as Newton did himself. In later years, positivists reproached all sides of the debates "for their foolishness in clothing the mathematical formalism [of physical theory] with the 'gay garment' of a physical interpretation," a concept that had lost substantive meaning.[12]

Newton's famous phrase "I frame no hypotheses" appears in this context: recognizing that he had been unable to discover the *physical* cause of gravity, he left the question open. He adds that "to us it is enough that gravity does really exist, and act according to the laws which we have explained, and abundantly serves to account for all the motions of the celestial bodies, and of our sea." But while agreeing that his proposals were so absurd that no serious scientist could accept them, he defended himself from the charge that he was reverting to the mysticism of the Aristotelians. His principles, he argued, were not occult: "their causes only are occult"; or, he hoped, were yet to be discovered in physical terms, meaning mechanical terms. To derive general principles inductively from phenomena, he continued, "and afterwards to tell us how the properties of actions of all corporeal things follow from those manifest principles, would be a very great step in philosophy, though the causes of these principles were not yet discovered."[13]

To paraphrase with regard to the contemporary analogue I mentioned, it "would be a very great step in science to account for mental aspects of the world in terms of manifest principles even if the causes of these principles were not yet discovered"—or to put the matter more appropriately, even if unification with other aspects of science had not been achieved. To learn more about mental aspects of the world—or chemical or electrical or other aspects—we should try to discover "manifest principles" that partially explain them, though their causes remain disconnected from what we take to be more fundamental aspects of science. The gap might have many reasons, among them, as has repeatedly been discovered, that the presumed reduction base was misconceived, including core physics.

Historians of science have recognized that Newton's reluctant intellectual moves set forth a new view of science in which the goal is not to seek ultimate explanations but to find the best theoretical account we can of the phenomena of experience and experiment. Newton's more limited goals were not entirely new. They have roots in an earlier scientific tradition that had abandoned the search for the "first springs of natural motions" and other natural phenomena, keeping to the more modest effort to develop the best theoretical account we can: what Richard Popkin calls the "constructive skepticism . . . formulated . . . in detail by Mersenne and Gassendi" and later in Hume's "mitigated skepticism." In this conception, Popkin continues, science proceeds by "doubting our abilities to find grounds for our knowledge, while accepting and increasing the knowledge itself" and recognizing that "the secrets of nature, of things-in-themselves, are forever hidden from us"—the "science without metaphysics . . . which was to have a great history in more recent times."[14]

As the impact of Newton's discoveries was slowly absorbed, such lowering of the goals of scientific inquiry became routine. Scientists abandoned the animating idea of the early scientific revolution: that the world will be intelligible to us. It is enough to construct intelligible explanatory theories, a radical difference. By the time we reach Russell's *Analysis of Matter*, he dismisses the very idea of an intelligible world as "absurd" and repeatedly places the word "intelligible" in quotes to highlight the absurdity of the quest. Qualms about action at a distance were "little more than a prejudice," he writes. "If all the world consisted of billiard balls, it would be what is called 'intelligible'—i.e., it would never surprise us sufficiently to make us realize that we do not understand it." But even without external surprise, we should recognize how little we understand the world, and we should also realize that it doesn't matter whether we can conceive of how the world works. In his classic introduction to quantum mechanics a few years later, Paul Dirac wrote that physical science no longer seeks to provide pictures of how the world works, that is, "a model functioning on essentially classical lines," but only seeks to provide a "way of looking at the fundamental laws which makes their self-consistency obvious." He was referring to the inconceivable conclusions of quantum physics but could just as readily have said that even the classical Newtonian models had abandoned the hope of rendering natural phenomena intelligible, the primary goal of the early modern scientific revolution, with its roots in common-sense understanding.[15]

It is useful to recognize how radical a shift it was to abandon the mechanical philosophy and with it any scientific relevance of our commonsense beliefs and conceptions, except as a starting point and spur for inquiry. The Galileo scholar Peter Machamer observes that by adopting the mechanical philosophy and initiating the modern scientific revolution, Galileo had "forged a new model of intelligibility for human understanding, [with] new criteria for coherent explanations of natural phenomena" based on the conception of the world as an elaborate machine. For Galileo, and leading figures in the early modern scientific revolution generally, true understanding requires a mechanical model, a device that an artisan could construct, hence intelligible to us. Thus Galileo rejected traditional theories of tides because we cannot "duplicate [them] by means of appropriate artificial devices."[16]

The model of intelligibility that reigned from Galileo through Newton and beyond has a corollary: when mechanism fails, understanding fails. The apparent inadequacies of mechanical explanation for cohesion, attraction, and other phenomena led Galileo finally to reject "the vain presumption of understanding everything." Worse yet, "there is not a single effect in nature . . . such that the most ingenious theorist can arrive at a complete understanding of it." Galileo was formulating a very strong version of what Daniel Stoljar calls "the ignorance hypothesis" in his careful inquiry into the contemporary study of philosophical problems relating to consciousness, concluding that their origins are epistemic and that they are effectively overcome by invoking the ignorance hypothesis—which for Galileo, Newton, Locke, Hume, and others was more than a hypothesis and extended far beyond the problem of consciousness, encompassing the truths of nature quite generally.[17]

Though much more optimistic than Galileo about the prospects for mechanical explanation, Descartes too recognized the limits of our cognitive reach. Rule 8 of the *Regulae* reads: "If in the series of subjects to be examined we come to a subject of which our intellect cannot gain a good enough intuition, we must stop there; and we must not examine the other matters that follow, but must refrain from futile toil." Specifically, Descartes speculated that the workings of *res cogitans* may lie beyond human understanding. He thought that we may not "have intelligence enough" to understand the workings of mind, in particular, the normal use of language, with its creative aspects his core example: the capacity of every human, but no beast-machine, to use language in ways appropriate to situations but not caused by them, and to formulate and express coherent thoughts without bound, perhaps "incited or inclined" to speak in certain ways by internal and external circumstances but not "compelled" to do so, as his followers put the matter.[18]

However, Descartes continued, even if the explanation of normal use of language and other forms of free and coherent choice of action lies beyond our cognitive grasp, that is no reason to question the authenticity of our experience. Quite generally, "free will" is "the noblest thing" we have, Descartes held: "there is nothing we comprehend more evidently and more perfectly," and "it would be absurd" to doubt something that "we comprehend intimately, and experience within ourselves" (that "the free actions of men [are] undetermined") merely because it conflicts with something else "which we know must be by its nature incomprehensible to us" ("divine preordination").[19]

Such thoughts about cognitive limits do not comport well with Descartes' occasional observation that human reason "is a universal instrument which can serve for all contingencies," whereas the organs of an animal or machine "have need of some special adaptation for any particular action." But let's put that aside and keep to the more reasonable conclusions about cognitive limits.

The creative use of language was a basis for what has been called the "epistemological argument" for mind-body dualism and also for the scientific inquiries of the Cartesians into the problem of "other minds"—much more sensible, I believe, than contemporary analogues, often based on misinterpretation of a famous paper of Turing's, a topic that I will put aside.[20]

Desmond Clarke is accurate, I think, in concluding that "Descartes identified the use of language as the critical property that distinguishes human beings from other members of the animal kingdom and [that] he developed this argument in support of the real distinction of mind and matter." I think he is also persuasive in interpreting the general Cartesian project as primarily "natural philosophy" (science), an attempt to press mechanical explanation to its limits, and in regarding the *Meditations* "not as the authoritative expression of Descartes' philosophy, but as an unsuccessful attempt to reconcile his theologically suspect natural philosophy with an orthodox expression of scholastic metaphysics."[21] In pursuing his natural science, Descartes tried to show that mechanical explanation reached very far but came to an impassable barrier in the face of such mental phenomena as the creative use of language. He therefore, quite properly, adopted the standard scientific procedure of seeking some new principles to account for such mental phenomena—a quest that lost one primary motivation when mechanical explanation was demonstrated to fail for everything.

Clarke argues that "Descartes' dualism was an expression of the extent of the theoretical gap between [Cartesian physics] and the descriptions of mental life that we formulate from the first-person perspective of our own thinking." The gap therefore results from Descartes' "impoverished concept of matter" and can be overcome by "including new theoretical entities in one's concept of matter."[22] Whether the latter speculation is correct or not, it does not quite capture the deficiencies of classical science from Galileo through Newton and beyond. The underlying concept of matter and motion—based on conceivability, intelligibility, and common-sense understanding—had to be abandoned, and science had to proceed on an entirely new course in investigating the simplest phenomena of motion, and all other aspects of the world, including mental life.

Despite the centrality of the creative use of language to Cartesian science, it was only one illustration of the general problems of will and of choice of appropriate action, which remain as mysterious to us as they were to seventeenth-century scientists, so it seems to me, despite sophisticated arguments to the contrary. The problems are hardly even on the scientific agenda. There has been very valuable work about how an organism executes a plan for integrated motor action—say, how a person reaches for a cup on the table. But no one even raises the question of why this plan is executed rather than some other one, apart from the very simplest organisms and special circumstances of motivation. Much the same is true even for visual perception. The cognitive neuroscientists Nancy Kanwisher and Paul Downing reviewed research on a problem posed in 1850 by Helmholtz: "even without moving our eyes, we can focus our attention on different objects at will, resulting in very different perceptual experiences of the same visual field." The phrase "at will" points to an area beyond serious empirical inquiry, still the mystery it was for Newton at the end of his life when he continued to seek some "subtle spirit" that lies hidden in all bodies and that might, without "absurdity," account for their properties of attraction and repulsion, along with the nature and effects of light, sensation, and the way "members of animal bodies move at the command of the will"—all comparable mysteries for Newton, perhaps even beyond our understanding.[23]

It has become standard practice in recent years to describe the problem of consciousness as "the hard problem," others being within our grasp, now or down the road. I think there are reasons for some skepticism, particularly when we recognize how sharply understanding declines beyond the simplest systems of nature. To illustrate with a few examples, a review article by Eric Kandel and Larry Squire on the current state of efforts aimed at "breaking down scientific barriers to the study of brain and mind" concludes that "the neuroscience of higher cognitive processes is only beginning." C. R. Gallistel points out that "we clearly do not understand how the nervous system computes" or even "the foundations of its ability to compute," even for "the small set of arithmetic and logical operations that are fundamental to any computation." Reviewing the remarkable computational capacities of insects, he concludes that it is a mistake to suppose that the nervous system does not carry out complex symbolic computations on grounds of "our inability, as yet to understand how the nervous system computes at the cellular and molecular level. . . . We do not know what processes belong to the basic instruction set of the nervous system—the modest number of elementary operations built into the hardware of any computing device." Semir Zeki, who is optimistic about the prospects for bringing the brain sciences to bear even on creativity in the visual arts, nevertheless reminds us that "how the brain combines the responses of specialized cells to indicate a continuous vertical line is a mystery that neurology has not yet solved," or even how one line is differentiated from others or from the visual surround. Basic traditional questions are not even on the research agenda, and even simple ones that might be within reach remain baffling.[24]

It is common to assert that "the mental is the neurophysiological at a higher level." To entertain the idea makes sense but, for the present, only as a guide to inquiry and without much confidence about what "the neurophysiological" will prove to be. Similarly, it is premature to hold that "it is empirically evident that states of consciousness are the necessary consequence of neuronal activity." Too little is understood about the functioning of the brain.[25]

History also suggests caution. In early modern science, the nature of motion was the "hard problem." "Springing or Elastic Motions" is the "hard rock in Philosophy," Sir William Petty observed, proposing ideas resembling those soon developed much more richly by Newton. The "hard problem" was that bodies that seem to our senses to be at rest are in a "violent" state, with "a strong endeavor to fly off or recede from one another," in Robert Boyle's words. The problem, he felt, is as obscure as "the Cause and Nature" of gravity, thus supporting his belief in "an intelligent Author or Disposer of Things." Even the skeptical Newtonian Voltaire argued that the ability of humans to "produce a movement" where there was none shows that "there is a God who gave movement" to matter, and "so far are we from conceiving what matter is" that we do not even know if there is any "solid matter in the universe." Locke relinquished to divine hands "the gravitation of matter towards matter, by ways, inconceivable to me." Kant rephrased the "hard problem," arguing that to reach his conclusions, Newton was compelled to tacitly "assume that all matter exercises this motive force [of universal attraction] simply as matter and by its essential nature"; by rejecting the assumption, he was "at variance with himself," caught in a contradiction. Newton therefore did not, as he claimed, really leave "the physicists full freedom to explain the possibility of such attraction as they might find good, without mixing up his propositions with their play of hypotheses." Rather, "The concept of matter is reduced to nothing but moving forces. . . . The attraction essential to all matter is an immediate action of one matter on another across empty space," a notion that would have been anathema to the great figures of seventeenth-century science, "such Masters, as the Great Huygenius, and the incomparable Mr. Newton," in Locke's words.[26]

The "hard problems" of the day were not solved; rather, they were abandoned as over time science turned to its more modest post-Newtonian course. Friedrich Lange, in his classic nineteenth-century history of materialism, observed that we have

> so accustomed ourselves to the abstract notion of forces, or rather to a notion hovering in a mystic obscurity between abstraction and con-

> crete comprehension, that we no longer find any difficulty in making one particle of matter act upon another without immediate contact . . . through void space without any material link. From such ideas the great mathematicians and physicists of the seventeenth century were far removed. They were all in so far genuine Materialists in the sense of ancient Materialism that they made immediate contact a condition of influence.

This transition over time is "one of the most important turning-points in the whole history of Materialism," depriving the doctrine of much significance, if any at all. Newton not only joined the great scientists of his day in regarding "the now prevailing theory of *actio in distans* . . . simply as absurd, [but] also felt himself obliged, in the year 1717, in the preface to the second edition of his 'Optics,' to protest expressly against [the] view" of his followers who "went so far as to declare gravity to be a fundamental force of matter," requiring no "further mechanical explanation from the collision of imponderable particles." Lange concludes that "the course of history has eliminated this unknown material cause [that so troubled Newton], and has placed the mathematical law itself in the rank of physical causes." Hence, "what Newton held to be so great an absurdity that no philosophic thinker could light upon it, is prized by posterity as Newton's great discovery of the harmony of the universe!" The conclusions are commonplace in the history of science. Fifty years ago, Alexander Koyrė observed that despite his unwillingness to accept the conclusion, Newton had demonstrated that "a purely materialistic pattern of nature is utterly impossible (and a purely materialistic or mechanistic physics, such as that of Lucretius or of Descartes, is utterly impossible, too)"; his mathematical physics required the "admission into the body of science of incomprehensible and inexplicable 'facts' imposed up on us by empiricism," by what is observed and our conclusions from these observations.[27]

George Coyne describes it as "paradoxical that the rise of materialism as a philosophy in the 17th and 18th centuries is attributed to the birth of modern science, when in reality matter as a workable concept had been eliminated from scientific discourse" with the collapse of the mechanical philosophy.[28] Also paradoxical is the influence of Gilbert Ryle's ridicule of the "ghost in the machine," quite apart from the accuracy of his rendition of the Cartesian concepts. It was the machine that Newton exorcised, leaving the ghost intact. The "hard problem" of the materialists disappeared, and there has been little noticeable progress in addressing other "hard problems" that seemed no less mysterious to Descartes, Newton, Locke, and other leading figures.

The third English edition of Lange's much expanded history of materialism appeared in 1925 with an introduction by Bertrand Russell, who shortly after published his *Analysis of Matter*. Developing his neutral monism, Russell carried further seventeenth- and eighteenth-century skepticism about matter and recognition of the plausibility (or, for some, necessity) of thinking matter. Russell held that there are "three grades of certainty. The highest grade belongs to my own percepts; the second grade to the percepts of other people; the third to events which are not percepts of anybody," constructions of the mind established in the course of efforts to make sense of what we perceive. "A piece of matter is a logical structure composed of [such] events," he therefore concluded. We know nothing of the "intrinsic character" of such mentally constructed entities, so there is "no ground for the view that percepts cannot be physical events." For science to be informative, it cannot be restricted to structural knowledge of such logical properties. Rather, "the world of physics [that we construct] must be, in some sense, continuous with the world of our perceptions, since it is the latter which supplies the evidence for the laws of physics." The percepts that are required for this task—perhaps just meter readings, Eddington had argued shortly before—"are not known to have any intrinsic character which physical events cannot have, since we do not know of any intrinsic character which could be incompatible with the logical properties that physics assigns to physical events." Accordingly, "what are called 'mental' events . . . are part of the material of the physical world." Physics itself seeks only to discover "the causal skeleton of the world, [while studying] percepts only in their cognitive aspect; their other aspects lie outside its purview"—though we recognize their existence, at the highest grade of certainty in fact.[29]

The basic conundrum recalls a classical dialogue between the intellect and the senses, in which the intellect says that color, sweetness, and the like are only convention while in reality there are only atoms and the void, and the senses reply: "Wretched mind, from us you are taking the evidence by which you would overthrow us? Your victory is your own fall."[30]

To illustrate his conclusion, Russell asks us to consider a blind physicist who knows the whole of physics but does not have "the knowledge which [sighted] men have" about, say, the quality of the color blue. In their review of related issues, Daniel Stoljar and Yujin Nagasawa call this the "knowledge intuition," as distinct from the "knowledge argument," presented in the resurrection of Russell's example by Frank Jackson: in this case the physicist (Mary) "learns everything there is to know about the physical nature of the world" while confined to a black-and-white room, but when released, she "will learn what it is like to see something red."[31]

There is a substantial literature seeking to evade the argument. One popular though contested proposal is that what Mary lacks is not the knowledge of the world that we have but rather a range of abilities, a species of "knowing how." That seems unhelpful, in part because there is an irreducible cognitive element in "knowing how," which goes beyond abilities, but also for the kinds of reasons that Hume discussed in connection with moral judgments. Since these, he observed, are unbounded in scope and applicable to new situations, they must be based on a finite array of general principles (which are, furthermore, part of our nature though they are beyond the "original instincts" shared with animals). The knowledge that *we* have but Mary lacks is a body of knowledge that does not fall within the knowing-how/knowing-that dichotomy: it is knowledge *of*—knowledge of rules and principles that yield unbounded capacities to act appropriately. All of this is for the most part unconscious and inaccessible to consciousness, as in the case of knowledge of the rules of language, vision, etc. Such conclusions have been rejected as a matter of principle by Quine, Searle, and many others, but not convincingly or even coherently, I think.[32]

Russell's knowledge intuition led him to conclude that physics has limits: experience in general lies "outside its purview" apart from cognitive aspects that provide empirical evidence, though along with other mental events, experience is "part of the material of the physical world," a phrase that seems to mean no more than "part of the world." We must have "an interpretation of physics which gives a due place to perceptions," Russell held, or it has no empirical basis. Jackson's knowledge argument leads him to the conclusion that "physicalism is false." Or in a later version, that to be valid "materialism [as] a metaphysical doctrine" must incorporate "the psychological story about our world"; the "story about our world told purely in physical terms [must] enable one to deduce the phenomenal nature of psychological states."[33] But that is uninformative until some clear concept of physicalism/materialism is offered. Classical interpretations having vanished, the notions of body, material, and physical are hardly more than honorific designations for what is more or less understood at some particular moment in time, with flexible boundaries and no guarantee that there will not be radical revision ahead, even at its core. If so, the knowledge argument only shows (with Russell) that humanly constructed physics has limits, or that Mary did not know all of physics (she had not drawn the right conclusions from Eddington's meter readings).

To resurrect something that resembles a "mind-body problem," it would be necessary to characterize *physicalism* (matter, etc.) in some post-Newtonian fashion or to argue that the problem arises even if the concepts are abandoned. Both approaches have been pursued. I will return to current examples. An alternative approach is to dismiss the mind-body problem and to approach the knowledge/intuition argument as a problem of the natural sciences. Rephrasing Russell's thought experiment, we might say that like all animals, we have internal capacities that reflexively provide us with what ethologists called an *Umwelt*, a world of experience, different for us and for bees—in fact, differing among humans, depending on what they understand. That's why radiology is a medical specialty. Galileo saw the moons of Jupiter through his primitive telescope, but those he sought to convince could only see the magnification of terrestrial objects and took his telescope to be a conjuring trick (at least if Paul Feyerabend's reconstruction of the history is correct). What I hear as noise is perceived as music by my teenage grandchildren, at a fairly primitive level of perceptual experience. And so on quite generally.

Being reflective creatures, unlike others, we go on to seek to gain a deeper understanding of the phenomena of experience. These exercises are called myth, or magic, or philosophy, or science. They reveal not only that the world of experience is itself highly intricate and variable, resulting from the interaction of many factors, but also that the modes of interpretation that intuitive common sense provides do not withstand analysis, so that the goals of science must be lowered in the manner recognized in post-Newtonian science. From this point of view, there is no objective science from a third-person perspective, just various first-person perspectives, matching closely enough among humans so that a large range of agreement can be reached, with diligence and cooperative inquiry. Being inquisitive and reflective creatures, if we can construct a degree of theoretical understanding in some domain, we try to unify it with other branches of inquiry, reduction being one possibility but not the only one.

We can anticipate that our quest might fail, for one reason, because our basically shared capacities of understanding and explanation have limits—a truism that is sometimes thoughtlessly derided as "mysterianism," though not by Descartes and Hume, among others. It could be that these innate cognitive capacities do not lead us beyond some understanding of Russell's causal skeleton of the world (and enough about perception to incorporate evidence within this mental construction), and it is an open question how much of that can be attained. In principle, the limits could become topics of empirical inquiry into the nature of what we might call "the science-forming faculty," another "mental organ." These are interesting topics, but the issues are distinct from the traditional mind-body problem, which evaporated after Newton, or from the question of how mental aspects of the world, including direct experience, relate to the brain, one of the many problems of unification that arise in the sciences.

In brief, if we are biological organisms, not angels, much of what we seek to understand might lie beyond our cognitive limits—maybe a true understanding of anything, as Galileo concluded and Newton in a certain sense demonstrated. That cognitive reach has limits is not only a truism but also a fortunate one. If there were no limits to human intelligence, it would lack internal structure and would therefore have no scope: we could achieve nothing by inquiry. The basic points were expressed clearly by Charles Sanders Peirce in his discussion of the need for innate endowment that "puts a limit upon admissible hypotheses" if knowledge is to be acquired.[34] Similarly, if a zygote had no further genetic instructions constraining its developmental path, it would at best grow into a creature formed solely by physical law, like a snowflake, nothing viable.

We might think of the natural sciences as a kind of chance convergence between our cognitive capacities and what is more or less true of the natural world. There is no reason to believe that humans can solve every problem they pose or even that they can formulate the right questions; they may simply lack the conceptual tools, just as rats cannot deal with a prime-number maze.

Russell's general conclusions seem to me on the right track. The formulation can be improved, I think, by simply dropping the words "matter" and "physical." Since the Newtonian revolution, we speak of the "physical" world much as we speak of the "real" truth: for emphasis, but adding nothing. We can distinguish various aspects of the world—say chemical, electrical, experiential, and the rest—and we can then inquire into their underlying principles and their relations with other systems, problems of unification.

Suppose we adopt the "mitigated skepticism" that was warranted after Newton, if not before. For the theory of mind, that means following Gassendi's advice in his *Objections*. He argued that Descartes had at most shown "the perception of the existence of mind, [but] fail[ed] to reveal its nature." It is necessary to proceed as we would in seeking to discover "a conception of Wine superior to the vulgar," by investigating how it is constituted and the laws that determine its functioning. Similarly, he urged Descartes, "it is incumbent on you, to examine yourself by a certain chemical-like labor, so that you can determine and demonstrate to us your internal substance"[35]—and that of others.

The theory of mind can be pursued in many ways, like other branches of science, with an eye to eventual unification, whatever form it may take, if any. That is the task that Hume undertook when he investigated what he called "the science of human nature," seeking "the secret springs and principles, by which the human mind is actuated in its operations," including those "parts of [our] knowledge" that are derived from "the original hand of nature," an enterprise he compared to Newton's: essentially what in contemporary literature is termed "naturalization of philosophy" or "epistemology naturalized." Gassendi's recommended course was in fact being pursued in the "cognitive revolution" of the seventeenth century by British neo-Platonists and continental philosophers of language and mind and has been taken up with renewed vigor in recent years, but I'll put that matter aside.[36]

Chemistry itself quite explicitly pursued this course. The eighteenth-century chemist Joseph Black recommended that "chemical affinity be received as a first principle, which we cannot explain any more than Newton could explain gravitation, and let us defer accounting for the laws of affinity, till we have established such a body of doctrine as he has established concerning the laws of gravitation." Being yet "very far from the knowledge of first principles," chemical science should be "analytical, like Newton's *Optics*, in the form of a general law, at the very end of our induction, as the reward of our labour." The course he outlined is the one that was actually followed, as chemistry established a rich body of doctrine, its "triumphs . . . built on no reductionist foundation but rather achieved in isolation from the newly emerging science of physics," as the historian of chemistry Arnold Thackray observes. Newton and his followers did attempt to "pursue the thoroughly Newtonian and reductionist task of uncovering the general mathematical laws which govern all chemical behavior" and to develop a principled science of chemical mechanisms based on physics and its concepts of interactions among "the ultimate permanent particles of matter." But the Newtonian program was undercut by Dalton's "astonishingly successful weight-quantification of chemical units," Thackray continues, shifting "the whole area of philosophical debate among chemists from that of chemical *mechanisms* (the *why*? of reaction) to that of chemical *units* (the *what*? and *how much*?)," a theory that "was profoundly antiphysicalist and anti-Newtonian in its rejection of the unity of matter, and its dismissal of short-range forces." "Dalton's ideas were chemically successful. Hence they have enjoyed the homage of history, unlike the philosophically more coherent, if less successful, reductionist schemes of the Newtonians."[37]

Adopting contemporary terminology, we might say that Dalton disregarded the explanatory gap between chemistry and physics by ignoring the underlying physics, much as post-Newtonian physicists disregarded the explanatory gap between Newtonian dynamics and the mechanical philosophy by ignoring (and in this case rejecting) the latter, even though it was self-evident to common-sense understanding. That has often been the course of science since, though not without controversy and sharp criticism often later recognized to have been seriously misguided.

Well into the twentieth century, the failure of the reduction of chemistry to physics was interpreted by prominent scientists as a critically important explanatory gap, showing that chemistry provides "merely classificatory symbols that summarized the observed course of a reaction," to quote Brock's standard history. Kekulé, whose structural chemistry was an important step toward the eventual unification of chemistry and physics, doubted that "absolute constitutions of organic molecules could ever be given"; his models and analysis of valency were to have an instrumental interpretation only, as calculating devices. Lavoisier before him believed that "the number and nature of elements [is] an unsolvable problem, capable of an infinity of solutions none of which probably accord with Nature." "It seems extremely probable that we know nothing at all about . . . [the] . . . indivisible atoms of which matter is composed"—and never will, he believed. Kekulé seems to be saying that there isn't a problem to be solved: the structural formulas are either useful or not, but there is no truth of the matter. Large parts of physics were understood the same way. Poincaré went so far as to say that we adopt the molecular theory of gases only because we are familiar with the game of billiards.

Boltzmann's scientific biographer speculates that he committed suicide because of his failure to convince the scientific community to regard his theoretical account of these matters as more than a calculating system—ironically, shortly after Einstein's work on Brownian motion and broader issues had convinced physicists of the reality of the entities he postulated. Bohr's model of the atom was also regarded as lacking "physical reality" by eminent scientists. In the 1920s, America's first Nobel Prize–winning chemist dismissed talk about the real nature of chemical bonds as metaphysical "twaddle": they are nothing more than "a very crude method of representing certain known facts about chemical reactions, a mode of representation" only, because the concept could not be reduced to physics. The rejection of that skepticism by a few leading scientists, whose views were condemned at the time as a conceptual absurdity, paved the way for the eventual unification of chemistry and physics, with Linus Pauling's quantum-theoretic account of the chemical bond seventy years ago.[38]

In 1927, Russell observed that chemical laws "cannot at present be reduced to physical laws,"[39] an observation that was found to be misleading: the words "at present" turned out to understate the matter. Chemical laws could not ever be reduced to physical laws, because the conception of physical laws was erroneous. The perceived explanatory gap was never filled. It was necessary, once again, to dismiss as irrelevant the notions of "conceivability" and "intelligibility of the world" in favor of the mitigated skepticism of methodological naturalism: seeking to increase our knowledge while keeping an open mind about the possibility of reduction.

There are fairly clear parallels to contemporary discussion of language and mind and some lessons that can be drawn. The study of insect symbolic representation, organization of motor behavior, mammalian vision, human language, moral judgment, and other topics is in each case well advised to follow Joseph Black's prescription. If these inquiries succeed in developing a "body of doctrine" that accounts for elements of insect navigation, or the rule that image motions are interpreted (if other rules permit) as rigid motions in three dimensions, or that displacement operations in language observe locality principles, and so on, that should be regarded as normal science, even if unification with neurophysiology has not been achieved—and might not be for a variety of possible reasons, among them that the expected "reduction base" is misconceived and has to be modified. Needless to say, the brain sciences are not as firmly established as basic physics was a century ago or as the mechanical philosophy was in Newton's day. It is also pointless to insist on doctrines about accessibility to consciousness: even if they could be given a coherent formulation, they would have no bearing on the "physical reality" of the rigidity principle or locality conditions. We should understand enough by now to dismiss the interpretation of theoretical accounts as no more than a way of "representing certain known facts about [behavior], a mode of representation" only—a critique commonly leveled against theories of higher mental faculties, though not insect computation, another illustration of the methodological dualism that is so prevalent in the critical discussion of inquiry into language and mind.[40]

It is also instructive to observe the reemergence of much earlier insights, though divorced from their grounding in the collapse of traditional physicalism. Thus we read today of the thesis of the new biology that "Things mental, indeed minds, are emergent properties of brains, [though] these emergences are . . . produced by principles that . . . we do not yet understand"—so writes the neuroscientist Vernon Mountcastle, formulating the guiding theme of a collection of essays reviewing the results of the Decade of the Brain that ended the twentieth century. The phrase "we do not *yet* understand" might well suffer the same fate as Russell's similar comment about chemistry seventy years earlier. Many other prominent scientists and philosophers have presented essentially the same thesis as an "astonishing hypothesis" of the new biology, a "radical" new idea in the philosophy of mind, "the bold assertion that mental phenomena are entirely natural and caused by the neurophysiological activities of the brain," opening the door to novel and promising inquiries, a rejection of Cartesian mind-body dualism, and so on.[41] In fact, all reiterate, in virtually the same words, formulations of centuries ago, after the traditional mind-body problem became unformulable with the disappearance of the only coherent notion of body (physical, material, etc.): for example, Joseph Priestley's conclusion that properties "termed mental" reduce somehow to "the organical structure of the brain,"[42] stated in different words by Hume, Darwin, and many others and almost inescapable, it would seem, after the collapse of the mechanical philosophy.

Priestley's important work was the culmination of a century of reflections on Locke's speculation and is their most elaborate development.[43] He made it clear that his conclusions about thinking matter followed directly from the collapse of any serious notion of *body*, or *matter*, or *physical*. He wrote that

> the principles of the Newtonian philosophy were no sooner known, than it was seen how few in comparison, of the phenomena of Nature were owing to solid matter, and how much to powers which were only supposed to accompany and surround the solid parts of matter. . . . Now when solidity had apparently so very little to do in the system, it is really a wonder that it did not occur to philosophers sooner . . . that there might be no such thing in Nature.

There is, then, no longer any reason to suppose that "the principle of thought or sensation [is] incompatible with matter," Priestley concluded. Accordingly, "the whole argument for an immaterial thinking principle in man, on this supposition, falls to the ground; matter, destitute of what has hitherto been called solidity, being no more incompatible with sensation and thought than that substance which without knowing anything farther about it, we have been used to call immaterial." The powers of sensation, perception, and thought reside in "a certain organized system of matter, [and] necessarily exist in, and depend upon, such a system." It is true that "we have a very imperfect idea of what the power of perception is" and that we may never attain a "clear idea," but "this very ignorance ought to make us cautious in asserting with what other properties it may, or may not, exist." Only a "precise and definite knowledge of the nature of perception and thought can authorize any person to affirm whether they may not belong to an extended substance which also has the properties of attraction and repulsion." Our ignorance provides no warrant for supposing that sensation and thought are incompatible with post-Newtonian matter. "In fact, there is the same reason to conclude, that the powers of sensation and thought are the necessary result of a particular organization, as that sound is the necessary result of a particular concussion of the air." And in a later discussion, "In my opinion there is just the same reason to conclude that the brain *thinks*, as that it is *white*, and *soft*."[44]

Priestley criticizes Locke for being hesitant in putting forth his speculation about thinking matter, since the conclusion follows so directly from "the universally accepted rules of philosophizing such as are laid down by Sir Isaac Newton." He urges that we abandon the methodological dualism that deters us from applying to thought and sensation the rules that we follow "in our inquiries into the causes of particular appearances in nature," and he expresses his hope "that when this is plainly pointed out the inconsistency of our conduct cannot fail to strike us and be the means of inducing" philosophers to apply the same maxim to investigation of mental aspects of the world that they do in other domains—a hope that has yet to be realized, I think.[45]

Priestley clearly "wished the disappearance of solid matter to signal an end to matter-spirit dualism," Thackray writes. And with it an end to any reason to question the thesis of thinking matter. In John Yolton's words, Priestley's conclusion was "not that all reduces to matter, but rather that the kind of matter on which the two-substance view is based does not exist," and "with the altered concept of matter, the more traditional ways of posing the question of the nature of thought and of its relations to the brain do not fit. We have to think of a complex organized biological system with properties the traditional doctrine would have called mental *and* physical."[46] Priestley's conclusions are essentially those reached by Eddington and Russell and developed in recent years particularly by Galen Strawson and Daniel Stoljar, in ways to which we will return.

Reviewing the development of Locke's suggestion in England through the eighteenth century, Yolton observes that "Priestley's fascinating suggestions were not taken up and extended; they were hardly even perceived as different from earlier versions of materialism. The issues raised by Locke's suggestion of thinking matter . . . played themselves out through the century, but no one gave the emerging view of man as one substance—foreshadowed by Priestley—a systematic articulation."[47] This conclusion remains largely true, even for simple organisms, if we interpret it as referring to the unification problem.

Having argued that the mind-body problem disappears when we follow the "principles of the Newtonian philosophy," Priestley turns to confronting efforts to reconstitute something that resembles the problem, even after one of its terms—body (matter, etc.)—no longer has a clear sense. The first is "the difficulty of conceiving how thought can arise from matter . . . an argument that derives all its force from our ignorance," he writes, and has no force unless there is a demonstration that they are "absolutely incompatible with one another." Priestley was not troubled by qualms arising from ignorance, rightly I think, any more than scientists should have been concerned about the irreducibility of the mysterious properties of matter and motion to the mechanical philosophy, or, in more modern times, about the inability to reduce chemistry to an inadequate physics until the 1930s, to take two significant moments from the history of science.

A common objection today is that such ideas invoke an unacceptable form of "radical emergence" unlike, for example, the emergence of liquids from molecules, where the properties of the liquid can in some reasonable sense be regarded as inhering in the molecules. In Nagel's phrase, "we can *see* how liquidity is the logical result of the molecules 'rolling around on each other' at the microscopic level," though "nothing comparable is to be expected in the case of neurons" and consciousness. Also taking liquidity as a paradigm, Galen Strawson argues extensively that the notion of emergence is intelligible only if we interpret it as "total dependence": if "some part or aspect of Y [hails] from somewhere else," then we cannot say that Y is "emergent from X." We can speak intelligibly about emergence of Y-phenomena from non-Y phenomena only if the non-Y phenomena at the very least are "somehow *intrinsically suited* to constituting" the X-phenomena; there must be "something about X's nature in virtue of which" they are "so suited." "It is built into the notion of emergence that emergence cannot be brute in the sense of there being no reason in the nature of things why the emerging thing is as it is." This is Strawson's *No-Radical Emergence Thesis*, from which he draws the panpsychic conclusion that "experiential reality cannot possibly emerge from wholly and utterly nonexperiential reality." The basic claim, which he highlights, is that "If it really is true that Y is emergent from X then it must be the case that Y is in some sense wholly dependent on X and X alone, so that all features of Y trace intelligibly back to X." Here "intelligible" is a metaphysical rather than an epistemic notion, meaning "intelligible to God": there must be an explanation in the nature of things, though we may not be able to attain it.[48]

Priestley, it seems, would reject Nagel's qualms while accepting Strawson's formulation, but without drawing the panpsychic conclusion. It should be noted that the molecule-liquid example, commonly used, is not a very telling one. We also cannot conceive of a liquid turning into two gases by electrolysis, and there is no intuitive sense in which the properties of water, bases, and acids inhere in hydrogen or oxygen or other atoms. Furthermore, the whole matter of conceivability seems to be irrelevant whether it is brought up in connection with the effects of motion that Newton and Locke found inconceivable, the irreducible principles of chemistry, or mind-brain relations. There is something about the nature of hydrogen and oxygen "in virtue of which they are intrinsically suited to constituting water," so the sciences discovered after long labors, providing reasons "in the nature of things why the emerging thing is as it is." What seemed "brute emergence" was assimilated into science as ordinary emergence—not, to be sure, of the liquidity variety, relying on conceivability. I see no strong reason why matters should necessarily be different in the case of experiential and nonexperiential reality, particularly given our ignorance of the latter, stressed from Newton and Locke to Priestley, developed by Russell, and arising again in recent discussion.

Priestley then considers the claim that mind "cannot be material because it is influenced by reasons." To this he responds that since "reasons, whatever they may be, do ultimately move matter, there is certainly much less difficulty in conceiving that they may do this in consequence of their being the affection of some material substance, than upon the hypothesis of their belonging to a substance that has no common property with matter"—not the way it would be put today but capturing essentially the point of the contemporary discussion leading some to revive panpsychism. But contrary to the contemporary revival,[49] Priestley rejects the conclusion that consciousness "cannot be annexed to the whole brain as a system, while the individual particles of which it consists are separately unconscious." That "a certain quantity of nervous system is necessary to such complex ideas and affections as belong to the human mind; and the idea of self, or the feeling that corresponds to the pronoun I," he argues, "is not essentially different from other complex ideas, that of our country for example." Similarly, it should not perplex us more than the fact that "life should be the property of an entirely animal system, and not the separate parts of it" or that sound cannot "result from the motion of a single particle" of air. We should recognize "that the term self denotes that substance which is the seat of that particular set of sensations and ideas of which those that are then recollected make a part, as distinct from other substances which are the seat of similar sets of sensations and ideas" and that "It is high time to abandon these random hypotheses, and to form our conclusions with respect to the faculties of the mind, as well as the properties and powers of matter, by an attentive observation of facts and cautious inferences from them," adopting the Newtonian style of inquiry while dismissing considerations of common-sense plausibility. That seems to be a reasonable stance.

Priestley urges that we also dismiss arguments based on "vulgar phraseology" and "vulgar apprehensions," as in the quest for an entity of the world picked out by the term *me* when I speak of "my body," with its hint of dualism. "According to this merely verbal argument," Priestley observes, "there ought to be something in man besides all the parts of which he consists," something beyond both soul and body, as when "a man says I devote my soul and body," the pronoun allegedly denoting something beyond body and spirit that "makes the devotion." In Rylean terms, phrases of common usage may be "systematically misleading expressions," a lively concern at the time, based on a centuries-old tradition of inquiry into the ways surface grammatical forms disguise actual meaning. Like Priestley, Thomas Reid argued that failure to attend "to the distinction between the operations of the mind and the objects of these operations" is a source of philosophical error, as in interpreting the phrase "I have an idea" on the model of "I have a diamond," when we should understand it to mean something like "I am thinking." In an earlier discussion, the encyclopedist César Chesneau du Marsais, using the same and many other examples, warned against the error of taking nouns to be "names of real objects that exist independently of our thought." The language, then, gives no license for supposing that such words as "idea," "concept," and "image" stand for "real objects," let alone "perceptible objects."[50] For similar reasons, Priestley argues that "Nothing surely can be inferred from such phraseology as ['my body'], which, after all, is only derived from vulgar apprehensions."

The need to resist arguments from "vulgar apprehensions" holds more broadly: for such phrases as "my thoughts," "my dreams," "my spirit," even "my self," which is different from myself (= me, even though in another sense, I may not be myself these days). When John thinks about himself, he is thinking about John, but not when he is thinking about his self; he can hurt himself but not his self (whatever role these curious entities play in our mental world). There's a difference between saying that his actions are betraying his true (authentic, former) self and that he's betraying himself, and "thine own self" indicates a more essential characteristic than "thyself." Inquiry into manifold questions like these, while entirely legitimate and perhaps enlightening, is concerned with the "operations of the mind," our modes of cognition and thought, and should not be misinterpreted as holding of the "real objects that exist independently of our thought." The latter is the concern of the natural sciences, and I take it also to be the prime concern of the tradition reviewed here.

The operations of the mind doubtless accommodate the thesis that "I am not identical to my body," a core assumption of substance dualism, Stephen Yablo proposes.[51] He suggests further that "substance dualism . . . has fallen strangely out of view," perhaps "because one no longer recognizes 'minds' as entities in their own right, or 'substances,'" though "*selves*—the things we refer to by use of 'I'—are surely substances, and it does little violence to the intention behind mind/body dualism to interpret it as a dualism of bodies and selves." In the tradition I am following here, it is *matter* that has lost its presumed status, and not "strangely." It is also by no means clear, as just noted, that by use of the first-person pronoun (as in "I pledge to devote my body and my soul"), or the name "John," we refer to *selves*. But truth or falsity aside, an argument would be needed to show that in using such words we refer (or even take ourselves to be referring) to real constituents of the world that exist independently of our modes of thought. An alternative, which seems to me more plausible, is that these topics belong not to natural science but rather to a branch of ethnoscience, a study of how people think about the world, a very different domain. For natural science, it seems hard to improve on Priestley's conclusion that Locke's suggestion was fundamentally accurate and that properties "termed mental" reduce to "the organical structure of the brain"—though in ways that are not understood, no great surprise when we consider the history of even the core hard sciences, like chemistry.

As noted above, with the collapse of the traditional notion of body (etc.), there are basically two ways to reconstitute some problem that resembles the traditional mind-body problem: either define *physical* or set the problem up in other terms, such as those that Priestley anticipated.

The first option is developed by Galen Strawson in an important series of publications.[52] Unlike many others, he does give a definition of "physical" so that it is possible to formulate a physical-nonphysical problem. The physical is "any sort of existent [that is] spatio-temporally (or at least temporally) located." The physical includes "experiential events" (more generally mental events) and permits formulation of the question of how experiential phenomena can be physical phenomena—a "mind-body problem" in a post-Newtonian version. Following Eddington and Russell and earlier antecedents, notably Priestley, Strawson concludes that "physical stuff has, in itself, 'a nature capable of manifesting itself as mental activity,' i.e., as experience or consciousness."

That much seems uncontroversial, given the definitions along with some straightforward facts. But Strawson intends to establish the much stronger thesis of *micropsychism* (which he identifies here with *panpsychism*): "at least some ultimates are intrinsically experience-involving." The crucial premise for that further conclusion, as Strawson makes explicit, is the No-Radical Emergence Thesis, already discussed, from which it follows that "experiential reality cannot possibly emerge from wholly and utterly non-experiential reality," a metaphysical issue, not an epistemic one. Strawson interprets Eddington's position to be *micropsychism*, citing his observation that it would be "rather silly to prefer to attach [thought] to something of a so-called 'concrete' nature inconsistent with thought, and then to wonder where the thought comes from," and that we have no knowledge "of the nature of atoms that renders it all incongruous that they should constitute a thinking object." This however appears to fall short of Strawson's micropsychism/panpsychism. Rather, Eddington seems to go no farther than Priestley's conception, writing that nothing in physics leads us to reject the conclusion that an "assemblage of atoms constituting a brain" can be "a thinking (conscious, experiencing) object." He does not, it seems, adopt the No-Radical Emergence Thesis that is required to carry the argument beyond to Strawson's conclusion. Russell too stops short of this critical step, and Priestley explicitly rejects it, regarding radical emergence as normal science. Textual interpretation aside, the issues seem fairly clearly drawn.

The second option is pursued by Daniel Stoljar, who has done some of the most careful work on physicalism and variants of the "mind-body problem." He does offer some answers to the question of what it means to say that something is *physical*[53]—a question that, he notes, has not received a great deal of attention in the literature, though "Without any understanding of what the physical is, we can have no serious understanding of what physicalism is." The answers he offers are not too convincing, I think he would agree, but he argues that it does not matter much: "we have many concepts that we understand without knowing how to analyze," and "the concept of the physical is one of the central concepts of human thought." The latter comment is correct, but only with regard to the common-sense concept of the mechanical philosophy, long ago undermined. The former is correct too, but it is not clear that we want to found a serious philosophical position on a concept that we think we understand intuitively but cannot analyze, particularly when a long history reveals that such common-sense understanding can often not withstand serious inquiry. But Stoljar's more fundamental reason for not being too concerned with characterizing the "physical" is different: the issues, he argues, should be shifted to epistemological terms, not seeking reduction to *the physical* but taking physicalism to be only the "background metaphysical assumption against which the problems of philosophy of mind are posed and discussed." Thus "when properly understood, the problems that philosophers of mind are interested in are not with the framework [itself], and to that extent are not metaphysical."

Stoljar suggests that "the problem mainly at issue in contemporary philosophy is distinct *both* from the mind-body problem as that problem is traditionally understood *and* from the problem as it is, or might be, pursued in the sciences"; a qualification, I think, is that the traditional problem, at least from Descartes through Priestley (taking his work to be the culmination of the post-Newtonian reaction to the traditional problem), can plausibly be construed as a problem within the sciences. The traditional questions "we may lump together under the heading 'metaphysics of mind,'" but contemporary philosophy Stoljar takes to be concerned with "epistemic principles" and, crucially, "*the logical problem of experience.*" It might be true that "the notion of the physical fails to meet minimal standards of clarity," he writes, but such matters "play only an illustrative or inessential role in the logical problem," which can be posed "even in the absence of . . . a reasonably definite conception of the physical."[54]

The logical problem arises from the assumption that (1) there are experiential truths, while it seems plausible to believe both that (2) every such truth is entailed by (or supervenes on) some nonexperiential truth and (3) not every experiential truth is entailed by (or supervenes on) some nonexperiential truth. Adopting (1) and (2) (with a qualification, see below), the crucial question is (3). As already discussed, following a tradition tracing back to Newton and Locke, Priestley sees no reason to accept thesis (3): our "very ignorance" of the properties of post-Newtonian *matter* cautions us not to take this step. In Russell's words (which Stoljar cites), experiential truths "are not known to have any intrinsic character which physical events cannot have, since we do not know of any intrinsic character which could be incompatible with the logical properties that physics assigns to physical events." From these perspectives, then, the logical problem does not arise.[55]

Stoljar's solution to the logical problem, the new "mind-body problem," is similar to the stance of Priestley and Russell, even if put somewhat differently. It is based on his "ignorance hypothesis, according to which we are ignorant of a type of experience-relevant nonexperiential truth," so that the "logical problem of experience" unravels on epistemic grounds.[56] He suggests elsewhere that "the radical view . . . that we are ignorant of the nature of the physical or non-experiential has the potential to completely transform philosophy of mind."[57] In Strawson's formulation, the (sensible) line of thought that was well understood up to a half century ago "disappeared almost completely from the philosophical mainstream [as] analytical philosophy acquired hyper-dualist intuitions even as it proclaimed its monism. With a few honorable exceptions it out-Descartesed Descartes (or 'Descartes' [that is, the constructed version]) in its certainty that we know enough about the physical to know that the experiential cannot be physical."[58]

The qualification with regard to (2) is that we cannot so easily assume that there are nonexperiential truths; in fact, the assumption may be "silly," as Eddington put it. Some physicists have reached such conclusions on quantum-theoretic grounds. The late John Wheeler argued that the "ultimates" may be just "bits of information," responses to queries posed by the investigator. "The actual events of quantum theory are experienced increments in knowledge" (H. P. Stapp).[59] Russell's three grades of certainty suggest other reasons for skepticism. At the very least, some caution is necessary about the legitimacy even of the formulation of the "logical problem."

Stoljar invokes the ignorance hypothesis in criticizing C. D. Broad's conclusions about the irreducibility of chemistry to physics, a close analogue to the Knowledge Argument, he observes. He concludes that Broad was unaware "that chemical facts follow from physical facts," namely the quantum-theoretic facts.[60] But putting the matter that way is somewhat misleading. What happened is that physics radically changed with the quantum-theoretic revolution, and with it the notion of "physical facts." A more appropriate formulation, I think, is to recognize that post Newton, the concept "physical facts" means nothing more than what the best current scientific theory postulates and hence should be seen as a rhetorical device of clarification adding no substantive content. The issue of physicalism cannot be so easily dispensed with. Like Marx's old mole, it keeps poking its nose out of the ground.

There are also lesser grades of mystery worth keeping in mind. One of particular interest to humans is the evolution of their cognitive capacities. On this topic, the evolutionary biologist Richard Lewontin has argued forcefully that we can learn very little, because evidence is inaccessible, at least in any terms understood by contemporary science.[61] For language, there are two fundamental questions in this regard: first, the evolution of the capacity to construct an infinite range of hierarchically structured expressions interpretable by our cognitive and sensorimotor systems; and second, the evolution of the atomic elements, roughly word-like, that enter into these computations. In both cases, the capacities appear to be specific to humans, perhaps even specific to language, apart from the natural laws they obey, which may have rather far-reaching consequences, recent work suggests. I think something can be said about the first of these questions, the evolution of the generative mechanisms. One conclusion that looks increasingly plausible is that externalization of language by means of the sensorimotor system is an ancillary process and also the locus of much of the variety and complexity of language. The evolution of atoms of computation, however, seems mired in mystery, whether we think of them as concepts or lexical items of language. In symbolic systems of other animals, symbols appear to be linked directly to mind-independent events. The symbols of human language are sharply different. Even in the simplest cases, there is no word-object relation, where objects are mind-independent entities. There is no reference relation, in the technical sense familiar from Frege and Peirce to contemporary externalists. Rather, it appears that we should adopt something like the approach of the seventeenth- and eighteenth-century cognitive revolution and the conclusions of Shaftesbury and Hume that the "peculiar nature belonging to" the linguistic elements used to refer is not something external and mind independent. Rather, their peculiar nature is a complex of perspectives involving Gestalt properties, cause and effect, "sympathy of parts" directed to a "common end," psychic continuity, and other such mental properties. In Hume's phrase, the "identity, which we ascribe" to vegetables, animal bodies, artifacts, or "the mind of man"—the array of individuating properties—is only a "fictitious one," established by our "cognoscitive powers," as they were termed by his seventeenth-century predecessors. That is no impediment to interaction, including the special case of communication, given largely shared cognoscitive powers. Rather, the semantic properties of words seem similar in this regard to their phonetic properties. No one is so deluded as to believe that there is a mind-independent object corresponding to the internal syllable [ba], some construction from the motion of molecules perhaps, which is selected when I say [ba] and when you hear it. But interaction proceeds nevertheless, always a more-or-less rather than a yes-or-no affair.[62]

There is a lot to say about these topics, but I will not pursue them here, merely commenting that in this case too there may be merit to Strawson's conclusion that "hyperdualist intuitions" should be abandoned along with the "certainty that we know enough about the physical to know that the experiential cannot be physical" and Stoljar's suggestion that "the radical view" might transform philosophy of mind and language, if taken seriously.

Returning finally to the core example of Cartesian science, human language, Gassendi's advice to seek a "chemical-like" understanding of its internal nature has been pursued with some success, but what concerned the Cartesians was something different: the creative use of language, what Humboldt later called "the infinite use of finite means," stressing *use*.[63]

There is interesting work on precepts for language use under particular conditions—notably the intent to be informative, as in neo-Gricean pragmatics—but it is not at all clear how far this extends to the normal use of language, and in any event, it does not approach the Cartesian question of creative use, which remains as much of a mystery now as it did centuries ago and may turn out to be one of those ultimate secrets that ever will remain in obscurity, impenetrable to human intelligence.

NOTES

1. David Hume, *The History of England*, VI, LXXI. John Locke, *An Essay Concerning Human Understanding*, book 4, chap. 3. Locke's reasons, of course, were not Hume's but relied on the boundaries of "the simple ideas we receive from sensation and reflection," which prevent us from comprehending the nature of body or mind (spirit).

2. R. Baillargeon, "Innate Ideas Revisited: For a Principle of Persistence in Infants' Physical Reasoning," *Perspectives on Psychological Science* 3 (2008): 2–13.

3. B. Cohen, *Revolution in Science* (Cambridge, Mass.: Harvard University Press, 1985), 155.

4. E. McMullin, *Newton on Matter and Activity* (Notre Dame, Ind.: University of Notre Dame Press, 1978), 52ff. He concludes that because of Newton's vacillation in use of the terms "mechanical," "spirit," and others, it is "misleading . . . to take Newton to be an exponent of the 'mechanical philosophy'" (73).

5. Locke, *An Essay Concerning Human Understanding*. Correspondence with Stillingfleet cited by Ben Lazare Mijuskovic, *The Achilles' Heel of Rationalist Arguments* (The Hague: Martinus Nijhoff, 1974), 73. On the development of "Locke's suggestion" through the eighteenth century, culminating in Joseph Priestley's important work (discussed below), see John Yolton, *Thinking Matter* (Minneapolis: University of Minnesota Press, 1983).

6. Pierre-Jean-George Cabanis, *On the Relations Between the Physical and Moral Aspects of Man*, vol. 1 (1802; repr. Baltimore, Md.: The Johns Hopkins University Press, 1981). Darwin cited

by V. S. Ramachandran and Sandra Blakeslee, *Phantoms in the Brain* (New York: William Morrow, 1998), 227.

7. *Principia*, General Scholium.

8. E. J. Dijksterhuis, *The Mechanization of the World Picture* (Oxford: Oxford University Press, 1961; Princeton, N.J.: Princeton University Press, 1986), 479–480.

9. Letter to Bentley, 1693. Cited in McMullin, *Newton on Matter and Activity*, 488. See chap. 3 of McMullin's volume for more detailed analysis.

10. Thomas Nagel, *Other Minds* (Oxford: Oxford University Press, 1995), 106.

11. For varying perspectives on the "explanatory gap," see essays in Galen Strawson et al., *Consciousness and Its Place in Nature* (Exeter: Imprint Academic, 2006).

12. T. Kuhn, *The Copernican Revolution* (New York: Random House, 1957), 259. Heinrich Hertz cited by McMullin, *Newton on Matter and Activity*, 124.

13. Dijksterhuis, *The Mechanization of the World Picture*, 489.

14. Joseph Glanvill; see John Henry, "Occult Qualities and the Experimental Philosophy," *History of Science* 24 (1986). R. Popkin, *The History of Skepticism from Erasmus to Spinoza* (Berkeley: University of California Press, 1979), 139–140, 213.

15. B. Russell, *Analysis of Matter* (London: Allen and Unwin, 1927; repr. New York: Dover, 1954), 18–19, 162. P. Dirac, *Principles of Quantum Mechanics* (Oxford: Oxford University Press, 1930), 10, brought to my attention by John Frampton.

16. P. Machamer, introduction and "Galileo's Machines, His Mathematics, and His Experiments," in P. Machamer, ed., *The Cambridge Companion to Galileo* (Cambridge: Cambridge University Press, 1998).

17. Ibid. Cited by Pietro Redondi, "From Galileo to Augustine." D. Stoljar, *Ignorance and Imagination* (Oxford: Oxford University Press, 2006). Recall that Newton hoped that there might be a scientific (that is, mechanical) solution to the problems of matter and motion.

18. On these topics, see my *Cartesian Linguistics* (New York: Harper & Row, 1966); third edition, edited by James McGilvray, with introduction, full translations, and quotes from updated scholarly editions (Cybereditions, forthcoming); and *Language and Mind* (New York: Harcourt Brace Jovanovich, 1968), chap. 1. Note that the concerns go far beyond indeterminacy of free action, as is particularly evident in the experimental programs by Cordemoy and others on "other minds"; see *Cartesian Linguistics*.

19. Letter to Queen Christina of Sweden, 1647; *Principles of Philosophy*. For discussion, see Tad Schmaltz, *Malebranche's Theory of the Soul* (Oxford: Oxford University Press, 1996), 204ff.

20. See my "Turing on the 'Imitation Game,'" in Stuart Schieber, ed., *The Turing Test* (Cambridge, Mass.: The MIT Press, 2004).

21. D. Clarke, *Descartes' Theory of Mind* (Oxford: Oxford University Press, 2003), 12. See also Descartes' 1641 letter to Mersenne on the goal of the *Meditations*, cited by Margaret Wilson, *Descartes* (1978), 2.

22. Clarke, *Descartes' Theory of Mind*, 258.

23. Kanwisher and Downing, "Separating the Wheat from the Chaff," *Science* 282 (October 2, 1998). Newton, *General Scholium*.

24. E. Kandel and L. Squire, "Neuroscience," *Science* 290 (2000): 1113–1120. C. R. Gallistel, "Neurons and Memory," in *Conversations in the Cognitive Neurosciences*, ed. M. Gazzaniga (Cambridge, Mass.: The MIT Press, 1997); "Symbolic Processes in the Insect Brain," in *Methods, Models, and Conceptual Issues: An Invitation to Cognitive Science*, ed. D. Scarborough and S. Sternberg, vol. 4 (Cambridge, Mass.: The MIT Press, 1998). S. Zeki, "Art and the Brain," *Daedalus* 127 (1998).

25. Nagel, *Other Minds*. For some cautionary notes on "sharp logical separation between the nervous system and the rest of the organism," see Charles Rockland, "The Nematode as a

Model Complex System," Working Paper (LIDS-WP-1865), Laboratory for Information and Decisions Systems, MIT (April 14, 1989): 30.

26. Henry, "Occult Qualities and the Experimental Philosophy." Alan Kors, "The Atheism of D'Holbach and Naigeon," in *Atheism from the Reformation to the Enlightenment*, ed. M. Hunger and D. Wootton (Oxford: Oxford University Press, 1992). Locke, *An Essay Concerning Human Understanding*; Yolton, *Thinking Matter*, 199. Voltaire, Kant, McMullin, *Newton on Matter and Activity*, 113, 122–123, from Kant's *Metaphysical Foundations of Natural Science*, 1786; Michael Friedman, "Kant and Newton: Why Gravity Is Essential to Matter," in *Philosophical Perspectives on Newtonian Science*, ed. P. Bricker and R. I. G. Hughes (Cambridge, Mass.: The MIT Press, 1990). Howard Stein, "On Locke, 'the Great Huygenius, and the incomparable Mr. Newton,'" in *Philosophical Perspectives on Newtonian Science*. Friedman argues that there is no contradiction between Newton and Kant because they do not mean the same thing by "essential," Kant having discarded Newton's metaphysics and making an epistemological point within his "Copernican revolution in metaphysics."

27. F. Lange, *Geschichte des Materialismus und Kritik seiner Bedeutung in der Gegenwart* (1865), translated as *The History of Materialism*, 3rd expanded ed. (London: Kegan Paul, 1925). A. Koyré, *From the Closed World to the Infinite Universe* (Baltimore, Md.: The Johns Hopkins University Press, 1958), 210.

28. G. Coyne, "The Scientific Venture and Materialism: False Premises," in *Space or Spaces as Paradigms of Mental Categories* (Fondazione Carlo Erba, 2000).

29. Russell, *Analysis of Matter*, chap. 37. Russell did not work out how percepts in their cognitive aspect were assimilated into the "causal skeleton of the world," leaving him open to a counterargument by mathematician Max Newman. See Russell's letter to Newman (April 24, 1928), in *The Autobiography of Bertrand Russell*, vol. 2: *1914–1944* (Boston: Little Brown, 1967).

30. Democritus, quoted by Erwin Schrödinger, *Nature and the Greeks* (Cambridge: Cambridge University Press, 1954), 89. Brought to my attention by Jean Bricmont.

31. D. Stoljar and Y. Nagasawa, introduction to Peter Ludlow, Yujin Nagasawa, and Daniel Stoljar, *There's Something About Mary* (Cambridge, Mass.: The MIT Press, 2004).

32. For Hume, see John Mikhail, *Rawls' Linguistic Analogy: A Study of the "Generative Grammar" Model of Moral Theory Described by John Rawls in* A Theory of Justice (Ph.D. diss., Ithaca, N.Y., Cornell University, 2000); *Moral Grammar: Rawls' Linguistic Analogy and the Cognitive Science of Moral Judgment* (Cambridge: Cambridge University Press, forthcoming); "Universal Moral Grammar: Theory, Evidence, and the Future," *Trends in Cognitive Sciences* (April 2007). On the irrelevance (and, as it is formulated, even incoherence) of the doctrine of "accessibility to consciousness, see my *Reflections on Language* (New York: Pantheon, 1975), *Rules and Representations* (New York: Columbia University Press, 1980), and *New Horizons in the Study of Language and Mind* (Cambridge: Cambridge University Press, 2000). On the rules of visual perception, inaccessible to consciousness in the interesting cases, see Donald Hoffman, *Visual Intelligence* (New York: Norton, 1998).

33. F. Jackson, "What Mary Didn't Know," "Postscript," in Ludlow et al., *There's Something About Mary*.

34. C. S. Peirce, "The Logic of Abduction," in *Peirce's Essays in the Philosophy of Science*, ed. V. Tomas (New York: Liberal Arts Press, 1957). See my *Language and Mind*, 90ff., for a discussion of his proposals and fallacies invoking natural selection that lead him to the ungrounded (and implausible) belief that our "guessing instinct" leads us to true theories.

35. Cited by Wilson, *Descartes*, 95.

36. Hume, *Inquiry*, 2.1. On dubious modern efforts to formulate what had been a reasonably clear project before the separation of philosophy from science, see my *New Horizons*, 79–80, 144–145, and generally chaps. 5–6 (reprinted from *Mind* [1995]: 104, 1–61).

37. Joseph Black and Robert Schofield, *Mechanism and Materialism* (Princeton, N.J.: Princeton University Press, 1970), 226; William Brock, *The Norton History of Chemistry* (New York: Norton, 1993), 271. A. Thackray, *Atoms and Powers* (Cambridge, Mass.: Harvard University Press, 1970), 37–38, 276–277.

38. Brock, *The Norton History of Chemistry*. For sources and further discussion, see my *New Horizons* and *Knowledge of Language* (New York: Praeger, 1986), 251–252. David Lindley, *Boltzmann's Atom* (New York: The Free Press, 2001). Some argue that even if quantum-theoretic unification succeeds, "in some sense the program of reduction of chemistry to [the new] physics fails," in part because of "practical issues of intractability." Maureen Christie and John Christie, "'Laws' and 'Theories' in Chemistry Do Not Obey the Rules," in *Of Minds and Molecules*, ed. Nalin Bhushan and Stuart Rosenfield (Oxford: Oxford University Press, 2000).

39. Russell, *Analysis of Matter*, 388.

40. See references in note 32, above. Sometimes misunderstanding and distortion reach the level of the surreal. For some startling examples, see my contribution to the Symposium on Margaret Boden, "Mind as Machine: A History of Cognitive Science, Oxford, 2006," *Artificial Intelligence* 171 (2007): 1094–1103 (Elsevier), http://www.sciencedirect.com. On "the rigidity rule and [Shimon] Ullman's theorem," see Hoffman, *Visual Intelligence*, 159. Needless to say, the rule is inaccessible to consciousness.

41. V. Mountcastle, in *Daedalus* (Spring 1988). For sources, see my *New Horizons*, chap. 5.

42. J. Priestley, "Materialism," from *Disquisitions Relating to Matter and Spirit* (1777). In John Passmore, ed., *Priestley's Writings on Philosophy, Science, and Politics* (New York: Collier-MacMillan, 1965).

43. Similar ideas appear pre-Newton, particularly in the *Objections to the Meditations*, where critics ask how Descartes can know, "without divine revelation . . . that God has not implanted in certain bodies a power or property enabling them to doubt, think, etc." Catherine Wilson, in Strawson et al., *Consciousness and Its Place in Nature*.

44. Priestley, "Materialism." Later discussion, Yolton, *Thinking Matter*, 113. Similar conclusions had been drawn by La Mettrie a generation earlier but in a different framework and without addressing the Cartesian arguments to which he is attempting to respond. The same is true of Gilbert Ryle and other modern attempts. For some discussion, see my *Cartesian Linguistics*.

45. For discussion and illustrations, see my *New Horizons*. Strawson, below, on "hyperdualism."

46. Thackray, *Atoms and Powers*, 190. Priestley's reasons for welcoming "this extreme development of the Newtonian position" were primarily theological, Thackray concludes. Yolton, *Thinking Matter*, 114.

47. Yolton, *Thinking Matter*, 125. See chaps. 5–6 for discussion. Yolton writes that "there was no British La Mettrie," but that exaggerates La Mettrie's contribution, I believe. See n. 44, above.

48. Nagel, *Other Minds*. Strawson, "Realistic Monism" and "Reply," in Strawson et al., *Consciousness and Its Place in Nature*. Printer's errors corrected. See essays in this volume for further discussion.

49. Priestley, "Materialism." Strawson and commentary.

50. See my *Aspects of the Theory of Syntax* (Cambridge, Mass.: The MIT Press, 1965), 199–200, and for much more extensive discussion, *Cartesian Linguistics*. On the accuracy of interpretations of the empiricist theory of ideas by Reid and others, see John Yolton, *Perceptual Acquaintance from Descartes to Reid* (Minneapolis: University of Minnesota Press, 1984), chap. 5.

51. S. Yablo, "The Real Distinction Between Mind and Body," *Canadian Journal of Philosophy*, suppl. vol. 16.

52. Quotes here from "Realistic Monism" and "Reply," in Strawson et al., *Consciousness and Its Place in Nature*.

53. Quotes in this paragraph from Stoljar, "Physicalism," in the *Stanford Encyclopedia of Philosophy* (2001).

54. Stoljar, *Ignorance and Imagination*.

55. Stoljar, *Ignorance and Imagination*, 17ff., chap. 2, 56–57, 104. Stoljar understands the "traditional problem" to be derived from the *Meditations* (45), hence not a problem of the sciences. But though a conventional reading, it is questionable, for reasons already discussed.

56. Stoljar, *Ignorance and Imagination*, chap. 4.

57. Stoljar, in Strawson et al., *Consciousness and Its Place in Nature*.

58. Strawson, "Realistic Monism," n. 21.

59. J. Wheeler, *At Home in the Universe* (American Institute of Physics, 1994), vol. 9 of *Masters of Modern Physics*. Stapp, in Strawson et al., *Consciousness and Its Place in Nature*.

60. Stoljar, *Ignorance and Imagination*, 139.

61. R. Lewontin, "The Evolution of Cognition: Questions We Will Never Answer," in *Methods, Models, and Conceptual Issues: An Invitation to Cognitive Science*.

62. Chomsky, *Cartesian Linguistics*, 94ff.; and McGilvray's introduction, on Cartesian and neo-Platonist conceptions of the role of "cognoscitive powers." For review and sources on referring, see *New Horizons*. For Shaftesbury, Hume, and forerunners, see Mijuskovic, *The Achilles' Heel of Rationalist Arguments*.

63. On misunderstandings about this matter, see my "A Note on the Creative Aspect of Language Use," *Philosophical Review* 41, no. 3 (July 1982).

2

THE GREAT SOUL OF POWER

SAID MEMORIAL LECTURE

NOAM CHOMSKY

It is a challenging task to select a few themes from the remarkable range of the work and life of Edward Said, who I was privileged to count as a treasured friend for many years. I will keep to two: the culture of empire and the responsibility of intellectuals—or, from a broader perspective, the culture of dominance generally and the responsibility of those with sufficient privilege and resources so that if they choose to enter the public arena, we call them "intellectuals."

The phrase "responsibility of intellectuals" conceals a crucial ambiguity: it blurs the distinction between "ought" and "is." In terms of "ought," their responsibility should be exactly the same as that of any decent human being, though greater: privilege confers opportunity, and opportunity confers moral responsibility. We rightly condemn the obedient intellectuals of brutal and violent states for their "conformist subservience to those in power"—I am borrowing the phrase from Hans Morgenthau, one of the founders of modern international-relations theory. He was, however, not referring to the commissar class of the totalitarian enemy but to Western intellectuals, whose crime is far greater, because they cannot plead fear but only cowardice and subordination to power. He was describing what *is*, not what *ought* to be. And regrettably, he is basically correct about what *is* and *has been* throughout much of history. It is noteworthy that he was writing in late 1970, after opposition to the Indochina wars had peaked, as had the most vocal dissidence of the educated classes.

The history of intellectuals is written by intellectuals, so not surprisingly, they are portrayed as defenders of right and justice, upholding the highest values and confronting power and evil with admirable courage and integrity. The record reveals a rather different picture. The term "intellectual" came into common usage with the Dreyfusards, the prototypical "engaged intellectuals." But they were a minority: most kept to conformist subservience to those in power. That has been the pattern back to the earliest recorded history. It was the man who "corrupted the youth of Athens" with "false gods" who drank the hemlock, not those who worshipped the true gods of the doctrinal system. A large part of the Bible is devoted to people who provided critical geopolitical analysis and condemned the crimes of state and immoral practices. They are called "prophets," a dubious translation of an obscure word. In contemporary terms, they were "dissident intellectuals." There is no need to review how they were treated: miserably, the norm for dissidents.

There were also intellectuals who were greatly respected in the era of the Prophets: the flatterers at the Court. Centuries later, they were condemned as "false prophets." The Gospels warn of "false prophets, who come to you in sheep's clothing, but inwardly are ravening wolves. By their fruits ye shall know them." That's correct: it is by their acts, not their lofty words, that we should know them, a good lesson to the present day.

The end of the last millennium was surely one of the low points in the generally dismal history of intellectuals. In the United States and Europe, respected figures were entranced by the "normative revolution" unfolding before our eyes, as U.S. foreign policy entered a "noble phase" with a "saintly glow." For the first time in history, a state was dedicated to "principles and values," acting from "altruism" alone. At last the "enlightened states" would undertake their "responsibility to protect" the suffering everywhere, led by the "idealistic New World bent on ending inhumanity." That is a small sample from the left-liberal end of the deluge, and deluge it was. The illustrations offered collapse under the slightest examination, and while the chorus of self-adulation was resounding, the idealistic New World and its European allies were conducting some of the most horrendous atrocities of those ugly years. But none of that matters in a well-disciplined intellectual culture, and those who dare sully the record with boring fact can quickly be dismissed as "anti-Americans" if not worse, as Edward Said knew well.

The jewel in the crown was the bombing of Serbia in 1999. Standard doctrine—to quote one respected source—is that NATO went to war "to stop the ethnic cleansing ordered by" Milošević against Kosovar Albanians, and the U.S.-led campaign "succeeded in stopping the violence." Close to 100 percent of the flood of commentary on the war repeats this story, reversing the chronology: the ethnic cleansing was the consequence of the NATO bombing, not its cause, and furthermore its anticipated consequence. The Milošević indictment, issued at the height of the bombing on the basis of U.S. and UK intelligence, refers only to crimes after the bombing, with a single exception two months earlier. The truth of the matter is demonstrated conclusively by a vast collection of detailed documents from the most impeccable sources: the U.S. State Department, the OSCE, NATO, a lengthy British parliamentary inquiry, and others, all in full agreement that the atrocities followed the bombing and admittedly were its anticipated consequence. The same documentary record shows that the prebombing period was ugly, though regrettably not in the least unusual, and not even close to the crimes that the United States and United Kingdom were implementing right at the same time. According to the British government, until shortly before the NATO bombing most of the atrocities were committed by KLA guerrillas attacking from across the border in an effort to elicit a harsh Serbian reaction, which could be used to bring about Western intervention. And the detailed Western documentation reveals that nothing changed up to the onset of the bombing, after which the anticipated atrocities did take place, in reaction.

There is much more, but what is interesting is the desperation to which the Western intellectual classes cling to the reversal of chronology and a flood of other lies about what happened. The single example I quoted illustrating the deluge is particularly interesting. It appears in the current issue of the journal of the American Academy of Political Science, in a laudatory review of a book that draws precisely the opposite conclusion, and from a highly authoritative source: the highest level of the Clinton administration. The conclusion of the book under review, articulating the thinking of administration planners, is that "it was Yugoslavia's resistance to the broader trends of political and economic reform—not the plight of Kosovar Albanians—that best explains NATO's war." As the words pass through the distorting prism of the intellectual culture, these conclusions are transmuted to the doctrine required for the chorus of self-adulation: NATO went to war "to stop the ethnic cleansing ordered by" Milošević against Kosovar Albanians, and the U.S.-led campaign "succeeded in stopping the violence." There is virtually nothing that can shake the dogmas required to uphold the nobility of state power, despite the occasional errors and failures that critics allow themselves to condemn.

The grip of imperial culture is sometimes utterly astounding. Even calls for genocide from the highest level of government elicit hardly a murmur. The prosecutors at The Hague were working hard to establish a charge of genocide against Milošević. Suppose they had come across a document in which he orders the Serbian armed forces to carry out a "massive bombing campaign: Anything that flies on anything that moves." The trial would have been over and Milošević sentenced to life imprisonment, if not worse. Actually they did find such a document, but from the wrong source: Henry Kissinger, conveying orders from the president to bomb Cambodia. Perhaps you can enlighten me, but I have never found such an explicit call for genocide in the archival record of any state. It was published in the *New York Times* two years ago, eliciting not a murmur of protest, even qualms. And as educated readers knew, the orders were implemented—though they could not have known how awesome was the scale, because that was kept quiet. Newly released Pentagon records, still unreported, reveal that the bombing of Cambodia was about five times the scale of the horrifying figures that were previously announced: close to three million tons of bombs, nearly half of all U.S. bombing of Indochina, making Cambodia the most heavily bombed country in history, by a wide margin.

All of this happens to be highly relevant to today's crises. The order for genocide in Cambodia was part of the last stages of the Vietnam war, when U.S. troops were being withdrawn to be replaced by airpower, just what is now being planned in Iraq.

The ability "not to see" what might conflict with the image of righteousness often reaches impressive heights. To mention another current example, last February the *New York Times* published an article by law professor Noah Feldman, who you will recall from his failed effort to impose a U.S.-written constitution on Iraq, part of the effort to ensure that the wrong people would not be elected, as he explained. Feldman was reviewing a collection of speeches of Osama bin Laden and described his descent to greater and greater evil, finally reaching the absolute lower depths, when he came to advocate "the perverse claim that since the United States is a democracy, all citizens bear responsibility for its government actions, and civilians are fair targets." The ultimate evil.

Two days later, the lead article in the *New York Times* announced that the United States and Israel were adopting Osama's "perverse claim," joining him in the lower depths of evil. The article reports that Palestinians bear responsibility for their government, so all must suffer for electing Hamas. They will be held hostage and punished until they elect a government favored by the imperial overlords. Detailed mechanisms are outlined, since implemented. The article also reports that Condoleezza Rice will visit the oil producers to ensure that they do not relieve the torture of the Palestinians. When we adopt Osama's perverse principle, it is not ultimate evil but has different names, like "promoting democracy" and noble pursuit of peace and justice.

All of this passed without notice, along with much more: crucially, the fact that Osama's perverse doctrine has been U.S.-UK policy as far back as we would like to go. Well-known cases include "making the economy scream" in Chile, when citizens of Latin America's oldest democracy elected the wrong person. That was the "soft track"; the "hard track" led to the imposition of the Pinochet regime on the first 9/11, as South Americans call it: September 11, 1973, far more hideous than the second one in 2001. Or to take a more recent example, Afghanistan, where after three weeks of bombing, the United States and United Kingdom announced a new war aim: to overthrow the Taliban. Admiral Sir Michael Boyce, chief of the British Defence Staff, announced that

U.S.-UK bombing will continue "until the people of the country themselves recognize that this is going to go on until they get the leadership changed," a particularly extreme version of Osama's "perverse claim," because the attackers were well aware that millions of people faced possible death from starvation if the bombing continued. We may also recall the murderous U.S.-UK sanctions against Iraq, killing hundreds of thousands of people and shattering the society, while "paralyzing all opposition to the discredited and moribund regime and giving it a new lease on life," as Iraqi dissident Kamil Mahdi wrote.

We also have to forget the most venerable current example: Cuba. Washington has been running campaigns of terror and economic strangulation against Cuba for over forty-five years. The reasons are frankly explained in the secret internal record from the start: The Eisenhower administration determined that "The Cuban people are responsible for the regime," so the United States has the right to cause them to suffer by economic strangulation. "Rising discomfort among hungry Cubans" will cause them to throw Castro out, President Kennedy advised, while also initiating a massive terror campaign that almost brought the world to destruction: the Cuban missile crisis. "Every possible means should be undertaken promptly to weaken the economic life of Cuba [in order to] bring about hunger, desperation and overthrow of the government," the State Department advised. The basic thinking has not changed since. After the collapse of the Soviet Union, Clinton Democrats used Cuba's desperate straits to tighten the vise, intensifying the blockade with the announced objective "to wreak havoc in Cuba" so that the people will suffer and overthrow the government—targeted for attack because of its "successful defiance" of U.S. policies going back to the Monroe doctrine 180 years ago, as we learn from records of the Kennedy-Johnson years.

Without continuing, adopting Osama's most perverse claim to punish Palestinians is no departure from the routine. But thanks to a good education, none of this can be perceived, even when the articles denouncing Osama and lauding ourselves for joining him in the depths of evil appear simultaneously. Such illustrations, easily multiplied, are real triumphs of imperial culture, fully shared in Europe.

Though the end-of-millennium chorus of self-adulation may well have set a new low in the annals of intellectual history, the norm is not very different. The prevailing truth was expressed by President John Adams two centuries ago: "Power always thinks it has a great soul and vast views beyond the comprehension of the weak." That is the deep root of the combination of savagery and self-righteousness that infects the imperial mentality—and in some measure, every structure of authority and domination.

We can add that reverence for that great soul is the normal stance of intellectual elites, who regularly add that they should hold the levers of control, or at least be close by. That doctrine holds across the narrow spectrum. It is the standard Leninist doctrine, shared by progressive thought in the West: the so-called action intellectuals of Kennedy's Camelot, for example. The leading public intellectual of twentieth-century America, the Wilsonian political analyst Walter Lippmann, explained in his essays on democracy that the "responsible" intellectuals who design and implement policy must "live free of the trampling and roar of [the] bewildered herd" as they labor selflessly for the common good, while the herd must be "put in its place": attending to private pursuits. One common expression of this prevailing view is that there are two categories of intellectuals: the "technocratic and policy-oriented intellectuals"—responsible, sober, constructive—and the "value-oriented intellectuals," a sinister grouping who pose a threat to democracy, as they "devote themselves to the derogation of leadership, the challenging of authority, and the unmasking of established institutions," even seeking to delegitimize the institutions responsible for the "indoctrination of the young" (the schools, colleges, churches, and so on). I am quoting from a study by the more progressive and humane wing of the international intellectual class, the liberal internationalists of the Trilateral Commission, from the United States, Europe, and Japan. The Carter administration was almost entirely drawn from their ranks, including the president. They were reflecting on the "crisis of democracy" that developed in the 1960s, when normally passive and apathetic sectors of the population, called "the special interests," sought to enter the political arena to advance their concerns. Those improper initiatives created what they called a "crisis of democracy," in which the proper functioning of the state was threatened by "excessive democracy." To overcome the crisis, the special interests must be restored to their proper function as passive observers, so that the "technocratic and value-oriented intellectuals" can do their constructive work undisturbed by the bewildered herd.

The disruptive special interests are women, the young, the elderly, workers, farmers, minorities, majorities—in short, the population. Only one specific interest is not mentioned in the study: the corporate sector. But that makes sense. They represent the "national interest," and naturally there can be no question of the national interest being protected by state power. I stress again that I am quoting from the liberal internationalist extreme of the spectrum. The business world and the right took a much harsher stand, and they implemented their decisions so as to beat back the dangerous civilizing and democratizing tide, with some success, though only partial. Those reactions have set their stamp on the contemporary era in a wide variety of ways, ranging from doctrinal management to global economic policies.

There have always been radical extremists who reject the prevailing premises. Some of them, centuries ago, went so far as to condemn the "merchants and manufacturers" who were the "principal architects of policy" in England and used their power to make sure that their own interests were "most peculiarly attended to," no matter how "grievous" the impact on others, primarily those in India and elsewhere who were subject to their "savage injustice," but also the population at home. According to these renegades, a primary task of the ruling economic class is to "delude and oppress the public" as they pursue the "vile maxim of the masters of mankind: all for ourselves, nothing for other people." Such radical extremists are either ignored or reviled. In the case of Adam Smith, whom I am quoting, ignored. He is revered but unread; his leading ideas are not only ignored but often simply falsified. To mention only one case of unusual contemporary relevance, everyone has heard his phrase "invisible hand," but few have taken the trouble to find the phrase in his classic *Wealth of Nations*. It appears just once, in an argument against what is now called "neoliberalism." Neoliberal principles, Smith recognized, would devastate England if the merchants and manufacturers who ruled the state in their own interest would invest abroad and rely on imports. But they will not do so, Smith suggested, because they will prefer to do business at home, so as if by an invisible hand, England will be saved from the ravages of neoliberalism. David Ricardo made similar observations, adding that he would be "sorry to see such feelings of the merchants and manufacturers weakened," in which case his theory of comparative advantage would collapse.

England, of course, did not rely on the invisible hand to ensure that it would develop. England relied on a powerful interventionist state, and its colonies were crushed by forceful imposition of free trade. The merchants and manufacturers who ruled England were unwilling even to toy with the idea of free trade until 1846, after 150 years of protectionism and violence had created far and away the most powerful industrial society in the world. They could therefore expect that the "playing field would be tilted" in their favor, to adapt the contemporary idiom, so that "free competition" would be acceptable. But they were careful to hedge their bets. They relied crucially on their protected markets in India and elsewhere, while also developing the most extraordinary narcotrafficking enterprise in human history to enable them to break into the China market. The United States followed the same path, as have other countries that have developed, in radical violation of the precepts of economic theory, again a pattern that persists to the present. But that is another topic.

For those who want to understand today's world and what is likely to lie ahead, it is of prime importance to look closely at the long-standing principles that are held to animate the decisions and actions of the powerful—bearing in mind that it is by their fruits that you shall know them, not the fine words. In today's world, that means primarily the United States. Though only one of three major power centers in economic and most other dimensions, it surpasses any power in history in its military dominance, which is rapidly expanding, and it can generally rely on the support of the second superpower, Europe, and also Japan, the second-largest industrial economy. There is a clear doctrine on the general contours of U.S. foreign policy. It reigns virtually without exception in Western journalism and almost all scholarship, even among critics of policies past and present. The major theme is "American exceptionalism": the thesis that the United States is unlike other great powers, past and present, because it has a "transcendent purpose": "the establishment of equality in freedom in America" and indeed throughout the world, since "the arena within which the United States must defend and promote its purpose has become world-wide."

The particular version of the thesis I have just quoted is particularly interesting because of its source: Hans Morgenthau, whom I quoted before, but that was after the Vietnam War, when he had shifted to a more critical phase in his thinking. His exposition of the "transcendent purpose" is in his book *The Purpose of American Politics*, written during the Kennedy years, another period of extreme self-adulation among responsible intellectuals. Morgenthau was the founder of the dominant tough-minded *realist* school of international affairs, which avoids sentimentality and keeps to the hard truths of state power. He was, incidentally, a very decent human being. He was one of the very few prominent scholars in these fields to oppose the Vietnam War on moral grounds, not on grounds of cost effectiveness, a stance that was extremely rare in intellectual circles, though not among the unwashed masses; by 1969, 70 percent of the public condemned the war as "not a mistake, but fundamentally wrong and immoral," words rarely seen or heard within the mainstream, across the spectrum.

Morgenthau was also a highly competent and honest scholar. While praising the "transcendent purpose" of America, he recognized that the historical record is radically inconsistent with it. But he explains that we should not be misled by that discrepancy. In his words, we should not "confound the abuse of reality with reality itself." Reality is the unachieved "national purpose" revealed by "the evidence of history as our minds reflect it." The actual historical record is merely the "abuse of reality," which is of only secondary interest—at least to those who are holding the clubs.

The principles continue to guide intellectual practice, including most scholarship. Just keeping to the present, the most extensive scholarly article on "the roots of the Bush doctrine" appears in the prestigious U.S. journal *International Security*. It opens with these words: "The promotion of democracy is central to the George W. Bush administration's prosecution of both the war on terrorism and its overall grand strategy." In Britain's leading journal of international affairs, the major article on the same topic extends the scope of the thesis. The author writes that "promoting democracy abroad" has been a primary goal ever since Woodrow Wilson endowed U.S. foreign policy with a "powerful idealist element," which gained "particular salience" under Reagan and has been taken up with "unprecedented forcefulness" under George W. Bush. Such declarations are close to uniform in scholarship. In journalism and intellectual commentary, they are taken to be the merest truisms. To be sure, there are critics who argue that it is important not to go too far in our idealism, because it can be harmful to our interests. To take one significant example, the veteran commentator of the *Washington Post*, David Ignatius, a former editor of the *International Herald Tribune*, warns that the "idealist-in-chief" of the Bush administration might be "too idealistic—his passion for the noble goals of the Iraq war might overwhelm the prudence and pragmatism that normally guide war planners." He was referring to Paul Wolfowitz, now free to pursue his passion for democracy and development as head of the World Bank. The many accolades to Wolfowitz at the time of his appointment scrupulously evaded his record, which is one of utter contempt for democracy and human rights, easily documented—but it is the abuse of history, hence irrelevant.

The scholarly and journalistic presentation of the reigning thesis also carefully evades empirical evidence, a wise decision, because it is overwhelmingly to the contrary. This is obliquely recognized in serious scholarship that focuses specifically on democracy promotion—Bush's "messianic mission," as it is described in the liberal press. The most prominent scholar-advocate of the cause of democracy promotion is Thomas Carothers, director of the Democracy and Rule of Law Project at the Carnegie Endowment. He identifies himself as a neo-Reaganite, agreeing with general scholarship that Wilsonian idealism took on particular "salience" under Reagan's leadership. A year after the invasion of Iraq, he published a book reviewing the record of democracy promotion by the United States since the end of the cold war. He finds what he calls "a strong line of continuity" running through all administrations: Bush I, Clinton, and Bush II: democracy is promoted by the U.S. government *if and only if* it conforms to strategic and economic interests. All administrations are "schizophrenic" in this regard, he concludes, with puzzling consistency. Again, actual history is the abuse of reality, so it can be ignored in responsible circles.

Carothers also wrote the standard scholarly work on democracy promotion in Latin America in the 1980s, in part from an insider's perspective. He was serving in the Reagan State Department in the programs of democracy promotion. Carothers regards these programs as sincere but a failure. Like Morgenthau, he is an honest scholar, and he points out that the failure of the programs was systematic. Where U.S. influence was least, in South America, progress toward democracy was greatest, despite Reagan's attempts to impede it by embracing right-wing dictators. Where U.S. influence was strongest, in the regions nearby, progress was least. The reasons, he explains, are that Washington would tolerate only "limited, top-down forms of democratic change that did not risk upsetting the traditional structures of power with which the United States has long been allied [in] quite undemocratic societies."

In short, the strong line of continuity goes back a decade earlier, to the Reagan years, when the "powerful idealist element" in traditional U.S. policy gained "particular salience," according to Western scholarship. Nonetheless, the dedication of our leaders to the principle is beyond question, and today we must believe that Bush is pursuing his messianic vision of creating a sovereign, democratic Iraq and bringing democracy everywhere, ignoring the overwhelming consistency of the record, the abuse of history. I am surprised to see the doctrine echoed even in the Arab world, where people surely should know better.

In fact, the strong line of continuity goes back much farther. Democracy promotion has always been proclaimed as a guiding vision, but it is not even controversial that the United States regularly overthrew parliamentary democracies, often installing or supporting brutal tyrannies: Iran, Guatemala, Brazil, Chile, a long list of others. There were cold war pretexts, but they regularly collapse on investigation. I will not insult your intelligence by recounting how Reagan brought democracy to Central America in the course of the "war on terror" that the Reagan administration declared when it took office in 1981—quickly becoming an extraordinary terrorist war that left hundreds of thousands of corpses in Central America and four countries in ruins.

The paradoxical character of policy is also recognized at the dovish extreme of the policy spectrum, where it elicits regret but is recognized to be unavoidable. The basic dilemma facing policymakers was expressed by Robert Pastor, a progressive Latin America scholar and President Carter's national security advisor for Latin America. He explains why the administration had to support the murderous and corrupt Somoza regime in Nicaragua, and when that proved impossible, to try at least to maintain the U.S.-trained National Guard even as it was massacring the population "with a brutality a nation usually reserves for its enemy," in his words, killing some forty thousand people. The reason was the familiar one: "The United States did not want to control Nicaragua or the other nations of the region," he writes, "but it also did not want developments to get out of control. It wanted Nicaraguans to act independently, *except* when doing so would affect U.S. interests adversely."

Once again we find the dominant operative principle, illustrated copiously throughout history: policy conforms to expressed ideals only if it also conforms to interests. The term "interests" does not refer to the interests of the domestic U.S. population but rather the "national interest"—the interests of the concentrations of power that dominate the domestic society. That truism is often derided by respectable opinion as a "conspiracy theory," or "Marxist," or some other epithet, but it is readily confirmed when subjected to inquiry. In a rare and unusually careful analysis of the domestic influences on U.S. foreign policy, published in the *American Political Science Review* a year ago, two prominent political scientists find, unsurprisingly, that the major influence on policy is "internationally oriented business corporations," though there is also a secondary effect of "experts," who, they point out, "may themselves be influenced by business." Public opinion, in contrast, has "little or no significant effect on government officials." As they note, the results should be welcome to "realists" such as the influential progressive public intellectual Walter Lippmann, who "considered public opinion to be ill-informed and capricious [and] warned that following public opinion would create a 'morbid derangement of the true functions of power' and produce policies 'deadly to the very survival of the state as a free society,'" in his words. The "realism" is scarcely concealed ideological preference. One will search in vain for evidence of the superior understanding and abilities of those who have the major influence on policy, apart from protecting their own interests—Adam Smith's neglected truism.

I will not tarry on how Wilsonian idealism and love of democracy was actually exercised, with devastating effects that remain until today, particularly in Haiti, once the richest colony in the world and the source of much of France's wealth, now decaying in misery and likely to become uninhabitable before too long, thanks to French brutality and avarice, which were then carried forward by Wilsonian idealism up through Clinton and Bush. And Haiti is far from the only case. We may recall that Wilson's high-minded passion for self-determination had a qualification: it did not apply to people "at a low stage of civilization," he explained, as in the Middle East, where these defective creatures must be given "friendly protection, guidance, and assistance" by the colonial powers that had tended to their needs in past years, in his words. Wilson's famous Fourteen Points held that in questions of sovereignty, "the interests of the populations concerned must have equal weight with the equitable claims of the government whose title is to be determined," the colonial ruler. Posturing aside, Wilson scarcely departed from Churchill's doctrine after World War II, when he advised that

> the government of the world must be entrusted to satisfied nations, who wished nothing more for themselves than what they had. If the world-government were in the hands of hungry nations, there would always be danger. But none of us had any reason to seek for anything more. The peace would be kept by peoples who lived in their own way and were not ambitious. Our power placed us above the rest. We were like rich men dwelling at peace within their habitations.

No sentimentalist, Churchill knew well how Britain's wealth and peace had been obtained. Speaking in secret to his cabinet colleagues on the eve of World War I, Churchill explained that

> we are not a young people with *an innocent record and* a scanty inheritance. We have engrossed to ourselves . . . an *altogether disproportionate* share of the wealth and traffic of the world. We have got all we want in territory, and our claim to be left in the unmolested enjoyment of vast

> and splendid possessions, *mainly acquired by violence, largely maintained by force*, often seems less reasonable to others than to us.

Churchill published these remarks a decade later but made sure to delete the offending passages, italicized above, which were only discovered fairly recently.

Churchill's sensible and realistic stance illustrates one of the many reasons for regarding the fabled "American exceptionalism" with some skepticism. The doctrine appears to be close to a historical universal, even including the worst monsters. Aggression and terror are almost invariably portrayed as self-defense and dedication to inspiring visions. The Japanese emperor Hirohito was merely repeating a broken record in his surrender declaration in August 1945 when he told his people that "We declared war on America and Britain out of Our sincere desire to ensure Japan's self-preservation and the stabilization of East Asia, it being far from Our thought either to infringe upon the sovereignty of other nations or to embark upon territorial aggrandizement." If Asians have a different picture, it shows that they are backward and uncivilized people—a leading source of tension in Asia right now. From Japan's perspective, Asians who worry about such ancient history as the Nanjing massacre, biological warfare, and other atrocities are "naughty children who are exercising all the privileges and rights of grown ups" and require "a stiff hand, an authoritative hand," to quote the description of Latin Americans by Secretary of State John Foster Dulles, though he advised President Eisenhower that to control the naughty children more effectively, it may be useful to "pat them a little bit and make them think that you are fond of them."

That stance extends worldwide and has recently been announced with regard to China. A few weeks ago, as President Hu Jintao was about to visit Washington, the respected commentator Frederick Kempe explained in the *Wall Street Journal* that "Americans aim to show Hu how his country can act as a 'responsible stakeholder,'" joining the United States and its allies in adherence to international law, principles of world order, and civilized behavior. The stance and commentary elicit no ridicule, just as Western intellectuals soberly observe the demand of the United States and United Kingdom that *Iran* end its interference in Iraq—rather like Hitler's condemnation of U.S.-UK interference in peaceful occupied Europe. Though the matter passed without notice in the media, we can be confident that shivers went up the spines of Washington planners when President Hu left Washington for Saudi Arabia, returning King Abdullah's visit to Beijing. And they surely watched with trepidation as Saudi Arabia became China's largest trade partner in West Asia and North Africa, with bilateral trade reaching $16 billion in 2005. China's influence is also growing in Washington's backyard, Latin America, which no longer can be regarded as a reliable source of oil and other resources.

While the law-abiding states of the West seek to civilize China, the most respected strategic analysts are calling on China to lead a coalition of peace-loving states to counter U.S. aggressive militarism, which they warn is driving the world toward "ultimate doom." They appeal to China because of all the nuclear powers it "has maintained by far the most restrained pattern of military deployment." China has also led the efforts at the United Nations to block the unilateral U.S. refusal, since the Clinton years, to preserve orbital space for peaceful purposes, now extended by the Bush administration to the doctrine of *ownership of space*, leaving every corner of the world subject to near-instantaneous lethal attack without the need for military bases. It is well understood that these moves, which are already eliciting the anticipated reaction among potential targets, pose a severe threat to the survival of the species. I happen to be quoting from the journal of the American Academy of Arts and Sciences, but the basic conclusions are widely shared among strategic analysts. Nevertheless, it is China that is to be portrayed to the public as another of those "naughty children" to whom we must teach manners.

The universal stance of "exceptionalism" extends even to the figures of the highest intelligence and moral integrity. Consider John Stuart Mill, who wrote the classic essay on humanitarian intervention that is presumably studied in every serious law school in the West. His essay raised the question of whether England should intervene in the ugly world or whether it should keep to its own business and let the barbarians carry out their savagery. His conclusion, nuanced and complex, was that England *should* intervene, even though by doing so, it will endure the "obloquy" and abuse of Europeans, who will "seek base motives" because they cannot comprehend that England is "novelty in the world," an angelic power that seeks nothing for itself and acts only for the benefit of others. Though England selflessly bears the cost of intervention, it shares the benefits of its labors with others equally. Mill's immediate concern was India. He was calling for the expansion of the occupation of India to several new provinces.

The timing of the article is revealing. The essay appeared in 1859, immediately after what British history calls the "Indian mutiny": the first rebellion in India, which Britain put down with extreme savagery. All of this was very well known in England. There were parliamentary debates and a huge controversy. There were people who opposed the crimes: Richard Cobden, a really committed liberal in the old-fashioned sense, and a few others. Mill knew all about it. He was corresponding secretary of the East India Company and was following it all closely. The purpose of the expansion of British power over India was to obtain a monopoly over opium so that England could break into the Chinese market, which British exporters could not penetrate because Chinese goods were comparable and they didn't want British goods. So the only way to sell them was to force the Chinese to become a nation of opium addicts at the point of a gun. Mill was writing right at the time of the Second

Opium War, which established Britain's extraordinary narcotrafficking enterprise, which I already mentioned, and did enable England, and later others, to subjugate China and break into its markets. The profits were an enormous boost to British capitalism, right during the period of much pious rhetoric about free trade. In the same essay, Mill also praised the civilizing mission of the French, then underway in North Africa under the orders of the French minister of war, who called for "exterminating the indigenous population," with little dissent among the engaged intellectuals.

Without proceeding, exceptionalism seems to be close to universal. I suspect if we had records from Genghis Khan, we might find the same thing. Nonetheless, it is the responsibility of intellectuals to recognize the doctrine as a driving force of U.S. policy, even if they sometimes criticize the idealism as excessive—as in the case of the "idealist-in-chief" in charge of the "noble war" in Iraq.

Perhaps I can end with some additional truisms. The great soul of power extends far beyond states. Slavery was defended with arguments similar to those of Mill: it was a selfless exercise of benevolence to poor people who needed the care of their masters, and if they were "naughty children," they sometimes needed the rod, or worse, for their own benefit. Some of the arguments of the slave owners were never really countered. To rephrase one anachronistically, suppose I buy a car and you rent one. A year later, which one is likely to be in better shape? Mine, surely, because I protect my capital investment, while you can discard yours and rent another one. Suppose now that I own workers and you rent them. Who is more benign? That argument had considerable force for working people who fought for the Union in the U.S. Civil War, under the banner that wage slavery is little different from chattel slavery and that it is an attack on fundamental human rights to reduce people to the level where they must rent themselves to survive—a perception so common that it was even a slogan of the Republican Party. That was 150 years ago to be sure; we have become more civilized since, and we are not supposed to see such social-economic arrangements as an attack on the most fundamental human rights.

The great soul of power extends far beyond, to every domain of life, from families to international affairs. And throughout, every form of authority and domination bears a severe burden of proof. It is not self-legitimating. And when it cannot bear the burden, as is commonly the case, it should be dismantled. That has been the guiding theme of the anarchist movements from their modern origins, adopting many of the principles of classical liberalism after it had been wrecked on the shoals of capitalism, as the anarchist historian Rudolf Rocker wrote. Keeping to the international arena, the contemporary system of nation-states was established with extreme violence and sadism, which for centuries made Europe the most savage region of the world, which ended in 1945, when it was recognized that the next time Europeans play the game of mutual slaughter, it will be the last. Eliminate that factor, and the thesis of "democratic peace" that is much prized in political science loses its core empirical support. There is good reason to believe that the culture of savagery that evolved in the course of establishing the state system was a significant factor in Europe's conquest of the world. The inhabitants of Asia and the Western Hemisphere were "appalled by the all-destructive fury of European warfare," the military historian Geoffrey Parker observes, enabling "the white peoples of the world [to] create and control" history's first "global hegemony." The British historian V. G. Kiernan comments aptly that "Europe's incessant wars" were responsible for "stimulating military science and spirit to a point where Europe would be crushingly superior to the rest when they did meet." Imposition of the European-style state system on the conquered lands was carried out with comparable brutality and lies at the root of indescribable horrors, including the conflicts that rage today.

One of the most healthy recent developments in Europe, I think, along with the federal arrangements and increased fluidity that the European Union has brought, is the devolution of state power, with the revival of traditional cultures and languages and a degree of regional autonomy. These developments lead some to envision a future Europe of the regions, with state authority decentralized. No one wants to reconstruct the Ottoman Empire, with its brutality and corruption, but that should not prevent us from recognizing that in some respects it had the right idea: leaving people alone to manage their own affairs, without strict borders and with substantial peaceful interaction at local and regional levels, a conception that should have particular resonance, and memories, in the complex societies of the Levant, I think—and has merits far beyond. To strike a proper balance between citizenship and common purpose on the one hand, and communal autonomy and cultural variety on the other, is no simple matter, and questions of democratic control of institutions of course extend to other spheres of life as well, but such questions should be high on the agenda of people who do not worship at the shrine of the Great Soul of Power, people who seek to save the world from the destructive forces that now literally threaten survival and who believe that a more civilized society can be envisioned and even brought into existence—the cause to which Edward devoted his life and work.

PART II

INTRODUCTIONS

3

CHOMSKY, FRANCE, REASON, POWER

JEAN BRICMONT AND JULIE FRANCK

> When I first read Noam Chomsky, it occurred to me that his marshalling of evidence, the volume of it, the relentlessness of it, was a little—how shall I put it?—insane. Even a quarter of the evidence he had compiled would have been enough to convince me. I used to wonder why he needed to do so much work. But now I understand that the magnitude and intensity of Chomsky's work is a barometer of the magnitude, scope, and relentlessness of the propaganda machine that he's up against.
>
> ARUNDHATI ROY, "The Loneliness of Noam Chomsky"

Among contemporary intellectuals, Noam Chomsky is probably the one most famous throughout the world and yet least known in France. His international reputation derives from his unremitting effort in the field of political analysis and commentary and from the revolution in the cognitive sciences he launched with his groundbreaking work in generative linguistics. If he is less well known in France, that is because he can't easily be situated on the grid (although its confines are not narrow) of French thought; one has to make an effort to step outside the usual categories in order to understand him. Chomsky is not and never has been a communist, Trotskyist, Maoist, third-world advocate, or even a Marxist in the strict sense of the term, but neither is he a liberal in the sense that Raymond Aron and Jean-François Revel are liberals. He is not close to Bergson, Sartre, Foucault, Althusser, Deleuze, Lacan, or Bourdieu, and he is not inspired by Nietzsche, Freud, or Heidegger. He is not a structuralist and has no sympathy for the *nouveaux philosophes* or postmodernism. He is not even really very close to most strands of French anarchism.

Chomsky is too original to attach himself to any school, but if we wish to locate his roots, we will find them in intellectual currents that are just about as little known in France today as Chomsky himself: Cartesian linguistics; Hume; Enlightenment materialism; the thought of Humboldt, Russell, and the logical positivists; anarchists like Kropotkin, Abad de Santillan, and Rocker; as well as the anti-Leninist current within Marxism based on "workers' councils" (Korsch, Mattick, Pannekoek).

Our purpose in this volume is to present an overview of Chomsky's thought to the French public and to pay homage to the importance of his oeuvre by highlighting the influence it has had in areas as diverse as theoretical linguistics, the study of the normal and pathological development of language, the philosophy of mind and the cognitive neurosciences, the analysis of ideology and power and of the media and the foreign policy of the United States, and in the discussion of freedom of expression, education, ethics, and political action. Over fifteen authors from various countries and different disciplines have helped us assemble this collection, which reflects the importance of Chomsky's work throughout the world and across disciplinary boundaries.

In this introduction, we supply the reader with an outline for understanding Chomsky's thought. The main axes of his scientific and political work are sketched out, and the reader is directed to the different chapters below for detailed and fully developed presentations. Our purpose is not to justify or certify Chomsky's ideas but to supply a backdrop against which to read them, emphasizing what it is that makes them original and what it is that makes them hard to accept in France.

A Classical Rationalist

Chomsky takes a scientific approach to the world; in other words, he adheres to the idea that what is accessible to the human being, apart from knowledge about everyday life, must be apprehended through the scientific method. Like the majority of the scientists and philosophers of science who have reflected on this question, Chomsky knows that the "scientific method" is not something that can be described by an algorithm or applied in automatic fashion. Nonetheless, there is a difference between a rational approach based on observation and logic and a religious approach that does not attempt to determine the true or false character of ideas on the basis of factual data or reasoning but rather trusts them implicitly because they emanate from an authority recognized as such. From this perspective, even nominally secular approaches can be religious, because the authority in question may, for example, be intellectual or political. For Chomsky, a good part of "Marxism" belongs to the history of organized religion, which is not at all to say that there are not interesting ideas to be found in Marx or his successors, but that the relation of his partisans to Marx's ideas has often been of a religious kind. Take the example of Louis Althusser, who left his mark on a whole generation of intellectuals, given that among his friends and students we find people as different as Michel Foucault, Jacques Derrida, Bernard-Henri Lévy, Alain Badiou, Régis Debray, and Dominique Lecourt: there is no doubt that Althusser's "reading" of Marx, in which one never encounters a number or a fact, was more religious than scientific in nature.

The religious attitude is not the only characteristic of the nonscientific approach to reality. Another feature is the excessive reliance on one's own intuition or introspection, as we see for example in the different schools of psychoanalysis.

Chomsky's stance vis-à-vis the scientific pretentions of various social and human sciences is well illustrated by the following anecdote: In 1965, during the Vietnam War, a conference was organized in order to bring together the views of social-science researchers and representatives of "various theological, philosophical, and humanist traditions" so as to "find solutions that are more consistent with fundamental human values than current American policy in Vietnam has turned out to be." Chomsky's reaction was blunt: "The only debatable issue, it seems to me, is whether it is more ridiculous to turn to experts in social theory for general well-confirmed propositions, or to the specialists in the great religions and philosophical systems for insights into fundamental human values." And he added: "If there is a body of theory, well tested and well verified, that applies to the conduct of foreign affairs or the resolution of domestic or international conflict, its existence has been kept a well-guarded secret."[1] Chomsky later noted that this remark applies to human affairs in general. What is important in his remark is the words "well tested and well verified" and the notion of "application to human affairs." Chomsky does not deny that disciplines such as history, economics, sociology, and psychology yield interesting results, but he does doubt that there exists among them a well-tested corpus of knowledge, sufficiently developed (i.e., comparable to the way physics is applied to engineering or biology to medicine, for example) to really solve human problems.

The Naturalization of Mind

One may ask what it is that justifies such skepticism. After all, doesn't Chomsky himself work in linguistics, and is that not a human science? Moreover, on what, then, does he base his political analyses? We shall return to the second question in the following sections, but to address the first, it should be noted that there are several differences between the perspective Chomsky adopts toward his own discipline and the one that often dominates in the human sciences. Chomsky sees his own practice of linguistics as part of psychology, and psychology as, ultimately, part of biology—a perspective sometimes called the naturalization of mind, which takes seriously the idea that the human mind is a natural object that must be studied using the same scientific methods used to study the rest of the natural world.

The first step in the scientific approach is perhaps to ask questions about phenomena that seem "obvious" and not to immediately accept ready-made answers: why do apples fall from trees; where do the various species come from; how do infants learn what they know, in particular their mother tongue? In the last case, the common-sense answer is that this knowledge comes from the family environment or from a "process of socialization" or "acculturation." A slightly more sophisticated version of this idea, advanced by empiricist thinkers, consists of regarding the mind at birth as a *tabula rasa*, upon which experience proceeds to "inscribe" what becomes the knowledge possessed by the child or the adult. But one may still wonder how this works exactly. How does the human mind go about transforming what is no more than noise and light (the sounds and images coming in from the environment) into ideas and concepts? Why does it do so when the brain of an ape or a cat, even though subjected to exactly the same sounds and images as an infant, does not? The reply might be that the infant, unlike the animals, possesses certain "general learning mechanisms" (which amounts to an admission that the *tabula* is not, in fact, *rasa* after all), but you might then pursue the line of reasoning further and ask how these mechanisms function. In particular, what meaning can we give to the idea that there exist "all-purpose" learning mechanisms that work equally well on absolutely any "content" to be learned? If the infant were placed in the right environment, would it be able to learn a computer language, for example? And if not (as seems likely), why not? What specific thing in its mind allows it to learn a natural language but not a computer language?

This is the type of question that Chomsky asks: in particular, what are the innate mental structures that allow the child, when immersed in a given linguistic environment, to construct the grammar of its mother tongue on the basis of data that are highly fragmented, not always correct, and not explicit—in other words, the things it hears around it? This argument from the *poverty of stimulus*, which is regarded as one of the main arguments in favor of the thesis that there exists an innate structure for language, is discussed by Cedric Boeckx and Norbert Hornstein in their contribution to this volume. According to Chomsky, the difficulty of accounting for language learning if we start from a *tabula rasa* is so great that it is necessary to postulate the existence of a specific organ for this purpose, capable of calculating internal representations on the basis of the information with which the outside world supplies it. The notion of a language organ is used here to designate a biological entity with a specific function and a structure that enables it to carry out that function (whether it be a particular area, or a network, in the brain). Just as we see with our eyes and hear with our ears, we handle language with a language organ capable of managing stimuli of a specifically linguistic kind.

This conception is radically different from that traditionally advanced by the current of thought flowing from empiricist philosophy. The empiricist tradition survives in experimental psychology in the behaviorist school (initially represented by scientists such as B. F. Skinner), which conceives of learning as an associative process. The prototypical illustration of this process is that of Pavlov's dogs learning to salivate in response to the sound of a bell when this sound occurs often enough in conjunction with food. This behavior illustrates the fact that the dog has created a mental representation arbitrarily linking the bell to food simply on the basis of their linkage in the environment. The infant is likewise thought to learn its language by associating the sounds it hears with the objects present in its immediate surroundings, to which it may assume that the sounds refer. Language is thus conceived as a stock of mental representations linking sounds and concepts. In this conception, a unique general device capable of encoding concomitant experiences is supposed to lie at the origin of the development of the different mental faculties, such as language, spatial representation, and arithmetic.

In his chapter, Randy Gallistel describes the revolution to which Chomsky's reconceptualization of learning led in the domain of cognitive neuroscience and illustrates it by using the example of the neurobiological modeling of the way bees learn the solar ephemerides. On the basis of his work, he establishes a parallel between the approaches to this function and to the language function. In both cases, learning mobilizes a genetically determined learning organ equipped with a certain number of parameters that must be set by experience. The structure of this organ is specific to the domain of stimuli that it is capable of handling. The language system is capable of transforming the linguistic information the infant receives into a representation of the particular grammar of the language to which it is exposed. The nervous system of the bee possesses a neural circuit capable of transforming the information with which its circadian clock supplies it into a representation of the sun's azimuth. Just as it is enough for the language system to be exposed to a limited number of phrases in order to set the parameters of the language on the basis of the internal innate structure that is its universal grammar, it is enough for the bee to observe the sun at a few different times of day in order to set the parameters of the innate function of the ephemerides.

The repercussions of Chomsky's naturalist approach in the study of spatial orientation in animals and infants are illustrated here by Elizabeth Spelke. Research carried out on rats shows that they are capable of associating a taste and a sensation of nausea and of associating an object and a physical blow but that they are completely incapable of associating a taste to a blow or an object to nausea. Rats therefore possess a system allowing them to apprehend the properties of edible things and another system allowing them to apprehend the mechanical properties of objects. These two systems are totally sealed off from each other. Spelke's research shows that infants too are equipped with independent systems for dealing with, for example, colors and geometrical information. However, unlike rats, children who have already acquired language show a capacity to combine information from these two systems.

So, according to Spelke, language allows humans to combine modular systems productively and thus endows them with a freedom of action unknown among the other animals.

This handful of examples of research in cognitive neuroscience illustrates the pertinence of the naturalist approach to mind advocated by Chomsky almost fifty years ago. At the time, his stance challenged the empiricist tradition going back to the seventeenth century and largely dominant in the human sciences. With his critique of Skinner's *Verbal Behavior*, Chomsky became one of the leading figures of the cognitive revolution. Today, conditioned reflexes have given way to the more sophisticated approach known as connectionism in the cognitive sciences, in which the mental faculties are modeled as statistical learning systems based on the general principles of association, memorization, generalization, and analogy. This approach to the human mind still holds a dominant place in psychology and the cognitive sciences in the French-speaking world.

The Scientific Study of the Language Faculty

For Chomsky, the grammar of a language must be a description of the intrinsic competence of the ideal speaker-hearer, in other words her competence independent of the hazards of her concrete performance, which is subject to a number of external factors that often cause her to produce incomplete and sometimes even incorrect sentences. This focus on the "internal language" emanating from the "language faculty" and reflecting its properties is what distinguishes the linguistic approach initiated by Chomsky from the study of "language" as a social object, which is how Ferdinand de Saussure envisaged it. It is also what drives him to attach linguistics to the cognitive sciences rather than to the humanities; in this respect, the fact that Chomsky's department of linguistics is part of the Massachusetts Institute of Technology is not insignificant. This approach to language, albeit predominant in North America and in certain regions of Europe, has very little purchase in France, where linguistics remains the study of language as medium of communication and a reflection of culture, subject to variation among individuals and cultures and often associated with literary study.

In their contribution to this volume, Cedric Boeckx and Norbert Hornstein distinguish three phases in the development of generative grammar begun by Chomsky at the end of the 1950s. During the first combinatory phase, Chomsky mounted a challenge to finite-state grammars as being incapable of accounting for nonlocal dependencies between the words in a sentence. Generative grammar would need to elaborate a formal system making it possible to engender the infinite set of well-formed sentences of the language and only those. The cognitive era gave a new impulse to this program by grounding it in a real program of biolinguistics: now the theory of grammar needed to account for how the infant is able, starting with the fragments of language it hears, to acquire an adult linguistic competence. Chomsky put forward the idea that the infant is born equipped with an array of universal principles and that these principles entail a certain number of unspecified parameters, which the infant must set through its experience of the particular language to which it is exposed. The aim of linguistics becomes that of identifying these principles, which are common to all languages, and the parameters of variation. Language acquisition is conceived, from this perspective, as a process of parameter setting. Last, the Minimalist Program rests on the hypothesis that the central computational system of human language, syntax, represents an optimal solution to the internal function of language, which is to link sounds to meanings. The aim is now to abstract the individual computational algorithms as much as possible, reducing them to a minimal set that explains the internal structure of language.

The guiding rationale of generative grammar throughout these different phases is the identification of the nucleus of properties that the several thousand languages produced by mankind have in common. This nucleus, which forms universal grammar, is what leads Chomsky to say that if a Martian landed on Earth, he would probably think that we all speak the same language. The most obvious of these properties is recursion, that is, the fact that a unit may contain an infinite number of units of the same type: a sentence, for example, may contain an infinite number of sentences (X thinks that Y thinks that Z thinks that . . .). Recursion allows the system to produce an infinite number of sentences with a finite number of words (even though we are incapable in practice of dealing with very long sentences; see the next section on language processes).

Another fundamental property of natural languages is structure dependence: the sentence cannot be reduced to a sequence of words related in purely linear fashion. For example, in order to form a question out of the sentence "the dog drinks the milk of the cat," it is necessary to know that "the milk of the cat" forms the entity that is the object of the question, which thus must be replaced by the interrogative word *what* ("what does the dog drink?"), and that this entity is not "the milk," which, if it were replaced by the interrogative *what*, would produce the incorrect sentence "what does the dog of the cat drink?" Another example of this structure dependence is agreement in number. In the sentence "the dog drinks the milk," the verb *drink* is in the singular because *the dog* is singular in number. But the same constraint applies in the sentence "the neighbors' dog that regularly bites passers-by drinks the milk." Thus it is not the proximity of the two words in the sentence that determines their agreement in number but rather their structural position: the verb agrees with the grammatical subject. The system must therefore be endowed with the capacity to identify structural entities behind the linearity of language.

In the example above, the constituent elements changed place to form an interrogative sentence. We could also form the sentences "it's the milk of the cat that the dog drinks," "the milk of the cat is drunk by the dog," "the milk of the cat that the dog drinks is fresh," and so on. Hence the constituent elements may be shifted around, but always in accordance with constraints that occur in all languages. One of these constraints is locality: movement cannot shift a constituent element to an excessively remote position. For example, it is not possible to move an interrogative element outside a relative clause: on the basis of the phrase "the cat pursued the dog [that was drinking its milk]," we cannot form a question about what the dog was drinking: "what did the cat pursue the dog [that was drinking]?" is incorrect.

The facts discussed so far bear on syntax, which is the domain of linguistics that deals with the possible combinations of words: generative grammar has made it possible to isolate universal principles that govern the functioning of syntax. But the influence of generative grammar extends to semantics as well, which is the domain of linguistics that studies the meaning of expressions. In his contribution, Gennaro Chierchia shows how the research paradigm of generative grammar has reformulated the question of the relation between language and thought posed by Whorf and Sapir. Semantics is no longer conceived as an arbitrary relation between knowledge of the world and language but as a set of universal representations linking the structures of denotation to syntactic representations. The author illustrates how the manner in which the different languages structure the domains of discourse is universal through his research on a property of nouns: their character of being "countable" (for example, "table") or "uncountable" (for example, "salt"). This property is rooted in the grammar of nouns: for example, uncountable nouns are harder to put into the plural ("the salts") or to combine with an indefinite article ("a salt"). For that matter, research shows that a three-month-old baby already categorizes the world into solid objects and substances, which suggests that it has an innate knowledge of the fact that the world is structured in this way. This universal distinction between objects and substances does not however entirely coincide with the grammatical distinction: the same noun can be countable in French ("*cheveu*") and uncountable in English ("hair"), just as two closely related nouns in the same language can either be countable ("carrot") or uncountable ("broccoli"). In other words, the uncountable/countable property of nouns correlates with the substance/object distinction in the real world, but it is formal: it cannot be reduced to this real-world distinction; it is specific to grammar. The author proposes a model that links the categorization of the real world and the categorization of grammar through universal principles of denotation.

The Scientific Study of Language Processing

Chomsky introduced an important distinction between language competence (internal language) and performance, that is, the processes put to work when we speak or understand our language. Whereas our competence theoretically permits us to generate or comprehend sentences implying infinite recursion ("the son of the son of the son of the son of the son . . . of John is born"), we are incapable of doing so in practice. The reason is that performance is subject to cognitive constraints (memory, attention span, etc.) that limit how competence is put into practice. Competence is often thought of as the object of linguistics and performance as the object of psycholinguistics.

Are the ways in which language is actually handled therefore of no interest for linguistics? The answer is no, as long as we hypothesize that competence is reflected in performance even though it does not entirely determine it. An illustration of the fact that this hypothesis is well grounded comes from the systematic analysis of slips of the tongue. It is the case that although speakers regularly alter and mix up words or phonemes, creating what are called in English "Spoonerisms" (for example, "tips of the slung"), and sometimes produce neologisms (for example, "hyposthesize"), they almost never produce ungrammatical sentences, which demonstrates the power of the grammatical machinery. Performance thus appears to be strongly constrained by grammar, and the study of it proves to be extremely fruitful for the understanding of competence.

As Yosef Grodzinsky shows in his contribution, it is also possible to study competence in the laboratory using various experimental techniques. This is the task of psycholinguistics: it comes down to isolating certain variables of competence in order to manipulate them experimentally while holding constant a maximum number of performance variables. Grodzinsky presents arguments from neurolinguistics suggesting that universal grammar is represented in the brain, exactly as is the visual system, for example. Research in aphasiology and recent cerebral imaging studies of healthy speakers both point toward a specific involvement of the Broca's area of the brain in the processing of syntactic movement.

Methodological Naturalism and Liberty

It is thus innate mental structures that fundamentally interest Chomsky; indeed, it is on the conception of the structures forming the hard core of human nature that the linguist and the political militant in him converge—at a very abstract level, as we shall see below. Just like our linguistic structures and the various cognitive functions with which we are endowed, our structures of moral and esthetic thought are part of the mental properties common to the human species and firmly planted in our universal biological heritage.

But in order to understand him clearly, it is important to emphasize that Chomsky operates a sharp distinction between the study of the cognitive functions and the study of the manner in which these intellectual tools are put to work in human action. Whereas it is possible to study mental faculties like language, spatial representation, or mathematical reasoning using the scientific approach, Chomsky is skeptical about the idea that one could really explain scientifically *why* an individual makes use of his linguistic competence to say one thing or another, his logical competence to think this or that, or his sense of social relations to act in a given fashion. So while Chomsky has contributed more than anyone else to the scientific approach to cognitive capacity, he rejects the possibility of examining questions relative to freedom of action, what we may call *intentionality*, following this approach.

In other words, Chomsky does not think that the scientific method makes it possible to answer all the questions human beings ask. He distinguishes between what he calls *problems* and *mysteries*. The former concern the *capacities* of a human being, the latter what she *does* with her capacities. The former are accessible to scientific study; the latter are questions to which we are unable to furnish an answer, on account of the limitations intrinsic to the biological constitution of our minds. To grasp the point, let us imagine a rat in a labyrinth, who, to escape, must turn right each time the number of intersections it has just passed is a prime number. The rat will obviously never learn the secret of the labyrinth, because the concept of "prime number" is inaccessible to it. Likewise it is possible—indeed, highly probable—that certain questions human beings ask themselves do have answers but that these answers are inaccessible to us because the concepts necessary to formulate them are not among the concepts that our innate mental capacities allow us to acquire. This is why Chomsky often says that literature teaches us more than science about certain questions, which does not mean that we ought to do science in a literary mode (however much that might please certain philosophers and psychoanalysts) but that literature offers us answers, vague and intuitive as they may be, to those questions to which the human mind is incapable of replying scientifically.

This stance leads Pierre Jacob, in his contribution, to discuss the limits of Chomsky's naturalism. As the author points out, naturalism in no way leads Chomsky to subscribe to the orthodox approach in the philosophy of science, which is called externalism and which regards intentionality as a property emanating from the interaction between the cognitive system and the environment rather than as a property intrinsic to the cognitive system. For Chomsky, the human being, unlike a machine, is free to choose; his intentional action is in a certain fashion undetermined. He may be incited to act in one way rather than another, but he can always choose to act differently. In her contribution, Elizabeth Spelke sets forth a possible approach to the scientific study of this freedom of action, which Chomsky thinks is inaccessible to our rational comprehension. On the basis of her recent work on the system of spatial representation, she suggests that language offers the human mind the possibility of combining, in a creative manner, the representations issuing from the different specialized cognitive systems and that the freedom of action felt by every human being might well flow from these combinations.

Chomsky's original stance, in which a naturalist and innatist approach to mental faculties rubs shoulders with the postulate that the human being has freedom of action in her environment, is strongly opposed to the idea that theories developed in the human sciences can, or could some day, "eliminate" the free and conscious subject by reducing her to language structures, to an effect of the unconscious, or to a pure product of her milieu or history. All these forms of reduction of freedom of action to a set of causal factors (whatever they may be) have had their hour of glory in one phase or another of French thought and certainly constitute an obstacle to the comprehension of Chomsky's thought.

A Nonrelativist Ethics

One of the best introductions to Chomsky's ethical stance can be found in his debate with Michel Foucault in 1971, filmed in a Dutch television studio.[2] For Foucault, whose position at the time was close to Marxist, or let's say Nietzschean-Marxist, "one makes [class] war to win, not because it is just." It is because the proletariat wants to seize power that it considers its war as just, not vice versa (51). For Foucault, "the idea of justice in itself is an idea which in effect has been invented and put to work in different types of societies as an instrument of a certain political and economic power or as a weapon against that power" (54).

Chomsky categorically rejects this outlook. As far as he is concerned, there exists "some sort of an absolute basis . . . ultimately residing in fundamental human qualities, in terms of which a 'real' notion of justice is grounded" (55). He adds that one must "give an argument that the social revolution that you're trying to achieve *is* in the ends of justice, *is* in the ends of realizing fundamental human needs, not merely in the ends of putting some other group into power, because they want it" (57). The rejection, or the bracketing, of the question of human nature naturally leads to "relativistic" attitudes like that of Foucault (and many other progressives), since if one eliminates moralities of the religious type because of their arbitrary character, the notion of justice can only be grounded on "fundamental human qualities"—in other words, on human nature. Since Chomsky allows that we do not know much about human nature, he also allows that our political attitudes are based more on our intuitions than on certainties.

It may seem surprising, from a political point of view, that Chomsky, adopting a firmly innatist attitude, devotes himself to the study of human nature, which he sees as a biological property: ideas of this kind are almost automatically linked, especially in France, to right-wing rather than left-wing thought. This is just one more obstacle to the understanding and acceptance of Chomsky's ideas. But on one hand, this stance is a consequence of the fact that he tries to understand the world as it is before trying to change it or pass any sort of value judgment. As he sometimes says, it is an objective analysis of the world that leads to radical conclusions, not any desire for change posited a priori, which then guides the analysis. On the other hand, in the interview published here, he explains in detail why he rejects the linkage of innatism to political conservatism. In fact, he thinks that it is the idea that the human being is infinitely malleable that is really "right-wing," because it legitimizes the action of the technocrats, the specialists in the human sciences, and more generally of what Bakunin called "the reign of *scientific intelligence*, the most aristocratic, despotic, arrogant and elitist of all regimes."[3] In addition, if the human being is infinitely malleable, why has it not been possible to convince slaves, in the inner sanctum of their moral conscience, that slavery is legitimate, or proletarians that capitalism is legitimate, or women that sexism is legitimate, or colonial peoples that colonialism is legitimate? On the contrary, it is solely in the name of fundamental human qualities, that is, of human nature, that one can oppose oppression in an intellectually coherent fashion or even give the term a meaning that is not entirely arbitrary.

In France, biological thought is also associated with racism. But from Chomsky's point of view, it points in precisely the opposite direction. He is interested in the *universal* properties of the human spirit, and he maintains that these are necessarily very rich (even though not immediately visible). That challenges the "empiricist" or "environmentalist" idea that reduces a person to his language, her religion, his culture, and that excludes in principle the idea of universal properties (which, let us note, can alone render possible communication between members of different cultures). In Chomsky's "biological" approach, to use a metaphor, there exists no "spirit of color," and this is a (modest) barrier *against* racism.

Finally, it is the importance that Chomsky accords to "fundamental human qualities" that, more than any other consideration (such as pragmatism), motivates his defense of freedom of expression. This defense caused him to be completely marginalized in France at the time of the "Faurisson affair." To this affair we shall return in detail in this volume, explaining exactly what took place and discussing with some precision the arguments for and against Chomsky's ideas on freedom of expression. We shall also review the numerous incoherences of Chomsky's principal French critic on this matter, Pierre Vidal-Naquet.

The Analysis of Ideology

On the conceptual plane, one of the principal contributions of Chomsky and Edward Herman to the comprehension of our societies comes from their analysis of ideological mechanisms, or what they call "brain-washing in freedom."[4] As the notion of ideology has given rise to so many learned analyses, it is interesting to observe how Chomsky addresses the matter. Here we will also find the answer to a query posed above: how does Chomsky combine his skepticism vis-à-vis the human sciences with the vast range of political analysis he has produced? When he analyzes American society or politics, Chomsky does not claim to be doing science, only to be exercising his common sense in light of a large quantity of information. This entails that he thinks his analyses are accessible to everyone, because they do not demand particular kinds of scientific knowledge—a stance diametrically opposed to that of many intellectuals in France, including ones on the left. On this, readers may consult the comparison sketched by Frédéric Delorca between Chomsky's views and those of Pierre Bourdieu.

In order to expound the vision Chomsky has of ideology, let us start with a banal observation: whenever some persons exercise their power over others, whether it be as proprietors, aristocrats, dictators, monarchs, bureaucrats, or colonists, they need a justifying ideology. And the abstract form of this justification is almost always the same: when X exercises power over Y, she does so "for Y's own good." In other words, power always presents itself as altruistic. To take a historical example, after the fall of Napoleon the emperors of Russia and Austria and the king of Prussia jointly signed a declaration in the context of their Holy Alliance. In it, Their Majesties claimed to base their rules of conduct "on the sublime truths contained in the eternal religion of Christ our Saviour" and on the principles "of our holy religion, precepts of justice, of charity, and of peace." They saw themselves "placed towards their subjects and their armies in the position of a father towards his children."[5] Today the language has changed, but the abstract structure is the same, meaning the presentation of power as altruistic. In the middle of the Vietnam War, the American historian Arthur Schlesinger described American policy in that country as "part of our general program of international goodwill."[6] At the war's end, a commentator in the *New York Times*, a liberal (in the American sense of the term) maintained that "for a quarter century, the United States has been trying to do good, encourage political liberty, and promote social justice in the Third World." But in this endeavor "we have been living beyond our moral resources and have fallen into hypocrisy and self-righteousness."[7] It is not easy to find holders of power expressing themselves in an overtly cynical manner; to do so, one probably must look toward those on the margins of society, like gang members or mobsters.

But the fact that the discourse of legitimation is awash in altruism is precisely what ought to arouse skepticism in the hearer. Remarkably, that is just what happens in daily life: altruistic statements are often greeted with skepticism, and people will advise you to pay attention to what others do, not what they say. It is exactly the reverse that often prevails in public life.

This comes about in part because the effect of ideology is constantly being reinforced by the existence of what Chomsky calls a secular priesthood,[8] which in our societies plays a role analogous to that played by priesthoods in traditional societies. Priests presented themselves as intermediaries between the human and the divine and legitimized the power of the ruling classes by interpreting the divine will in suitable fashion. As a perquisite of that, they assured themselves a relatively privileged social position and some protection against the secular power. The contemporary secular priesthood is composed of opinion shapers, philosophers with a media presence, and an army of faculty members and journalists. One of the most common mechanisms of ideological reinforcement is to focus discussion on the means employed to attain the putatively altruistic ends to which those in power lay claim rather than to ask whether the ends proclaimed are the ones really being pursued or whether the agent claiming to pursue such ends is really entitled to do so. To take a current example, it will be asked whether the United States has sufficient resources and intelligence to impose democracy in the Middle East or whether the eventual cost (the war) is not too high. This debate only reinforces the idea that the intentions proclaimed (liberating the peoples, propagating democracy) are the real ones and that the less edifying consequences, like control of oil or the reinforcement of American hegemony (globally) and Israeli hegemony (locally), are no more than collateral effects of a high-minded undertaking.

In order to grasp how ideology functions in our societies, it is helpful to emphasize the difference between this functioning and what takes place in more autocratic societies, where the behavior of individuals is controlled mainly by fear. In a society in which people can freely demonstrate and vote, control of "hearts and minds" has to be a lot deeper and more constant. Hence the importance of debates that confine the discussion within the narrow limits of the effectiveness of means and thus eclipse debates on the nature and the legitimacy of the ends. In an autocratic society, such debates would be forbidden. In ours, not only are they not forbidden, they are extremely useful. The intellectual left that strives for respectability plays a large part in this process of legitimation, because it is generally this left that focuses the debate on the first kind of question (the effectiveness of the means) and marginalizes the second kind (the nature and the legitimacy of the ends). An analysis of power that takes account of its mechanisms of legitimation and doesn't take official declarations at face value is a normal procedure when it is applied to the Roman, Napoleonic, or Soviet empires. But in discussing our own contemporary societies, we no longer follow this normal procedure. The other mechanism frequently employed by the respectable left is the ritual denunciation of "totalitarian" systems of indoctrination, usually with a quasi-religious reference to Orwell, particularly when it comes to the aspects of these systems that are most specific and the most remote from our own. This makes it possible to encourage the notion that universal mechanisms like the control and manipulation of minds by the powers in place are part of the fabric of every society except our own.

It is evident that such a radical questioning of our system of thought, our media, our teaching profession, and the social role of the dominant intelligentsia, including much of the intelligentsia of "the left," does not earn Chomsky a lot of sympathy from that quarter, which no doubt explains why his most hostile critics often come from the ranks of the "liberals" and "progressives." They tend to denigrate Chomsky's analysis by characterizing it as a "conspiracy theory," whereas Chomsky and Herman explain in detail why the behavior of the media is exactly what one would expect in a society in which journalists are "free" but work for large private corporations highly dependent on advertising revenue and on the information supplied to them by governments or other powerful institutions and have a more or less spontaneous ideological bias in favor of their own country and sociopolitical system. Serge Halimi and Arnaud Rindel dismantle in detail the accusation that Chomsky lapses into "conspiracy theory"; Pierre Guerlain shows how Chomsky's critique of the world of the intellectuals is systematically ignored or distorted by the intellectuals themselves.

An Anarchism of Sorts

What would a "just society" look like for Chomsky? This is not something he writes about at length, but he does think that the fundamental human quality to be developed is the opportunity for everyone to do free and creative work, with the hard but necessary labor being shared as equally as possible. For that matter, he thinks that the hard work could be considerably reduced if a full scientific and technological effort were made to do so, an effort that is not made today precisely because there exists abundant manpower that has no other choice but to carry out these tasks. Evidently, such a reorientation of society is incompatible with a society based on the hunt for maximum profit for a small number of people, itself the result of the private property of the means of production and the "free market," which inevitably engender what he calls "private tyrannies." For Chomsky, in an anarchist society control of production must be collective, but in a decentralized manner and making maximum usage of science and technology. This, to use a classic formulation, would be "order minus power" but not disorder.[9]

In this sense, Chomsky is close to the emancipatory perspective of Marx but also of various libertarian currents, which makes him, as he sometimes says, "an anarchist of sorts." Only "of sorts," because there exist numerous movements calling themselves anarchist, or so called by others, with which he does not identify: individualism or subjectivism, including such variants as "sexual liberation"; the antiscience and antitechnology movements; and gratuitous violence, provocation, and the tendency to want "everything right now." Chomsky's own definition of anarchism is a good summary of his own position: "Anarchism is a tendency in the history of human thought and action which seeks to identify coercive, authoritarian, and hierarchic structures of all kinds and to challenge their legitimacy—and if they cannot justify their legitimacy, which is quite commonly the case, to work to undermine them and expand the scope of freedom."[10]

Thus understood, anarchism, far from having disappeared, is alive and well. An antianarchist cartoon from the beginning of the twentieth century illustrates this idea: it shows men doing the dishes. Anarchists are always derided—but not always with the same arguments. Anarchism as a political movement may indeed not have made enormous progress, but the ideas of freedom and autonomy for which anarchists fought have. Think of the world as it was in the years leading up to World War I: hundreds of millions of men and women living under the European colonial yoke; millions of young people getting ready to slaughter one another in the name of countries in the government of which they had little say; the majority of women unable to vote, obtain an abortion, or enjoy any autonomy with respect to their husbands; little or no trade unionism or social protection; children regularly beaten at school and at home; personal and sexual relations dominated by rigidity and hypocrisy; and, as the glowing halo over this vale of tears, the burden of religion. Not all of this has changed, of course, but the twentieth century, contrary to what is often stated, especially in postmodern circles, has been a century of immense progress, over the course of which "coercive, authoritarian, and hierarchic structures of all kinds" have been radically challenged.

The alterglobalist movement, which Chomsky discusses in the interview published here, is for that matter essentially anarchist in the broad sense of the term, both as regards its aims and its functioning. Of course, only the future will tell us if it will attain the goals it has set itself better than the Marxist-inspired movement did in the twentieth century. The reminiscences about Chomsky of Susan George, a protagonist of the alterglobalist movement to which Chomsky is close, are included in this volume. Readers are also referred to Larry Portis's analysis of the origin of French "resistance" to libertarian ideas and an account of Chomsky's ideas on pedagogy by Normand Baillargeon.

Still, Chomsky devotes relatively little time to describing what an alternative to the existing system might be like, in part because he thinks that the alternative is constructed step by step over the course of the struggles of the present but primarily because, ever since the Vietnam War, he has been working urgently, under pressure continuously imposed by the policies of the American government in Indochina, Central America, and the Middle East, and by its mad rush to acquire arms (today the military expenditures of the United States exceed those of the rest of the world combined).[11]

Choices and Political Priorities

A final aspect of Chomsky's work that unsettles France is some choices he has made and the political priorities these choices reflect. To explain this, let us take the examples of Bosnia and Iraq. The conflict in Bosnia spurred a massive mobilization of French intellectuals and even the creation of an electoral list ("Sarajevo") for the European elections of 1994. And, in May 2006, a large number of artists "mobilized" in support of the cancellation of a play by Peter Handke from the program of the Comédie Française; the Austrian author was "guilty" of having been present at the burial of Slobodan Milošević and of having delivered a speech judged "denialist" on that occasion.[12]

But what about Iraq? The town of Falluja, attacked by the marines in 2004, was something like a Guernica without a Picasso. A town with 300,000 inhabitants lacking water, electricity, and food was emptied of its inhabitants, who were subsequently parked in camps. Then came a methodical bombardment and the retaking of the town quarter by quarter. When a hospital was occupied, the *New York Times* justified it by stating that it was being used as a propaganda center by inflating the number of victims.[13] Exactly: how many victims were there? Nobody knows; no count of (Iraqi) bodies was made. When estimates are published, even by the most reputable scientific journals, they are denounced as exaggerations.[14] Finally, the inhabitants can reenter their wrecked town via American checkpoints and set about rebuilding it, guarded by soldiers and biometric controls.

How many protests did that provoke? How many demonstrations in front of American embassies? How many petitions calling on our governments to demand that the United States withdraw? Which popular organizations are showing the same concern for these victims they showed for the victims of Hurricane Katrina? How many newspaper editorials are denouncing these crimes? Who among the supporters of "civil society" and nonviolence recall that Falluja's troubles began shortly after the invasion, when its inhabitants demonstrated peacefully and the Americans fired on the crowd, killing sixteen persons? It is not just Falluja: there is also Najaf, Al Kaïm, Haditha, Samarra, Bakouba, Hit, Bouhriz, Tal Afar—and the longer this war goes on, the longer the list will grow.

The difference in the reactions to Bosnia and Iraq is easy to understand: in the case of Bosnia, there was almost no one to defend or even state the Serbian point of view. So people could adopt a heroic posture of "resistance to fascism" with no risk. On the contrary, by appealing to the United States and Europe to intervene, they were able to side with the strong, giving still more reinforcement to powers responsible for far worse crimes than anything the Serbs could be accused of.[15] But to openly denounce American policy in Iraq is to expose oneself to a whole series of charges from the media and dominant intellectuals: anti-Americanism or support for terrorism or Islamism. Bertrand Russell said that true courage lay in denouncing the crimes of one's own side, not those of the enemy. Chomsky would no doubt add that it is not just a question of courage but one of sincerity: people who are really concerned about human suffering will begin by concentrating on the instances of it they can influence most easily: the ones that result from the actions of their own governments. But this elementary moral stance often provokes hysterical reactions on the part of those who specialize in the denunciation of "the crimes of the other side." And, unhappily, there are a lot of those in France today.

In fact, one can trace the divorce, on the political plane, between Chomsky and the dominant intelligentsia in France, including the intelligentsia of the left, to the end of the Vietnam War. While the war was going on, his writings enjoyed a certain readership among those who were opposed to it, even in France. But there was already an implicit misunderstanding back then. In the French anticolonial, anti-imperialist, and Marxist movements (and their adversaries), the dominant mentality was that of "taking sides." You had to choose your side: for the United States or for the Soviet Union (or China), for the West or for nationalist revolution in the third world.[16] Chomsky's attitude was fundamentally different: his libertarian stance makes the idea of taking sides totally foreign to him. Not that he is an intellectual "above the fray" (there are few intellectuals more committed than he), but his commitment is exclusively in favor of ideas like truth, justice, and moral principles and not to any particular historical or social camp.[17] His opposition to the war was not based on the idea that the Vietnamese revolution was going to offer a radiant future to the peoples of Indochina but simply on the observation that the American aggression in Vietnam, far from being motivated by the defense of democracy, was aimed at preventing any sort of independent development in Indochina (or anywhere else in the third world) and that, in consequence, it could only have catastrophic consequences.

Since Chomsky's writings are extraordinarily rigorous and well documented, he supplied opponents of the Vietnam War with valuable intellectual tools, and the difference in attitude between him and his supporters in France barely registered. For that matter, at the end of this war the left (in the broad sense of the term) was stronger in worldwide terms than it had been since the defeat of fascism. A counteroffensive was inevitable. The occasion for it was supplied when the boat people began to flee from a Vietnam exhausted and devastated by the war, and even more when the Khmer

Rouges committed enormous massacres in Cambodia. A mechanism of culpabilization of all those who had opposed the war—or opposed imperialism in general—was set in motion, in which blame for these events (and for other tragedies that occurred in postcolonial countries) was laid at their door. As Chomsky notes, it would be just as absurd to blame Soviet citizens opposed to the invasion of Afghanistan by their country for the atrocities committed there by the Afghan rebels after the withdrawal of the Soviet troops.[18] They might very well reply that the invasion led to a disaster, and that in opposing it they had done all that they could to prevent a catastrophe, the responsibility for which lay with those who had unleashed the invasion, not the opponents. But the counterpart of this virtually banal argument is almost inaudible in the Western camp.[19] In France, the partisan spirit had led a good many opponents of the colonial wars to delude themselves about the likelihood of a shining future for the decolonialized societies, which made the process of culpabilization particularly effective. And this was precisely the epoch at which the dominant French intelligentsia was effecting a *grand tournant* (a major change of direction) that was to lead it to take its distance from Marxism, from the so-called socialist countries, and from third-world revolution. With the influence of the *nouveaux philosophes* dominant in the little world of the media intellectuals, the majority of them lined up in a position of virtually systematic support of Western policy with respect to the third world—a position that, albeit nominally liberal, is accompanied by practices of intimidation and diabolization of the adversary not very different from the Stalinism of days gone by. This explains why a large number of French intellectuals, especially the ones from the generation of 1968, became more and more politically passive, more so even than the rest of Europe, first in the struggle against the euro-missiles and then against the Gulf War, and ended up frankly pro-war on the occasion of the Western interventions in Bosnia and Kosovo.

For his part, never having had any illusions of which to be disabused, Chomsky had no struggle to give up. In consequence, he remained at the forefront of the struggle against the military interventions and the embargos that, from Central America to Iraq and Kosovo, have caused hundreds of thousands of deaths. But in the eyes of all who had effected the *grand tournant*, Chomsky passed for a bizarre and dangerous anachronism.

Hope for the Future?

When one reads *Une imposture française*[20] or other works on "BHL" (Bernard-Henri Lévy) and becomes aware of the degree of corruption that characterizes the portion of the French intelligentsia that enjoys media visibility, one is forced to admit that the severe diagnosis made by Chomsky some time ago has been more than confirmed:

> French intellectual life has, in my opinion, been turned into something cheap and meretricious by the "star" system. It is something like Hollywood. Thus we go from one absurdity to another—Stalinism, existentialism, structuralism, Lacan, Derrida—some of them obscene (Stalinism), some simply infantile and ridiculous (Lacan and Derrida). What is striking, however, is the pomposity and self-importance at each stage.[21]

Chomsky emphasizes, however, that comments in this vein apply only to the portion of the French intelligentsia overexposed in the media, and that among the remainder of it are to be found many honest and lucid individuals—something to which the many French social movements in which Chomsky's ideas are gaining ground today bear witness.

NOTES

1. *The Chomsky Reader*, edited by James Peck (New York: Pantheon Books, 1987), 71–72.

2. Noam Chomsky and Michel Foucault, *The Chomsky-Foucault Debate: On Human Nature* (New York: The New Press, 2006). The page numbers given in parentheses in the text refer to this English translation.

3. Cited by Noam Chomsky in Chomsky, *Reflections on Language* (London: Temple Smith, 1976), 133. American imprints of this book include New York: Pantheon Books, 1975; and New York: The New Free Press, 1998. For this quotation from Bakunin, Chomsky cites Sam Dolgoff, *Bakunin on Anarchy* (New York: Knopf, 1972). In *Reflections on Language*, Chomsky also explains why the linkage between innatism and conservatism does not hold.

4. Edward S. Herman and Noam Chomsky, *Manufacturing Consent: The Political Economy of the Mass Media*, with a new introduction (1988; repr. New York: Pantheon, 2002).

5. Bertrand Russell, *Freedom Versus Organization, 1814–1914* (1934; repr. New York: Norton, 1962), 33.

6. *New York Times* (February 6, 1966).

7. William V. Shannon, *New York Times* (September 28, 1974); cited by Noam Chomsky in *"Human Rights" and American Foreign Policy* (Nottingham: Spokesman Books, 1978), 2–3.

8. Taking up an expression of Isaiah Berlin in "The Bent Twig: On the Rise of Nationalism," in *The Crooked Timber of Humanity: Chapters in the History of Ideas* (New York: Vintage Books, 1992), 240, which evidently referred to the communist intelligentsia in the socialist countries.

9. Chomsky explains his view on all these questions in the interview published in this volume.

10. Cited in French translation by Normand Baillargeon in *L'Ordre moins le pouvoir: Histoire et actualité de l'anarchisme*, 3rd ed. (Montréal: Lux impression, 2004), 26. This book offers a good exposition of anarchism from a point of view close to that of Chomsky. The English text is from http://www.zmag.org/chomsky/links.cfm#Anarchism. Neither the book by Baillargeon nor the Web site supplies the source of this quotation.

11. See *The Defense Monitor* 35 (March–April 2006).

12. See Olivier Py, "À plus tard, Peter Handke," *Le Monde* (May 10, 2006).

13. "After the Marines first invaded the city in April, inflated civilian casualty figures from Falluja General Hospital inflamed opinion throughout the country, driving up the political costs of the conflict and ultimately forcing the American occupation authority to order a with-

drawal." Robert Worth, "Sides in Falluja Fight for Hearts and Minds," *New York Times* (November 17, 2004).

14. See Les Robert et al., "Mortality Before and After the 2003 Invasion of Iraq: Cluster Sample Survey," *The Lancet* 364 (November 20, 2004). They assess the number of civilians killed in the wake of the American aggression at 100,000. In November 2006, the same journal published a new survey estimating the number of deaths since the aggression began at 655,000.

15. See, for example, William Blum, *Rogue State: A Guide to the World's Only Superpower*, 3rd ed. (Monroe, Me.: Common Courage Press, 2005); and Blum, *Killing Hope: U.S. Military and CIA Interventions Since World War II*, 2nd ed. (Monroe, Me.: Common Courage Press, 2004), for details on the American wars and interventions that have caused millions of victims, principally in the third world, since 1945.

16. This attitude reached well beyond politics; Louis Althusser, whose influence on a whole generation of intellectuals was enormous, described his "materialist" approach as a *prise de partie* in philosophy. See his *Lenine et la philosophie* (Paris: Maspero, 1982). Bernard-Henri Lévy for his part takes the view that one ought to think in the same way one makes war. See the final section of his *La Pureté dangereuse* (Paris: Grasset, 1994).

17. A stance that, for that matter, Marxists have often mistakenly denounced as "idealist."

18. See Noam Chomsky, *Power and Prospects: Reflections on Human Nature and the Social Order* (Cambridge, Mass.: South End Press, 1996).

19. This comparison is adduced only to demonstrate the absurdity of the dominant reasoning in the West. The two situations were nonetheless very different, the Soviet intervention in Afghanistan having been motivated in part by considerations of territorial security totally absent in the case of the American aggression in Vietnam.

20. Nicolas Beau and Olivier Toscer, *Une imposture française* (Paris: Les arènes, 2006).

21. Noam Chomsky, *Language and Politics*, ed. C. P. Otero, expanded 2nd ed. (Oakland, Calif.: AK Press, 2004), 285. In the first edition, *Language and Politics*, ed. C. P. Otero (Montreal: Black Rose Books, 1988), this quotation appears on 310–311.

4
AN INTERVIEW WITH NOAM CHOMSKY

JEAN BRICMONT

Politics

JB *The motivation behind most questions in this part is this: I feel that an increasing number of people realize that the current social arrangements are deeply unjust and that the main answer from the mainstream is the famous TINA. So, even though I know that you will not want to describe an alternative in detail, I think it is important to give concrete answers to the various forms of TINA. Hence, below, I try to formulate the case for TINA as sharply as I can.*

Q1. First there is the issue of human nature. Since you hold that humans cannot be explained except by assuming that their mind is highly structured innately, why are you not more pessimistic about the possibility of political change?

NC Pessimism about political change would follow from the assumption that existing sociopolitical arrangements conform to human innate capacities so closely that they cannot be changed, consistent with these capacities. If we make different assumptions about innate human nature, or if we simply recognize that so little is understood that no firm conclusions can be drawn, then there is ample room for optimism about political change. In fact, very little of relevance is understood, so I do not personally think that the question arises in a serious way. Optimism and pessimism are subjective reactions to situations and problems that are understood only in a superficial way. If that is true—and I think it is—then the reasonable stance is optimism, meaning commitment to change things for the better, in the hope that it is possible, which we have no reason to doubt. As I've occasionally said, we face a kind of "Pascal's wager": assume that nothing is possible, and the worst will come; assume that things can be improved, and perhaps they can. Given these choices, it is clear which one we should accept, whatever our subjective judgments may be (on extremely weak grounds).

Those committed to existing sociopolitical arrangements do sometimes hold that existing social systems somehow conform to human nature in ways that sharply different alternatives would not. But those who search for a serious argument will be sorely disappointed. It is hardly even clear what is being asserted. What do we mean by "existing social systems"? For almost all of human history, existing social systems were small hunter-gatherer groups. There have been no relevant evolutionary changes since, so from a narrow point of view, that's the best source of information about human nature: *very narrow*, because there is quite strong evidence that capacities may be latent for very long periods. What they are only comes to light as circumstances change, though they are all part of human nature, a fact with obvious relevance to the possibilities for social change. The next largest component is peasant societies of varying kinds. TINA is just the last instant of human history. In fact, the state-capitalist market systems of which TINA is a curious distortion were themselves imposed very recently, with plenty of force and violence and over strong resistance. There is classic work (such as Karl Polanyi's *Great Transformation*, along with ample additional scholarship) that regards this sharp break from tradition as truly revolutionary and largely coerced, over fierce resistance. We can say the same about the modern nation-state system, instituted in Europe through centuries of vicious savagery and imposed on the world with extreme brutality. The violence and savagery may reflect the fact that the system is so "unnatural." It radically deviates from the structures that had developed over long eras in traditional societies, in Europe as well.

Without proceeding beyond even these superficialities, what can we learn from history about conformity of social systems to human nature? Answer: precious little. What is learned is largely in the eye of the beholder. The heroes of those who laud TINA—Adam Smith and David Hume, for example—considered sympathy to be a basic principle of human nature and the foundation of decent human relations. "To feel much more for others and little for ourselves" is, for Smith, "the perfection of human nature" and the basis for a harmonious life. Smith regarded equality as an obvious desideratum. He argued that under conditions of perfect liberty, markets would tend toward perfect equality (*Wealth of Nations*, book 1, chap. 10)—an obvious good, in his view, a conception traceable back to the first major work on political organization, Aristotle's *Poli-*

tics. The founder of what is now called "sociobiology" or "evolutionary psychology"—the natural historian and anarchist Peter Kropotkin—concluded from his investigations of animals and human life and society that "mutual aid" was a primary factor in evolution, which tended naturally toward communist anarchism (in his book *Mutual Aid*). Of course, Kropotkin is not considered the founding figure of the field and is usually dismissed if mentioned at all, because his quasi-Darwinian speculations led to unwanted conclusions. But although a great deal has been learned in the century since, it would be hard to argue that current speculations about the matter are much more firmly grounded.

It is sometimes assumed that if the mind is highly structured, it does not allow for choice, change, and creativity. There is no basis for that belief. Creativity presupposes fixed structure; even I could be a creative artist if random noises qualified as poetry or music, and the fact that the structure can (in part) be chosen does not entail that "anything goes." That much is familiar since the aesthetic theorizing of the Enlightenment and the romantic period. The same is true in other domains. There is no doubt that fixed human nature sets bounds on societies that can function at all successfully, just as it sets bounds on what humans can come to understand about the world, what kinds of artistic traditions they can create and explore, and so on. But we scarcely have an idea of what these bounds are or, for that matter, what is their basis in human biology.

JB *I am not suggesting that your scientific ideas (about human nature) should be determined by your political preferences, but I observe that people who have the kind of nativist views that you hold tend to be conservative, while people who look forward toward social change tend to be more "environmentalist."*

NC That's not so obvious. Think of Kropotkin again. He was certainly a "nativist" but argued for radical social change, claiming that the communist anarchism he advocated conformed to essential human nature. Or take the place where I've spent most of my adult life, Cambridge, Massachusetts, one of the centers of thought and debate about these issues for half a century. No one, of course, believes in a "blank slate"; that is incoherent. But the most influential and respected advocates of the more extreme "empty-organism theories"—B. F. Skinner, W. V. Quine, and Nelson Goodman—happened to be very far to the right politically, while those who took the strongest positions on innate structure, primarily a few graduate students fifty years ago, were very far to the left end of the spectrum. I was one, Eric Lenneberg another—and both of us were heavily influenced by the work of Konrad Lorenz, who had been a Nazi sympathizer. So what can one conclude from that about nativist views and political attitudes?

Many Marxists also professed an extreme empty-organism stand, even arguing that there is no human nature apart from human history, but it is hard to see how one can make sense of that doctrine. It is also hard to see how one can even make sense of Marx's own views in these terms: for example, such concepts as "alienation" and the intrinsic human need for creative work under one's own direction and control, ideas that Marx drew from the Enlightenment-romantic intellectual environment, with its rich texture of libertarian and innatist conceptions.

Perhaps one can argue that rather generally, conceptions of human malleability have often been attractive to intellectuals across the political spectrum. Individuals have their own reasons, but a tendency in that direction would hardly be surprising, given the general role of intellectuals as managers: doctrinal, political, economic. Doctrines of human malleability remove a moral barrier to control and manipulation and hence are naturally attractive to managers. But at best these are tendencies.

I am not much convinced by the way the issues are commonly presented. For example, best-selling popular-science books written right here in Cambridge commonly attribute blank-slate doctrines generally to the left (Stephen Pinker's *Blank Slate*, for example). The left can then be discredited by its alleged association with ridiculous positions. That proceeds though everyone concerned with these issues in Cambridge knows that right here the correlation did not hold. Something closer to the opposite was true, if one wants to pursue this (in my view, rather pointless) endeavor.

JB *The reasons are fairly obvious. Consider the following factors, which seem, from an observation of history, to be part of human nature, if there is any such thing: first, a readiness to hold irrational beliefs. Even the decline (in Europe) of institutionalized religions has not brought more rationality. Look, for example, at the successes of pseudosciences, alternative medicines, etc. But if people aren't more rational, what hope is there that they'll manage their affairs in a sensible way?*

NC Can we really say "if there is any such thing"? No one doubts that our biological endowment ("human nature") determines that we have arms rather than wings, or a mammalian rather than an insect visual system, or . . . fill in the blanks, for all aspects of our being "below the neck," metaphorically speaking. It's conceivable that the human brain somehow escapes the bounds of biology and logic and is some nonnatural kind of organ, in which case there would be no human nature in the domain of higher mental faculties. But that belief is hardly one that we would take seriously. If humans are part of the natural world, then their cognitive, aesthetic, moral, and other faculties are firmly grounded in their biological nature. And to the limited extent that anything is understood about these topics, that is what we find.

As for the rest, I don't think we can conclude much. Humans are prone to irrational beliefs and are also committed to rational thought—otherwise they could hardly survive. I don't think that the prevalence of pseudoscience in contemporary Western culture proves much, for one reason, because only a tiny minority has the opportunity to gain any appreciation of the marvelous intellectual achievements of the natural sciences and mathematics. For those who do not—almost everyone—it may seem to be just more mystery, like the mysteries admired and advocated by the Grand Inquisitor of Dostoyevsky's *Brothers Karamazov*, in accord with his conception of human nature. We have no sensible way to measure to what extent people are rational by nature, or even in practice. I also do not know of any reason to believe that people who are generally committed to rationality in their intellectual lives "manage their affairs in a sensible way": mathematicians and logicians, for example. I doubt that one would want to seriously advance that empirical hypothesis.

JB *Second, the willingness to follow a leader: if a radical social change should occur, presumably after some socioeconomic disaster, wouldn't one expect people to follow some great leader, as they did, for example, in Russia after 1917?*

NC Sometimes people are willing to follow a leader, sometimes not. Sometimes they rebel, with courage and dedication, facing severe punishment for their refusal to submit to authority. I also do not think we should take for granted that radical social change requires socioeconomic disaster. As for Russia in 1917, I would tend to draw rather different conclusions. The factory councils, Soviets, peasant organizations and communities, constituent assembly, and other popular initiatives were not abandoned by their participants because they preferred a Maximal Leader; they were destroyed, with considerable force and violence, and discipline was imposed from above. One can inquire into the historical and other factors that led to that outcome, but I do not see any reason to attribute it to human nature.

JB *Third, the willingness to internalize oppression: you quote approvingly S. Biko's phrase about the most powerful tool in the hands of the oppressor being the mind of the oppressed. But why should one expect that to change (if we accept a nativist view)? Such an internalization of oppression seems so widespread.*

NC So it does. And so does revolt against dominance and oppression seem widespread, despite the great costs it typically carries. If the nativist view is that of Rousseau (in his more libertarian work, deeply suffused by nativist conceptions, some drawn straight from the Cartesian tradition) or Wilhelm von Humboldt and other founders of classical liberalism, or the left libertarian (anarchist) traditions that developed from these roots, then it suggests quite different conclusions about a willingness to internalize oppression. One can draw on history and current practice to support virtually any conclusion one wants in these domains, and the contributions of the sciences remain very thin.

JB *Fourth, selfishness: people look out first of all for themselves. Is there any way to design an alternative social order that does not rely on the implausible assumption that people suddenly become altruistic?*

NC Why is it implausible? Suppose someone is walking down a street, happens to be hungry, comes across a starving child holding a piece of bread, and doesn't see a policeman. Would the natural instinct be to steal the bread from the child? If that were to happen, we would regard it as pathological. Suppose dolphins are beached on the falling tide. Hundreds of people rush out and labor under harsh and difficult conditions to try to rescue them. Can we explain that on the basis of selfishness—or even sophisticated theories of kin selection and reciprocal altruism? I don't think that history or experience demonstrate that Adam Smith and David Hume—among the heroes of the contemporary chorus of praise for selfishness—were wrong in assuming that sympathy and concern for the welfare of others are basic features of human nature. The belief that selfishness is a predominant human instinct is very convenient for the rich and powerful, who hope to dismantle the social institutions that have been developed on the basis of sympathy, solidarity, and mutual aid. More barbaric elements of rich and powerful sectors—those now at the helm in Washington, for example, or the TINA enthusiasts elsewhere—are intent on demolishing Social Security, health programs, schools, in fact, all the achievements of popular struggle that serve public needs and marginally diminish their own wealth and power. And for them, it is most convenient to concoct fanciful theories about how selfishness is the core of human nature, so that it is wrong ("evil," in the favored contemporary terminology) to care whether the disabled widow on the other side of town has food and medical care or whether the child across the street has an opportunity for a decent education. Do we have some evidence to support these convenient doctrines? Not to my knowledge.

JB *Fifth, inequality: again, in a nativist view, some if not most inequalities (for example, in intellectual abilities) must be innate. But if people are both selfish and not equal, what can one hope for, except some combination of the rule of law and of market regulation—namely, the present social order? An environmentalist would say (or at least hope) that new social conditions will mold the human mind in a different direction, toward more solidarity or lucidity. But from your point of view, this answer is excluded. On the other hand, the above arguments are commonplace among conservatives. So, why don't you join their ranks?*

NC I don't join their ranks because the arguments are merely claims, not arguments, and claims that have little credibility as far as I can see. It is also not easy to make much sense of these unargued claims. If the present social order is the only possible one, given human nature, then how do we account for that fact that it did not exist through almost all of human history and was imposed quite recently and with considerable force and coercion, in England and elsewhere? One might just as well ask why "conservatives" don't join the ranks of libertarian socialists, on the grounds that people really are innately sympathetic and considerate of others, as

Hume, Smith, and other TINA heroes held. And I think reasonable people can agree that inequality of ability to solve problems in mathematics, or to crush someone's head with a blow, do not lead to any specific conclusions about how society should be organized.

Let's be clear about the very little that we do know about these matters.

First, everyone takes for granted that environment influences development: for arms and legs, the visual system, or any other relevant property of an organism. Only confirmed dualists might believe that human intellectual, moral, aesthetic, and other faculties are somehow exempt from these principles of nature. Just to add a historical note, the two cofounders of the theory of evolution, Darwin and Alfred Russel Wallace, had a famous disagreement about the origin of human "man's intellectual and moral nature." Wallace argued, contrary to Darwin, that natural selection was insufficient to account for their emergence in the evolutionary record and that they are based on some principle of nature alongside gravitation, cohesion, and other forces without which the material universe could not exist. But he did not doubt that human nature fell within the natural world and its laws, and the questions he raised, though differently formulated today, have by no means disappeared.

Second, there is no empirical evidence to support the claim that among the numerous characteristics that every normal human being shares, selfishness and cruelty overwhelmingly dominate, suppressing sympathy, compassion, solidarity, mutual aid, and other traits and tendencies. Again, with equal (that is, zero) logic, one might as well ask why conservatives don't join the ranks of communitarian anarchists of the Kropotkin variety. Or we could argue that societies must necessarily be founded on torture, slavery, brutal oppression and abuse of women, genocide, etc. given their prevalence in human history, which proves that they reflect intrinsic human nature, and therefore we should not try to overcome them, defying intrinsic human nature. I presume there are few who would take that stance. And for good reasons. There is nothing in history, science, or logic to suggest that the particular forms of social organization that have taken shape at one or another historical moment are somehow a necessary reflection of fundamental human nature—a belief that could hold for insects but is utterly senseless in the case of humans, surely. We can say that these social arrangements—say, those just mentioned—are rooted in some aspects of human nature, or we can hold—as seems to me more sensible—that they are severe pathologies of particular forms of human organization that have come to exist for one or another reasons and should be overcome by constructive social change. What we cannot do is claim scientific support for our judgments about these matters.

Comparative evidence is also very limited. Chimpanzees and Bonobos are about the same evolutionary distance from humans. Chimpanzees are very aggressive; Bonobos are sometimes described by those who have studied them most carefully as being like hippies of the 1960s, acting in accord with the slogan "make love not war." We can draw few if any lessons about humans that bear on the forms of social organization that we should seek.

JB *Q2. The R word (or words). You are sometimes accused by other anarchists of being reformist. What do you respond to that?*

NC If a "reformist" is someone who cares about the conditions of life for suffering people and works to improve them, then anyone who is even worth talking to is a "reformist." To be concrete, "reformists" in this sense support measures to improve safety in the workplace, to ensure adequate food and health care and potable water for everyone, and so on. To "accuse" someone of being a reformist in this sense is to accuse him or her of being a minimally decent human being—a rather odd sort of accusation. Of course, we sometimes observe that people who claim to be reformists in this sense—that is, minimally decent people—may be adopting this posture to preserve repression and dominance. But that is an entirely separate matter.

JB *In actual fact, what do you look forward to? Revolution?*

NC Revolution is a means, not an end. No one looks forward to a particular choice of means. If we are committed to certain goals, whatever they are, we would seek to attain them peacefully, by persuasion and consensus, if possible—at least if we are sane and accept the most minimal moral standards. That is true no matter how revolutionary our goals. There cannot possibly be any general formula as to whether, or when, other means are necessary and appropriate. Even the most ardent revolutionary agrees to that—again, assuming sanity and commitment to the most elementary moral standards.

JB *But isn't that passé? Isn't that a means of social change that belongs to a period where the majority of people were far more destitute than they are now, at least in the West?*

NC That is unsubstantiated dogma, as far as I am aware. To borrow some of Marx's terminology, one could just as well argue that as people's "animal needs" are more fully satisfied they become more able to devote thought and energy to achieving their "human needs"—which would require radical changes in social organization and human relations. Note again that if this (rather plausible) assumption holds, it still leaves entirely open the choice of means, for which there cannot possibly be any fixed formula.

JB *On the other hand, isn't reformism always co-opted by the system?*

NC I am not sure what that means. Take, say, slavery. Overcoming slavery is "reformism." Was it "co-opted by the system"? In some sense it was;

in the United States, for example, postslavery conditions, laws, and practices sustained a legacy of servitude that has still not been completely eliminated. But do we therefore deny that eliminating slavery was a tremendous achievement, and that the "reformist" efforts to overcome the legacy have not registered significant achievements as well? Surely that is the wrong conclusion. Let us proceed. At the time of the U.S. Civil War, northern workers understood "wage slavery"—as they regarded it, without the dubious benefit of radical intellectuals—to be not very different from chattel slavery, and they sought to overcome that as well. That struggle has barely been joined even a century and a half later, but should it be abandoned on the grounds that existing power structures would, naturally, make every effort to co-opt any progress that is made? That makes no sense either.

If by "co-optation" one means that some aspects of an unjust order always survive human progress, that's doubtless true. No one seriously expects to move to utopia in one step—or ever. Progress in human affairs is rather like mountain climbing. You see a peak, struggle to reach it, and discover that beyond it there are others of which you may not even have been aware. Undeniably, throughout history there have been real achievements in extending human rights and freedom and relieving suffering and oppression. Each victory provides the opportunity to explore our inner nature more fully and thoughtfully and to perceive forms of injustice and violence of which we may not even have been aware. Furthermore, as in mountain climbing, there will often be falls, sometimes precipitous ones, and the task has to be carried forward from where one finds oneself. Why should we suppose that this process has come to an end, or ever will?

JB *Look at what the social democrats have supported since they have accepted to participate in governments (around World War I): colonialism, including several wars (Algeria, Vietnam), and now they have become essentially neoliberals. Or look at the Greens; it is a relatively new movement and it took them less time than the socialists to integrate into the system. They also support de facto the neoliberal agenda as well as the NATO war against Yugoslavia. Similar remarks could be made about former revolutionary movements like the ANC or the Sandinistas. Why doesn't that make you pessimistic?*

NC I think that is a very misleading view of history. First, each case has to be looked at in itself. They are sharply different. Second, the fact that perfection isn't achieved does not entail that nothing is achieved. To take the first example, social democracy achieved many important gains; surely it is unnecessary to debate that. And the fact that people who struggled for these goals and partially attained them nevertheless continue to tolerate or even strongly support other severe abuses does not negate the gains. Nor does anything general follow from the fact that history is not a process of permanent progress toward rights and justice, without regression. Neither pessimism nor optimism follows from the fact that there has been quite a lot of progress in advancing freedom and human rights over the past centuries, even over the past few years, along with regression and inability to achieve more—so far.

JB *Do you envisage any form of organization (for example, like those of the anarchist movements) that would avoid those pitfalls? And if so, why do you think so?*

NC If by "avoiding pitfalls" one means a guarantee of endless progress without regression, it is not a serious goal. We can oppose torture, slavery, exploitation, oppression, violence, and other abuses without subjecting ourselves to the illusion that the choices are either utopia tomorrow or no progress at all. That makes no sense: it is a form of capitulation to power and abusive authority. No one can dream of a form of organization that is guaranteed to "avoid pitfalls" in this sense. On the other hand, there is good reason to believe that more democratic and participatory forms of organization can overcome the pitfalls that are inherent to the Central Committee or the tyrannical workplace or other forms of illegitimate hierarchy and domination. Can we be sure of this? Certainly not. There is no certainty outside of mathematics—and, to be precise, not even there.

JB *Q3. What is your view of violence? When is it legitimate?*

NC Only pure pacifists can answer this question. They can say that violence is never legitimate. I do not think that is a moral position and do not hold it. Apart from pure pacifists, there can be no general answer to the question. It depends on all sorts of circumstances, and in anything as complex as human affairs—in fact, even in far simpler systems—these cannot be spelled out in advance and subjected to abstract formulas. The world doesn't work like that. We can, sensibly, say that violence is a last resort, but that truism does not provide answers for many of the dilemmas of human life, from personal affairs to international relations. They have to be investigated in their own terms.

JB *For example, how do you react to the tactics of the Black Bloc?*

NC There is, first of all, a question about what was done by the Black Bloc and what should be attributed to police provocateurs. There is credible evidence that there was police provocation, under the guise of "Black Bloc," and it would hardly be surprising. The authorities welcome these tactics as a way of defaming dissent and protest, suppressing its actual content, and demoralizing those engaged in constructive action. Therefore, they are very likely to incite it. We know of many such cases. Anyone who has been involved in activist movements will know of them from personal experience. It was such a familiar lesson of the 1960s that care had to be taken to identify likely provocateurs—which is often not very difficult; the FBI is not very imaginative—and exclude them from sensitive discussions and decisions. There are many important historical examples. To take one with very significant current ramifications, the CIA-run coup that overthrew Iran's conservative parliamentary government in 1953, restoring the Shah, began with CIA-organized violence by mobs they recruited who pretended to be supporting Prime Minister Mossadegh. That is a sensible tactic for barbarians.

That pretty much explains my reaction to the tactics of the Black Bloc, at a very general level. Doubtless there is variation, but in general, I think that the state authorities know what they are doing when they instigate violence and fanaticism among activists. In general, I think the tactics are tactically and morally wrong, and I was therefore not surprised to learn that some of the actions may have been instigated by the state authorities.

JB *Since they are depicted as anarchists by almost everyone in the media, don't you fear that this will again give anarchism a bad name, like the tactic of the propagande par le fait in the nineteenth century?*

NC The actions are not right or wrong because the media depict them as "anarchist." We should not be interested in labels, nor should we determine what is right or wrong by practices of the media. That is true whether we have in mind state-capitalist media or those of fascist, Stalinist, theocratic, or other societies. Those who organized rural health clinics in the U.S.-backed Latin American neo-Nazi National Security State were commonly depicted by the authorities and their servants as "communists." We do not therefore conclude that they were communists—whatever that is supposed to mean. The notorious School of the Americas, which trains Latin American security forces, publicly takes pride in the claim that the U.S. Army "helped defeat liberation theology," therefore saving the civilized world from the Gulag. We do not therefore condemn the priests and nuns and lay workers who adopted "the preferential option for the poor" and suffered bitterly for that crime—far more so than East European dissidents, though that fact cannot be perceived in elite Western intellectual circles.

As for the *propagande par le fait*, we ask whether it was right or wrong, not whether those who carried out the actions called themselves "anarchists" or whether others exploited their actions to defame anarchism.

JB *On the other hand, what if there was no Black Bloc? Wouldn't the media simply ignore the protests?*

NC Actions are not undertaken in order to gain media attention. The World Social Forum is barely reported on in the United States, and the few reports verge on the ridiculous. The 100,000 participants in the last WSF meetings do not therefore conclude that their efforts were a waste of time. As for the media, it would be better to have them ignore the protests than to adopt tactics that offer the media an opportunity to defame the protests and suppress their valid objectives. Again, I presume that would be the point of police provocation if—as I have been informed—it occurred. But that aside, the goal is not to make sure that the media take notice. Rather, protests are part of an ongoing process of education, organization, resistance, and construction of alternatives. They are successful if they contribute to these ends. One can naturally expect defamation from institutions that support existing structures of power and domination.

JB *Q4. You often dismiss out of hand the notion that socialism failed (in Eastern Europe). But isn't that too easy? Not only was it regarded as a form of socialism by most leftists (hence, the present discouragement), but also, resource allocations have to be done by one method or another. If not by planning, then by some form of market. Or do you have another suggestion? But planning failed, as we saw in Eastern Europe, and the market leads to inequalities, as anyone can observe.*

NC It is easy, but I don't think it is "too easy." A lot depends on how we understand "socialism." I understand socialism to involve, at a very minimum, democratic control of production and other aspects of life. If so, socialism was not even tried in Eastern Europe any more than it was under Nazi National Socialism. Nor was there even an intention to institute socialism on the part of the Bolshevik leadership in 1917–1918 or their successors. One may argue, perhaps, that Lenin and Trotsky followed the only course they could under the circumstances, but it was about as remote from socialism as one can imagine. Leftists should, in my opinion, have welcomed the collapse of the Soviet tyranny as a victory for socialism, just as the collapse of fascism was: it eliminates some of the barriers to socialism.

Of course, no one owns the word "socialism." One can choose to use the term to refer to a system of central management by dictatorial authorities, ruling a "labor army" that follows orders with strict discipline (in Leninist terminology). In accord with that usage, socialism was tried and, thankfully, collapsed.

It is worth bearing in mind that the world's two major propaganda systems agreed that the brutal and tyrannical system instituted by Lenin and Trotsky and turned into a monstrosity by Stalin was "socialism." Western doctrinal systems delighted in that absurd and outrageous usage in order to defame authentic socialism. The Russian propaganda system enthusiastically adopted the same usage in an effort to exploit the sympathy and commitment to authentic socialist ideals for their own purposes. The convergence of the two major doctrinal systems may induce people to adopt their usage, thus helping advance their common purpose of undermining authentic socialism. We can choose to be slaves to propaganda if we like, but it is a choice, not a necessity.

I also don't know what it means to say that "planning failed." The Soviet system was an utter monstrosity, but as an economic model of development, by what standards did it fail? Certainly that was not the view of Western leaders, who were deeply concerned by the economic successes of Soviet planning and feared that it would inspire others to imitate it. That was true well into the 1960s, as we know from the declassified record. In the 1960s, the Soviet system began to stagnate and decline, for complex reasons that did not reduce to failures of central planning in narrow economic terms. The claim that the system "failed" is often based on comparisons of Eastern and Western societies, but that is about as sensible as saying that the kindergartens in Cambridge, Massachusetts, fail, as demonstrated by the fact that graduates of MIT know more quantum physics than graduates of kindergarten. The last time Eastern and Western Europe were more or less on a par economically was probably the fifteenth century. From about that time, the disparity grew until Eastern Europe extricated itself from the quasi-colonial relations that had developed and undertook a course of brutal and miserable independent development. If we compare Eastern Europe to societies that were more or less similar a century ago but were subject to Western development models—say, if we compare Russia to Brazil or Bulgaria to Guatemala—it would be rather difficult to conclude that the Soviet system "failed." True, no two cases are identical, but a close look at realistic comparisons only enhances these conclusions—which is hardly a compliment to Bolshevism, though it does tell us quite a lot about Western practices and operative ideals. That is why realistic comparisons are rarely undertaken and why the very idea arouses horror and outrage.

Are there better alternatives to dictatorial central planning, or the monstrous systems imposed on the third world, or state-capitalist systems in which planning is carried out primarily by a network of basically totalitarian institutions linked to one another and to powerful states in intricate ways and generally unaccountable to the public? I don't see any reason to doubt that there are preferable alternatives, just as in the eighteenth century it was possible to envision and realize significant and far preferable alternatives to feudalism—leaving, as always, plenty of fundamental injustices to overcome. Democratic control of the workplace and community, which can take many forms, is the bare beginning of preferable alternatives.

JB *Q5. Sometimes you relate your ideas to those of classical liberalism, including those of such strange bedfellows (for a leftist) as Adam Smith.*

NC There is an unfortunate tendency to assign individuals (for example, Smith) or ideas (classical liberalism) to abstract categories that we must either accept or reject, love or hate. That makes no sense. There is much of value in classical liberalism, and I, at least, am not inclined to reject

Adam Smith's commitment to equality or his sharp critique of the division of labor because its pernicious effects would not be tolerated in any civilized society, or reject his arguments against core principles of what is now called "neoliberalism" (which, he hoped, could be avoided as if by an "invisible hand," his only use of the phrase in *Wealth of Nations*) and other such ideas merely because one does not accept everything he stood for. By such arguments, contemporary physicists should not recognize Newton's achievements, on the grounds that many of his ideas and beliefs are understood to have been completely wrong and often pretty strange.

JB *However, you also say that when pure free-market ideas have been applied, they have quickly been abandoned, because they would have led to the total collapse of society.*

NC I don't claim any originality for that observation. It is a virtual commonplace, developed, for example, in such classic works as Polanyi's *Great Transformation*, which I already mentioned. There is good reason to accept the logic of Polanyi's argument that free markets would lead to the collapse of society, and the limited evidence of history supports the conclusions and also reveals that business leaders understood them very well and quickly sought measures to protect themselves from the ravages of the market. So did working people, in other ways, sometimes called "reformist," which achieved many successes in improving the quality of human life: implementing "reformist" measures advocated by Marx and Engels, for example. We should, however, recognize that the notion of "pure free-market ideas" has a large element of myth in the first place. Markets in any form are instituted, often by force, and are historical residues in ways that impose severe distortions. And the rich and powerful have never tolerated them for themselves, though they delight in imposing them on their subjects. Even during the classical moment of modern "laissez faire"—Britain's experiment in the latter part of the nineteenth century, which did not long endure—the powerful British state carried out vast programs of interference with free markets, for example, by creating the greatest narcotrafficking empire in history, on which the whole imperial system crucially relied, and by taking advantage of the controlled Indian and East African markets for its exports, and so on. Same with other very limited experiments.

JB *But why isn't the same true for anarchist ideas? If there were no government, no courts, no police whatsoever, why wouldn't there be a war of all against all? Do you have an optimistic view of human nature, and if so, based on what? And if some form of government has to be maintained, in what sense are you an anarchist?*

NC It depends which anarchist ideas you have in mind. The major mass popular anarchist movements and leading anarchist figures envisioned (and sometimes partially established) highly organized societies, based on free associations of many kinds, interacting through federal structures, and so on. Sometimes these plans were developed in enormous detail, as in Abad de Santillan's models for revolutionary Spain. Would they lead to a war of all against all? Or to fruitful cooperation among people who are enjoying new forms of freedom and are exploring the opportunities they provide? We cannot know the answers to these questions with any confidence, just as no one could know in the eighteenth century whether a society could exist in which voting rights were widely distributed. Social change always involves experimentation and requires open-mindedness. As for human nature, so little is understood about it that speculation is rather idle. Optimism and pessimism are reflections of one's own state of mind and personal preferences, not grounded in any solid knowledge or understanding. Pessimism, for example, is a convenient choice for those who are seeking, perhaps unconsciously, some way to disengage themselves from the difficult struggle to extend the realms of freedom and justice. But suppose that some day we discover compelling evidence that for a society to survive in ways that satisfy legitimate human needs, some form of government has to be maintained. If so, then no sensible person will reject the conclusion on the grounds that they would like to call themselves "anarchists." That would be a form of self-obsession that borders on lunacy. It's not very useful to wave banners and shout slogans. We have to try to discover what forms of interaction and organization will foster freedom, justice, self-realization, and other values that we hold.

JB *Q6. What about the relationship of forces? Nowadays, enormous power (think of the nuclear arsenals) is concentrated, in actual fact, in the hands of very few people. Considering the brutality with which the ruling classes have reacted in the past (the Paris Commune, the Spanish Revolution) when their power was challenged, what would prevent them from crushing any new movement aiming at radical social change?*

NC That seems to me a misreading of history and of the present situation. No doubt power systems will try to sustain themselves, sometimes resorting to violence. Sometimes they succeed. They often fail, and struggles for justice and freedom can then continue from a higher plane. Suppose that a mass popular movement were to develop in the United States calling for worker's control of industry—or something less dramatic, such as a national health program of the sort that is now called "politically impossible" (even though a large majority of the population favor it), because it is opposed by financial capital and the pharmaceutical corporations. Would nuclear weapons be of any use in undermining and destroying those movements? Or even armies and police, which can easily disintegrate when confronted with genuine popular movements? Armies and police consist of people, after all.

In fact, the hold on power is fragile, and those who hold power know that very well: we learn this from their internal documents. Even in the United States, where the power of business is unusually strong, business leaders, sometimes joined by elite intellectuals, regularly express deep concerns about "the hazard facing industrialists" in "the newly realized political power of the masses," the "crisis of democracy" (that is, *too much* democracy), and the need to wage and win "the everlasting battle for the minds of men" and "indoctrinate citizens with the capitalist story" until "they are able to play back the story with remarkable fidelity." And they devote enormous efforts to control the "great beast," as Alexander Hamilton called the people. They believe, rightly I think, that there is much truth to David Hume's observation in his *First Principles of Government* that "as Force is always on the side of the governed, the governors have nothing to support them but opinion. 'Tis therefore, on opinion only that government is founded; and this maxim extends to the most despotic and most military governments, as well as to the most free and most popular"—in fact, more so for the more free and popular governments. That is why we find that as rights were won by popular struggle, rulers turned more and more to sophisticated modes of propaganda. It's quite natural that the modern techniques of control of thought and attitudes were forged primarily in the more free societies, Britain and the United States.

JB *Q7: Worse, what about the control of the media over our ways of thinking? Aren't we moving toward a society where the mass of the population is undereducated and brainwashed by the infotainment industry? Isn't the latter much more dangerous than the lies of the New York Times (that you like to expose), which is after all an elite newspaper? Consider the Gulf War or the war in Kosovo. There were only limited protests against them. But since the infotainment industry actually attracts the people, satisfying their desire for distractions, what can be done about that?*

NC I pay attention to the *New York Times* and other elite journals for several reasons. One is that they set the agenda within which others operate. Another is that they are part of the dominant intellectual culture, a matter that interests me a great deal. Undoubtedly the "infotainment industry" is huge. And, as its leaders are kind enough to tell us, they are dedicated to establishing "off-job control" as a counterpart to the "on-job control" of the Taylorist systems designed to turn working people into unthinking and obedient robots, to focus people's attention on the "superficial things of life such as fashionable consumption," and to induce among the general population to a "philosophy of futility." It is no doubt important to emphasize all of this, and there is very good work on it, which I have often cited. I am happy to leave that work to others, who do it very well,

I think. One reason is that I do not know much about it and have neither the interest or resources to learn more: about TV, for example. In contrast, critical analysis of the intellectual culture and the elite agenda-setting media functioning within it is, not surprisingly, a highly unpopular venture among intellectual elites and is rarely undertaken seriously. But I would not try to convince others to adopt my priorities, though I think they are reasonable ones, for me at least.

Let us take a look at the two examples you bring up. The Kosovo war was largely fostered by elite intellectuals, who continue to propagate the most extraordinary fabrications about it, evading massive documentation from impeccable Western sources that demonstrates, at a level of confidence rarely attained in the study of history, that the standard claims and justifications are completely untenable. The general population is presented with a residue of this, and few have the time, energy, or resources to carry out a personal research project to find out the truth of the matter. For the general population, it was never much of an issue: remote, bloodless (for the United States), and framed as a humanitarian exercise to put an end to ethnic cleansing—which, as was known but concealed (and still largely is), was the anticipated consequence of the bombing. It is an enlightening case that demonstrates the crucial importance for a critical analysis of the elite intellectual culture and its media.

The first Gulf War was also in important respects an intellectual's war, fostered by the elite media. Just to illustrate, as the bombing began, about two-thirds of the American public favored Iraqi withdrawal from Kuwait in the context of a regional conference on security measures, including the Arab-Israeli conflict. The figures would undoubtedly have been much higher had people been permitted to know that Iraq had just made such an offer, regarded by State Department specialists as "serious" and "negotiable" but instantly rejected by Washington. The agenda-setting media and the elite intellectual culture kept the facts hidden and largely still do; there was, literally, only one daily newspaper in the United States, *Long Island Newsday*, which reported the facts accurately throughout, from August 1990 to January 1991. So the public was deluded, but not by the infotainment industry.

Nevertheless, the first Gulf War was the occasion of the greatest popular antiwar protests in American history, to my knowledge—demonstrations of hundreds of thousands of people organized even before the war began. And the second Gulf War elicited even more massive protests, well before it was officially launched—actions without any historical precedent. The contrast to the 1960s is dramatic. Protest against the Vietnam War at first was so slight that few even knew about it. There was scarcely any detectable protest or opposition for years after the attack against South Vietnam began in full force. By the time protest reached a significant scale, the renowned French military historian and Indochina specialist Bernard Fall was warning that "Vietnam as a cultural and historic entity . . . is threatened with extinction [as] the countryside literally dies under the blows of the largest military machine ever unleashed on an area of this size." And even then, protest focused on the attack against the North, even though it was South Vietnam that always bore the brunt of the U.S. assault. To this day, the public is unaware of the horrifying impact of the chemical warfare that was initiated by John F. Kennedy in 1962 against South Vietnam: the North was spared this particular atrocity, which still severely affects newborn children in the South. There is plenty of evidence available, including scientific studies by prominent U.S. and Canadian investigators. But one has to search hard to find them, or to discover the shocking accounts by the few journalists (notably, the fine Israeli journalist Amnon Kapeliouk) and health workers who have sought to investigate. Again, the responsibility lies crucially on the elite intellectual culture and the agenda-setting media, not the infotainment industry.

It is for Europeans to consider comparisons to their own behavior: for example, comparisons to protest in France against the murderous French wars in Indochina, or against France's role in horrendous atrocities in Algeria just in the last decade, or the British reaction to their shocking crimes in Kenya in the 1950s, or much else in the past and current history of European violence throughout the world. It is true that there is often protest against the crimes of others, but that is always much easier and more appealing than looking in the mirror.

The vast increase in popular opposition to aggression among the general U.S. population since the 1960s is well known to government planners and of great concern to them. When any administration comes into office, it commissions a review of the international situation from the intelligence agencies. They are of course secret, sometimes declassified decades later. But parts were leaked when Bush Senior came into office in 1989, dealing with conflicts against "much weaker enemies"—that is, any enemy that might be confronted. Intelligence advised that the United States must win "decisively and rapidly" or else popular support will erode, because it is so thin. It is no longer the 1960s, when the public would tolerate years of war with vast devastation and slaughter before any protest developed. And when the business classes turned against the war and intellectual elites and the elite media began to as well, it was mostly on very narrow grounds of cost and failure, though for the public it was a moral issue. Unlike intellectual elites (and the elite media), by 1969 about 70 percent of the public regarded the war as "fundamentally wrong and immoral," not "a mistake," figures that have remained pretty stable until the present, despite the lack of articulate support and the elite consensus that it was at worst a "mistake" that became too costly.

I think the questions are posed in a form that is unfair to the general public. I do not know of any evidence that they are any more subordinated by propaganda assaults than they are by elite intellectuals, and I suspect that the opposite may be the case, for good reasons. And while those in the faculty clubs and editorial offices may not spend as much time watching sitcoms and sports events as the "great beast" does, they have other methods of "distraction" from uncomfortable realities.

JB *Q8. You sometimes say, when discussing violence, that you do not hold absolutist views on that or on anything else. But isn't freedom of speech a counterexample? Don't you hold absolutist views on that? You sometimes say that there are only two possible attitudes on freedom of speech: either the totalitarian one or your own—the libertarian one. But why not adopt an intermediate view, like the one that exists in France or Germany, where some extreme and horrendous views are banned but where everything else is allowed? You surely cannot use a slippery slope argument and claim that such a mild censorship will spill over to other views. After all, in those countries a huge variety of political and philosophical views still exist (a variety perhaps even more diverse than in the United States).*

NC I certainly do not hold absolutist views on freedom of speech or anything else. There are imaginable occasions, sometimes real ones, when infringement on freedom of speech is legitimate, in my view, and I have always been completely explicit about that. Thus, I have argued, and believe, that the U.S. Supreme Court finally reached an appropriate characterization of the limits of freedom of speech in 1969—reiterating Enlightenment views, those of Jeremy Bentham for example. That is an achievement well beyond anything in Europe or elsewhere, to my knowledge. That was, incidentally, in a case involving the Ku Klux Klan, which advocates "extreme and horrendous views" that have taken a terrible toll in the past and cause great harm today. While of course despising their views, the Supreme Court nevertheless upheld their right to freedom of speech. In contrast, the call for suppressing views that one regards as "extreme and horrendous" is quite normal. Even Hitler and Stalin were happy to grant freedom of speech to views of which they approved. The test of honest commitment to freedom of speech is the aphorism attributed to Voltaire.

The "intermediate view," if taken seriously, would virtually shut down the media, the universities, and the elite intellectual culture in general. Very clearly, the "intermediate view" would ban support for massive ongoing Western or Western-backed atrocities, a far more serious matter than denial of even the most horrendous past crimes or crimes of others. That it is a far more serious matter should be too obvious even to debate, at least among people who accept the most elementary moral standards. And it therefore follows, by simple logic, that if the "intermediate view" grants the state the right to ban what state authorities determine to be "extreme and horrendous views," then the state should ban even more extreme and horrendous views, such as support for massive ongoing Western or Western-backed atrocities—that is, close down the state propaganda systems, the media, the universities, and the elite intellectual culture in general. The cases brought up earlier are a few of a great many examples.

But the "intermediate view" is never taken seriously. It grants the state the power to silence "extreme and horrendous views" that one does not like but not those that one finds tolerable, like direct support for massive ongoing crimes. Still worse, many European intellectuals are even willing to grant the state the authority to be the official custodian of History, to determine "Historical Truth" and punish deviation from it, thus taking their stand alongside of Goebbels and Zhdanov. They may try to convince themselves that they are opposing Nazism and Stalinism, but the reality is that they are supporting some of its outrageous doctrines—highly selectively of course. The "slippery slope" argument never arises, because of the extreme selectivity of these judgments. No one who advocates these principles, disgraceful in my view, would dream of using them to silence advocacy of the terrible crimes of their own states and corporations, or apologetics for these crimes, or the routine denial of them that disfigures the intellectual culture. These are weapons against enemies and not to be applied to oneself.

I do not, incidentally, agree with the belief of many European intellectuals that there is a "huge variety of political and philosophical views" in Europe, more so than in the United States. I do not believe that these self-serving and comforting views can withstand analysis.

JB *Q9. You are strongly opposed to any form of "popular dictatorship." But isn't that wishful thinking? Look at the Paris Commune, Allende, the Sandinistas, and many other examples. If a popular revolution isn't strong enough, it will be crushed by internal subversion and external pressure. In fact, the political thought of Lenin is to a large extent based on a reflection on the failure of the Paris Commune. Isn't it preferable, in the end, to have some form of limitation of political freedoms, as in Cuba, and be able to preserve some minimal level of subsistence and some health care for the majority of the population, rather than having them sink into utmost poverty, as elsewhere in Latin America? Given the meaninglessness of most elections today, on what human basis can one consider such games as more important than health care and food?*

NC I understand the examples quite differently. Lenin did not institute a "popular dictatorship" but rather a dictatorship of the Central Committee, ultimately the Maximal Leader, over the population—much as Rosa Luxemburg and (before he joined in) Trotsky had predicted in earlier years. Lenin and Trotsky moved very quickly to suppress the popular revolution and to demolish its achievements and the institutions of popular power that were taking shape in the months before the Bolshevik take-over, which was more of a coup or a counterrevolution than a revolution, in my opinion. They led the subversion and destruction. Again, one may argue that they had no choice: I do not happen to believe that, but it is an entirely distinct argument. The same is true in other cases. A strong case can be made that the communist-led destruction of the popular libertarian movements in Spain did not contribute to the defense of the Republic against Franco but was a significant factor in its collapse and that other modes of popular struggle, such as those advocated by Camilo Berneri before he was murdered, would have had more chances of success, though they would of course have been anathema to communist totalitarians and the Western democracies. The same holds in other cases, including those mentioned, if we look closely. It is quite possible that even under the best of circumstances, the Sandinistas could not have withstood the vicious U.S. terrorist war, supported by many Europeans who like to posture about their courageous opposition to terror and oppression when carried out by official enemies, mimicking the commissars. But anyone familiar with what was happening in Nicaragua knows full well that their authoritarian practices contributed to undermining popular support that might have overcome the terror, and these were also exploited effectively by supporters of Washington's terrorist war to undermine solidarity movements in the United States and Europe. These were all matters that were discussed quite frankly and openly during those years by opponents of Washington's terrorist war. I had plenty of experience with it myself, in Nicaragua particularly. And there are many other examples that can easily be given by those who have been involved in solidarity work, internationally or domestically.

I also do not agree with the implicit assumption that "limitation of political freedom" contributes to preserving subsistence and health care in Cuba. On the contrary, I think it serves to undermine these achievements, which are quite impressive, certainly by the standards of U.S. and European client states. One needs an argument to demonstrate that elections are somehow in conflict with health care and food, as presupposed in these comments: I know of no such argument and think that political freedom and social welfare are mutually supportive, not contradictory. The belief that the Paris Commune, Allende, and the Sandinistas were defeated by virtue of allowing internal freedom flies in the face of the facts, in my opinion. The assumptions that underlie the previous comments should be challenged throughout, I believe. They cannot simply be presupposed without argument and will not, I think, withstand serious analysis. Since I do not see any reason to accept the assumptions, I cannot respond to the questions.

I also do not agree that elections in Western-style democracies are "meaningless." The spectrum is far too narrow and too restricted by concentrations of private power. And the neoliberal onslaught against democracy—its primary thrust—has imposed even narrower limits on functioning democracy, as intended. But it does not follow that politics has become a "meaningless game" or that the attack on democracy cannot be beaten back. Electoral politics has in the past achieved gains in human welfare that are by no means insignificant, as the great mass of the population understands very well. And there is no reason why elections in Western countries cannot rise to the level of, say, Brazil, under far more difficult conditions. Or why we must surrender the freedom and privileges that have been won by centuries of struggle just because there are serious obstacles to making use of them.

JB *Q10. You are often regarded as one of the "thinkers" behind the antiglobalization movement. So, how do you answer the following: what does the movement want? There are all these demonstrations, but there is no leadership, no specific political demands, no party running for elections in democratic countries. So, what can one do to satisfy those involved in the movement?*

NC There may be some marginal "antiglobalization" movement, but if so, I am not part of it and know very little about it. It would surely be a very odd stance for the left or the labor movements. The development of a true "international" has been a primary goal of the left and labor movements since their modern origins, and the same is true of other popular movements, for example, the rapidly growing peasant movements. The Via Campesina aims to be an international organization linked to others. The dominant propaganda systems naturally pretend that the form of investor-rights globalization that they prefer is "globalization" and that efforts to construct other forms of international integration, which privilege the rights of people over those of private concentrations of power, are some primitive and irrationalist form of "antiglobalization." But as usual, it is an error to succumb to the machinations of the propaganda of concentrated power. The issue is not "globalization" but rather something quite different: what form should international integration, economic and other, assume, and what should be its priorities? On these matters, the global-justice movements (as they should more properly be called) are remarkably explicit and detailed about what they want. Of course one would not find this out from the elite media, but the merest look at the programs for the meetings of the World Social Forums in Porto Alegre, for example, would reveal that they cover a wide range of issues of great human concern, and attendance at the sessions would reveal serious discussion often with quite definite programs. The same is true of parallel conferences, such as those of the Via Campesina. Even on the specific issues of the WTO, there are very definite demands with regard to TRIMS, TRIPS, GATS, subsidies, debt, programs of independent development, and much else. And the concerns of these various movements and their proposals go far beyond that, reaching to a host of other issues.

JB *Isn't the movement contradictory? It appears to be strong and worldwide, but only because it joins together many protesters with very different agendas. To take only one example, the South wants less protectionism from the North for its exports, but if that were implemented, there would be huge job losses in the North, which wouldn't please the unions there. For its development, the South needs more globalization (at least of a certain type), not less.*

NC It is true that the international media present the issues in those terms, and as is almost always true even for the most vulgar propaganda, there is an element of truth lurking within. But the reality is quite different. To see why, one has to go beyond very abstract formulations and turn to concrete examples. To take one of the most important ones, consider the North American Free Trade Agreement, NAFTA, which came into force in 1994. The propaganda image is that the needs of U.S. workers were in conflict with Mexico's call for less protectionism, and that U.S. labor opposed NAFTA for these reasons. At the dissident extreme of the elite media (which strongly supported NAFTA), Anthony Lewis of the *New York Times* berated the "backward, unenlightened" labor movement for its "crude threatening tactics" in opposition to NAFTA, motivated by "fear of change and fear of foreigners."

The reality was radically different, as one discovers by looking at the official position of the U.S. labor movement and the stands of other popular movements. True, it takes some work to find this out. The official position of the U.S. labor movement was suppressed in the media and still is. The same is true of the very similar analysis and proposals of Congress's own research bureau, the Office of Technology Assessment: also suppressed. One could learn about them from dissident literature (I wrote about them, for example, in Z magazine and a book that appeared in 1994, *World Orders Old and New*). But few are aware of their analyses and proposals. The labor movement and OTA were both in favor of a NAFTA, but not the executive version to which attention was restricted. Both argued that the executive version of NAFTA would be harmful to working people in all three countries, as proved to be the case, as documented at some length in reports of highly respected mainstream organizations that have also been suppressed in quite illuminating ways, which I and others have discussed, for example, at the World Social Forum. Both OTA and the labor movement presented specific plans for a very different form of a NAFTA, with compensatory funding and other devices that would permit close economic integration but in ways that would benefit, not harm, the mass of the population in the three participating countries, in contrast to the executive version, which, they argued (apparently correctly) would benefit narrow sectors of wealth and power. That aside, NAFTA (like the

WTO rules) is far from a "free-trade agreement." A look at the facts suggests that the questions should be rather significantly reformulated.

We may also recall that the foundation of free trade according to the unread patron saint, Adam Smith, is "free circulation of labor," and, as I already mentioned, Smith clearly perceived the dangers of free circulation of capital and sought to show that it would be prevented "as if by an invisible hand." NAFTA and the WTO are based on the opposite principle: the primary concern is free circulation of capital, and free circulation of labor is not even considered, at least if it infringes on elite interests. Looking more closely, we find that when NAFTA was instituted in 1994, the Clinton administration also initiated Operation Gatekeeper, militarizing the Mexican-U.S. border. Like most borders, the U.S.-Mexico border is artificial, the result of conquest. It had been fairly porous, in both directions, pre-NAFTA. But the advocates of NAFTA recognized that it would lead to an "economic miracle," benefiting corporations and the rich while harming the population, and they wanted to block the anticipated "free circulation of labor" from Mexico to the United States. And so it continues.

The propaganda images are very far from the reality, and there are very concrete proposals for alternative forms of international integration that might benefit the mass of the population generally. That is not to say that there will be no conflicts within national societies or among them. Of course there will. That is simply part of human life. But if that shows that "the movement is contradictory," then so is every human institution and endeavor. Furthermore, there is no "movement"; rather, there are many different intersecting and overlapping movements, to which individuals may associate themselves in a great many complex ways. And once again, there are very few who are opposed to "more globalization," unless we restrict the term "globalization" to the specific version favored by concentrations of power and called "globalization" by their doctrinal systems and media.

JB *A year ago, the World Bank released a study (I have to find it) showing that growth actually helps the poor. So why oppose economic growth? Isn't it selfish, to put it mildly, for Western elites to oppose trade—hence growth—at the expense of the poor?*

NC I presume you are referring to the study by Dollar and Kraay, which has been sharply criticized for analytic errors by a number of international economists, including the Harvard economist Dani Rodrik. If so, the issue is not whether growth helps the poor (a question that cannot be asked, unless "growth" is characterized more carefully) but narrower ones, specifically whether lowering trade barriers and more trade helps the poor by promoting growth. There may be a relation between growth and trade, but it is at best obscure, and as the leading figures of modern growth theory observe, even if there is a relation, no one knows what the causal direction may be (as argued the Nobel laureate Robert Solow, the founder of the modern theory, in a recent interview in the economics journal *Challenge*). Some noted economic historians have observed that one might conclude from the historical evidence that protectionism increases trade and have even suggested how this might be explained: protectionism has tended to increase growth, and growth increases trade (Paul Bairoch). Rodrik, for one, has argued that the remarkable growth of exports of the East Asian "Tigers" is the result of output growth, not conversely, and it is widely recognized that the output growth did not result from adopting neoliberal maxims—quite the contrary. In fact, serious international economists are well aware that the international economy is very poorly understood and do not offer sweeping judgments about such matters.

Who opposes economic growth and trade? Keynes? Herman Daly? Third-world peasants and northern environmentalists who share concerns about environmental destruction? It is true that they give serious arguments against unconstrained "growth" and "trade." We should, however, recognize that as commonly used, these terms have a far-reaching ideological component. The measures of growth exclude "externalities" and would be very different if questions of major human significance were included: sustainability, effects on the environment, depletion of resources, production for use or for profit, the vast transfer of costs to consumers that is associated with increased productivity (another highly ideological notion), and much else. Or take trade. After NAFTA, "trade" between the United States and Mexico increased, as measured by the highly ideological standards adopted. However, the component of "trade" that is intrafirm—that is, internal to huge command economies that are basically totalitarian in character—increased from about half to two-thirds. I rather doubt that Adam Smith would have called that "trade," any more than it is "trade" when General Motors transfers parts from Illinois to Indiana for assembly and then sends the product to New York for sale. As for authentic trade, we know very little, because it is not measured, but it would not be a great surprise to learn that it declined after NAFTA.

A strong case can be made that the most powerful opponents of growth, and even trade, are the advocates of neoliberalism. Among those who followed the rules, growth rates significantly declined under the neoliberal regime. There was substantial growth elsewhere, but as in the past, rejection of the neoliberal rules appears to have been a significant factor. And as international economist David Felix pointed out, the apparent growth in trade is partly illusory (even under the weird concept of "trade" just discussed): the ratio of growth of trade to growth of GDP increased, but that is largely attributable to the slowdown in economic growth, which many prominent economists attribute to core principles of neoliberalism, such as elimination of the capital controls and regulation of currencies that were part of the Bretton Woods system that was dismantled thirty years ago in favor of neoliberal measures.

Rather generally, I cannot respond to the questions as posed because, again, they are based on presuppositions that I do not see any reason to accept—insofar as I even understand them.

JB *What about the antitechnological bias of that movement? Do you really believe that, say, GMOs are a serious health risk? Isn't that just the usual technophobia that has existed all along? Moreover, on what grounds can privileged Westerners oppose such technologies that are embraced by poor countries, such as China?*

NC Why should I believe that GMOs are a serious health risk? The question presupposes that the global-justice movements are totalitarian organizations, in which participants must share beliefs and commitments, perhaps on pain of expulsion. As a matter of fact, I do not know of evidence that GMOs pose serious health risks or health risks at all. But I also believe that people have the right to adopt the "precautionary principle" if they choose and suspend judgment until they are convinced that I am right about this. To call that "technophobia" is like saying that it is "technophobia" to call for testing of pharmaceuticals before they are approved for public use. There are questions of judgment about uncertain matters in all such cases. I do not know which Westerners demand that we must somehow prevent Chinese peasants from growing GMOs if they choose to do so, so I cannot respond to the last question. If the question is whether privileged Westerners, say in Europe, choose to observe the precautionary principle for themselves in this case, contrary to my judgment, then one must ask them, not me, what their grounds are. My own judgment happens to be different, but they have the right to make their decisions whether I happen to agree with them or not.

JB *Q11. Isn't it the case that all form of self-organizations following anarchist principles eventually collapsed (think of various communes in the 1960s and the 1970s, but there were previous experiments). Again, given your strongly nativist views (which may be correct—that is not the issue), why would you expect this to change?*

NC By a similar argument, one could have held in the eighteenth century that efforts to institute political democracy or overcome war or slavery or to protect women's rights or anything else had always collapsed, so why should we even try to promote peace and justice and human rights? Surely that is not much of an argument. Furthermore, it has nothing whatsoever to do with nativism unless we attribute to the nativist the view that the environment (in this case, a complex array of power structures, etc.) has no effect on what human beings do. But that is not nativism but insanity. I also happen to disagree with the historical observation.

There are no fixed "anarchist principles," a kind of catechism to which one must pledge immutable allegiance. Anarchism, at least as I understand it (with ample justification, I believe, but that is a separate matter), is a tendency in human thought and action that seeks to identify structures of authority and domination, to challenge them to justify themselves, and when they cannot do so (as is commonly the case), to work to overcome them. Far from having "collapsed," it is alive and well, and the source of much of the—very real—progress of the past centuries, including very recently, since the 1960s and 1970s. Forms of oppression and injustice that were barely even perceived, let alone opposed, in the recent past are not considered tolerable today. That's success, not failure. I do think it provides some (weak) evidence about fundamental human nature and about what we discover about ourselves when we break barriers to self-recognition, but these are different matters.

JB *Q12. You have argued that the Kosovo war had the effect of strengthening the power of Milošević. But that was at best true in the very short run, as subsequent events showed.*

NC I do not agree with the factual statements. Subsequent events, I believe, have tended to support the judgment of democracy and human-rights activists in Belgrade that the Kosovo war delayed the overthrow of the Milošević regime. I have written about this elsewhere, as have others, and do not understand why one should simply assume that the assessment of leading Serbian democracy activists, or that of the electoral victor, was incorrect. I do not think we can so easily dismiss the judgment of the director of the Belgrade Center for Human Rights and many other anti-Milošević activists that the bombing of Serbia erased "the results of ten years of hard work" by the democratic opposition. Perhaps the Western propaganda that is presupposed in the question is correct, but more than assertion is required.

JB *In fact, it can be argued, and it would certainly be admitted privately by the people now in power in Belgrade, that the United States gave a lot of help to the pro-Western forces in Serbia, first by weakening the country under Milošević through sanctions and bombing and then by massively funding the opposition during the electoral campaign. A similar argument can be made for the Soviet bloc, which was weakened through ideological campaigns and the arms race. But that was mostly thanks to the hardliner forces in the West. Insofar as you consider the fall of communism as a victory for the human spirit, shouldn't you at least credit those forces for that victory?*

NC That view probably is held by the people who gained power in Belgrade, undermining the longstanding efforts of the democratic opposition, and who are dedicated to neoliberal programs for the usual reasons. But the word "admitted" goes too far, because it presupposes that their judgment must be correct, and that requires evidence and argument. Conformity with Western propaganda and the stance of those who gained power does not suffice. The way the question is formulated also presupposes that the bombing and sanctions weakened Milošević, not the opposition to him, but that cannot simply be presupposed: it has to be argued, and I know of no convincing argument to show that activists of the democratic opposition (or Koštunica, for that matter) are incorrect in their judgments.

As for the collapse of the Soviet bloc, it is true that the far right in the United States takes the position described in the question, offering no serious argument, but again, we cannot simply assume that the doctrinal system—in this case, its right-wing jingoist component—is necessarily correct. Respected conservative scholarship holds that these claims are baseless; Raymond Garthoff, to mention only one prominent and highly authoritative example, has reviewed these claims carefully and argued on the basis of research in Russian archives and other material that they are without substance.

However, a good argument can be made that the Kennedy administration did contribute significantly toward the weakening of the Soviet Union by, first, rejecting Khrushchev's proposals for a mutual reduction of offensive military forces and responding to his sharp unilateral reduction by a huge increase in U.S. military spending in order to increase substantially the already enormous U.S. military advantages and, second, by its actions during the missile crisis, further undermining Khrushchev by flatly refusing to announce publicly that the United States would not invade Cuba and that the United States would withdraw Jupiter nuclear missiles from Turkey (which were being replaced by much more lethal Polaris submarines). These moves ended Khrushchev's attempts to shift resources from military to economic development. He was removed, his reformist initiatives ended, and Russia turned to a massive arms buildup of its own in a (hopeless) effort to match the United States. That led to a renewed arms race, an increased threat of nuclear destruction, and the stagnation of the Soviet economy. Exactly as Khrushchev had argued internally (so we know from released archives), Russia could not possibly compete in an arms race with the far richer U.S. economy. A plausible judgment is that the kinds of reforms that Gorbachev attempted far too late might have been initiated in the 1960s and that there might have been a much more constructive transition of Russian society and economy, avoiding the catastrophic consequences of the 1990s and many other tragedies in the intervening years.

Of course we cannot judge with confidence what might have happened, but this is a more plausible interpretation than the standard right-wing jingoist version about the effects of the arms race and hard-line ideological campaigns. And these, we should not allow ourselves to forget, were largely devised and applied in order to support vast atrocities and repression in Latin America and much of the rest of the world, under the guise of "combating communism," taking a terrible toll that is of little concern to U.S. intellectuals or their European counterparts, with the usual margin of exceptions.

Science and Philosophy

JB *I know you often reject "polysyllabic" words like philosophy, but at least two philosophers, McGinn and McGilvray, and maybe more, have written books supposedly inspired by your philosophy. So, let me ask you some questions on your "philosophical" views.*

NC For clarification, I have no objection to the word "philosophy" or to philosophy. The "polysyllabic words" I dislike are those that are devised, as far as I can see, to obscure what can be said simply, creating a false impression of profundity. I again claim no originality for this position. It is a leading theme of twentieth-century Anglo-American philosophy and of its eighteenth-century philosophical antecedents (including the French and Scottish Enlightenment), with which I have a good deal of sympathy.

JB *Again, I know that some of my questions may seem somewhat strange, but I think that they are in some sense natural (from the point of view of your readers).*

Q1. You often sound like a realist, a down-to-earth scientist, maybe even the last positivist (see question 5, below). On the other hand, your nativist views have a strongly constructivist aspect. Sometimes you suggest that experience affects very little our view of the world, which mostly develops itself by internal mechanisms. You seem to hold a biological version of Kantianism. How do you reconcile both views? If our view of the world is so much given a priori, then how can it be objective? Or is it only a mental construction with little connection to reality? Or, if there is a coincidence between the two, do you regard it as some sort of accident? If so, how do you account for it? Surely, you don't believe in some preestablished harmony? Could it instead be explained by natural selection?

NC I have never suggested that experience has little effect on our view of the world. Rather, I adopt the standard view of the biological sciences with regard to organisms, humans among them. Our so-called mental states, like other states of the organism, are determined by an interaction of genetically determined factors, experience, and the operation of extraorganic laws of nature. We seek to unravel the interaction and its effects, without a priori demands. I know of no reasons to suppose that human "higher mental faculties" are somehow outside of nature. There is a kind of "biological version of Kant," developed explicitly, in virtually those words, by Konrad Lorenz, and I think the formulation makes some sense, but for other reasons. The loose similarity to Kant results from the fact that Kant was drawing from earlier views to which there is a much closer similarity: those of the Cartesians and the seventeenth-century neo-Platonists, particularly. That is discussed in some of the classical work on the history of philosophy, by Arthur Lovejoy for example. I have written about this some years ago. James McGilvray, in the book you mention, goes into these matters further.

There is nothing to reconcile, any more than we have to reconcile our understanding of the mammalian visual system with the standard conception of the "down-to-earth scientist" that there is a real world out there, one independent of our perceptions (and independent of the quite different perceptions of a bee or a pigeon in the same fixed outside world).

There is a traditional and important question about why it is that our interpretations of the external world are often more or less realistic. We can ask the same question about bees and pigeons. In the case of humans, there is a further question: why do our intellectual endeavors in the sciences often find that our intuitive, common-sense interpretations have gone far astray, and why can that realization lead to a deeper understanding of the real, external world that we assume does exist? These are all serious questions—about bees, pigeons, and other organisms, and in apparently unique ways, about humans and their conscious intellectual endeavors. They have troubled many philosophers over the ages, rightly. In the post-Darwinian period, it became conventional to adopt the formula that it is the result of natural selection. That is unproblematic, as long as we recognize how little we are saying when we merely intone the formula without adding the work that is necessary for it to be significant. In effect, mere repetition of the slogan says that organisms could not survive long enough to reproduce if they were so poorly adapted to their environment that they could not survive long enough to reproduce. That is undoubtedly true but not very informative. It is at this point that serious questions arise, and they are hard ones.

Some influential philosophers, among them Charles Sanders Peirce, held that the intellectual capacities put to work in the sciences lead toward truth as a result of natural selection, but that is quite unconvincing: the ability to solve problems in quantum mechanics or more primitive stages of the sciences was not, as far as we know, a factor in human evolution. So we are left with more unsolved questions of science to add to the huge list that already exists—fortunately, for those who enjoy and appreciate the quest for knowledge and understanding.

JB *Q2. When discussing mind and body, you sometimes say that physics has shown that "nothing is a machine." But isn't your own view of the mind quite mechanical? People build up the grammar of their native tongue, given some fragmentary data, according to some innate rules. Presumably, they do the same with other mental properties, and maybe with their ethical views as well. All this sounds quite mechanical to me—de la Mettrie and his mind machine (to whom you sometimes refer favorably) would be quite happy.*

NC The question conflates quite different notions of what is a "machine." La Mettrie appears to have adopted the standard conception of the so-called mechanical philosophy: a machine is something with gears and levers, like a complex clock, with interactions through direct contact. But as Newton had already demonstrated, much to his dismay and disbelief, that is not the way the world works. That slowly came to be understood after Newton, though not without much resistance, even among the most prominent scientists.

If by "mechanical" one means operating in accord with deterministic physical principles, then the growth of an organism (as far as we know), including the acquisition of language and other cognitive (ethical, etc.) systems, is mechanical—as is the solar system, though it is not a machine in the sense of Galileo, Descartes, Newton, and other pioneers of the modern scientific revolution. But that is pretty much beside the point. La Mettrie, for one, utterly failed to appreciate or to address the major problems that Descartes and his followers raised about the apparent limitations of machines, which had nothing to do with the acquisition of language and other "mental systems" but rather with their use, an entirely different matter. It is reflections on the use of language that led Descartes and his followers (and some predecessors as well, such as Juan Huarte) to question whether humans fall within the category of machines. Of course, as Newton showed, nothing falls within this category in the sense of "machine" that was taken for granted in the early modern scientific revolution and to which La Mettrie apparently still adhered. But that leaves unanswered the nontrivial questions about choice of action that troubled the Cartesians and others, with the ordinary use of language serving for them as a dramatic and unusually clear (but not unique) example.

Descartes would not have regarded acquisition of language (or other cognitive faculties) as problematic and never proposed that they offered a challenge to the mechanical philosophy. The use of language is quite a different matter.

JB *Q3. You often dismiss the mind-body problem by saying that we don't have a fixed concept of body. True enough, but why does it matter? We do have a fixed concept of body at a certain scale: fluid mechanics did not change radically with the advent of quantum mechanics, and planetary motion can still be, in the main, accounted for by Newton's laws. So, if further discoveries have to be made about quarks or strings, they are very unlikely to affect our views of the cells and neural connections of which the brain is made. But, even apart from that, the traditional way to pose the mind-body problem is to say that, no matter how much you know about cells and their connections, you are not going to see the emergence of sensations—say, pain—from that, unless you know from personal experience what those sensations are. The sensations are simply not part of your physical descriptions, because the latter are quantitative, while the former possess an irreducible qualitative aspect.*

NC Whether it "matters" depends on where one's interests lie. Those who are concerned with the mind-body problem have to begin by telling us what they mean by "body." Galileo, Descartes, and Newton could do so, but post-Newtonians slowly came to realize that what Newton regarded as an "absurdity" that no reasonable person can take seriously is in fact true: there are no bodies (machines, physical world) in the sense that was taken for granted by Newton and his predecessors, and nothing has replaced that notion. That is a remarkable moment in the history of science, with many consequences, including a significant change in the very standards of scientific intelligibility.

As was recognized by the leading historian of science in the nineteenth century, Friedrich Lange, that realization, which slowly came to prevail, marks the end of materialism in any interesting sense. No significant mind-body problem remains. We are left with the position expressed quite lucidly by late eighteenth-century scientists such as Joseph Priestley, who concluded that those properties "termed mental" are the result of the "organical structure of the brain." That should have ended the discussion of the mind-body problem, recognized to be fictitious, leaving the huge and fascinating scientific problem of how the organical structure of the brain yields the properties termed mental. Unfortunately, the understanding of these issues has regressed substantially over the past two centuries, so that today leading philosophers and scientists reiterate Priestley's observations, often in virtually the same words, presenting them as an "astonishing hypothesis," "a radical new idea in philosophy of mind," "a bold hypothesis," "the dramatic leading idea of current neurophysiology," or other such pronouncements. Once we recognize that we are repeating truisms familiar two centuries ago, we can put all that aside and turn to the hard problems.

A current version is that the "hard problem" is the special role of sensations (consciousness, etc.). But that is not the traditional problem of mind. Descartes, for example, regarded free choice of action, appropriate to situations but not caused by them or by internal states, as the true criterion of mind, and his followers sought to devise experimental tests that would enable us to judge whether some other creature or object has a mind like ours. These might be regarded as precursors of the Turing test, but they were far more serious. The tests developed by Cordemoy, for example, were rather like the litmus test for acidity: attempts to determine whether some object has a particular property, in Cartesian science, the property of mind. As noted, that question disappeared with Newton, or it should have. In contrast, the Turing test, which many contemporary philosophers and cognitive scientists regard as of crucial significance to their discipline, is not proposed within the framework of the sciences: it is not an effort to discover some aspect of the real world but rather is a terminological proposal. Turing himself was quite clear about this, but others have seriously misunderstood the matter, I believe.

The problems that concerned Descartes in his ruminations about mind remain as obscure as they were in his day, I believe (though others disagree). Like other problems that seem too hard to address, they were abandoned, not solved. As for sensations and consciousness, one task is to formulate a clear problem that can be addressed. It tells us nothing to say that sensations have an "irreducible qualitative aspect" for us—and presumably for bees and pigeons too (even though we can't ask them). It does not help much to ask "what is it like to be me" or "what is it like to be a bat"? A question becomes meaningful to the extent that we can think of possible answers: otherwise it is an expression with an interrogative form but not a question that can be addressed. And in this case, it is hard, if even possible, to formulate a coherent answer, even a wrong one.

As any scientist knows, formulating a question that can be seriously addressed is no trivial matter, and the success in doing so sometimes opens up completely new branches of scientific inquiry. The transition from perplexity to meaningful question is often a considerable feat of the intellect and imagination. For the core problems of mind, such as those of Descartes, or the contemporary preferences, there is mostly perplexity.

JB *Q4. You have been frequently challenged (for example, on Znet) by supporters of nonviral theories of AIDS. In your responses, you appear like a true believer in official science. However, when you discuss economics, you are rather dismissive of that science, especially of the experts of the World Bank and IMF. Isn't that a display of politically motivated double standards again? You don't like the conclusions of the economists, especially those who are for free trade, while you are indifferent to the real source of AIDS. Since you are neither an economist nor a biologist, what makes you trust some experts and not others?*

NC I am not at all dismissive of economics. If I happen to be interested in the conclusions drawn by the World Bank and IMF, I look at their assumptions, the underlying theories, and the real-world phenomena closely enough so that I think I can draw some sensible conclusions. As to whether I succeed in that or not—that's for others to decide. I don't happen to be interested in the nonviral theories of AIDS and do not see much reason to become interested in them. But if I were, I would undertake the same course. In fact, I have looked at some of the papers that are alleged to undermine standard theories and have discussed the matter with outstanding scientists and medical practitioners, and I have found no reason to question their judgment that the nonviral theories are not well grounded. Not seeing any reason to pursue the matter further, I have left it at that, as in innumerable other areas.

I do not see any double standard at all, beyond the fact that no human being can conceivably master every discipline and no sane human being would even try. I do not follow the rest of the question. It says "linguist" on my door, but does that mean I am an expert on Hittite? It says "biologist" on the door of a friend who works on large molecules. Does that mean he is an expert on AIDS? It says "economist" on the door of some other friend. Does that mean he knows anything about how military spending has served as a cover for much of the "new economy" or what the effects of neoliberalism are in Brazil? We investigate what we find interesting and important. Beyond that, we rely on those we have some reason to believe are fairly trustworthy, just as I don't take a graduate course in civil engineering before I decide whether or not to drive over a bridge.

JB *Q5. You often plead for methodological monism, rejecting the adoption of nonscientific methods when studying humans "above the neck." In that sense, you qualify as the last positivist. But isn't that too simple? There is a vast school of thought, including many subdisciplines, that can be loosely called hermeneutical. This school was always strong in continental Europe and now is being developed in the United States, with the demise of positivism and the rise of postmodernism. Their main claim is that there are many situations, of profound human importance, that cannot be studied by ordinary scientific methods. For example, how do people feel, subjectively, in certain situations (of oppression, for example), or what motivates their reactions, etc. Do you reject that whole approach as unscientific? Isn't then your attitude pretty insensitive, given the human importance of such questions?*

NC I do not know what is meant here by "positivism." If it is the positivism of Comte, its demise took place long ago, at least in domains I know anything about. If it is the "logical positivism" of Carnap, that has been hopelessly misunderstood and I know nothing about its demise in its actual form. If the question is whether one should adopt the methods of rational inquiry in studying humans "above the neck," then yes, I'm a positivist in that sense, and so is anyone who believes in rational inquiry (by definition). If the question presupposes some distinction between rational inquiry and the methods of science, I cannot answer until someone explains the distinction to me. As for what is meant by "positivism" in the postmodern and hermeneutic literature, I can say nothing, because what they seem to have in mind, to the (limited) extent that I understand it, has little resemblance to anything familiar to me or to what has been called "positivism" at least in the intellectual traditions I am more or less familiar with.

If those in the hermeneutic and postmodern traditions have something interesting or instructive to say about the vast domains in which rational (scientific) inquiry tells us nothing much, I would be the first to cheer. But it is "too simple"—to borrow your phrase—to say in a deep voice with a somber look that we are pondering the problem of "how do people feel, subjectively, in certain situations (of oppression, for example), or what motivates their reactions," etc. I ponder that too, and I am afraid I learn more about it from literature than from hermeneutic or postmodern tomes. Perhaps that's my intellectual limitation. It could well be.

I cannot fail, however, to be struck by a certain oddity. If I open the pages of a professional journal in the sciences or mathematics, I usually understand very little. If I happen to be interested in learning more, I know what course to follow and sometimes have done so. Usually the task is far too arduous and remote from my primary concerns. If, nevertheless, I happen to be interested in understanding more, I know just what to do: I can ask someone who concentrates in these areas to explain to me what I want to know at a level corresponding to my interest and understanding. And quite commonly they can do so to my satisfaction, at least if they have a serious grasp of their own field. My experience happens to be quite different when I try to read hermeneutic or postmodern literature. Sometimes I think I can understand it, but in those cases I do not see why it is not put much more simply (which may, of course, mean that I am not grasping the contents). But when I am told, or simply see, that I do not understand, I do not know how to proceed to gain a better understanding, and the practitioners of these arts do not seem to be able to explain to me what is being said, or what insights have been reached, in the way that this is commonly done in the sciences or mathematics or musical composition or other domains.

There might be various reasons for these differences. Perhaps these enterprises have reached hitherto unknown levels of depth. Perhaps I simply have a missing gene. Or other possibilities come to mind. I have no position on these matters, but I see absolutely no basis for the claim that by failing to understand work that no one can make comprehensible to me I am being "insensitive" or showing that I lack interest in questions of "human importance."

I have occasionally agreed, reluctantly, to take part in published discussion of these issues, with prominent figures, in some cases close friends. I am afraid they left me as baffled as when I began. I cannot really go beyond what is already in print.

JB *Q6. There is a new intellectual trend in the social sciences, loosely called "Darwinism," whose supporters have been very disappointed by your attitude. On the one hand, they were grateful to you for having brought back the issue of human nature into the intellectual debate. They applauded your criticism of behaviorism, for example. But you seem to refuse to take the next step, which is to admit that human nature has been shaped by evolution and that the only known mechanism that drives evolution is natural selection.*

NC I do not "admit" but *insist* that human nature has been shaped by evolution. What alternative is there? That it is created by some divinity? But I do not "admit" that "the only known mechanism that drives evolution is natural selection" any more than I "admit" anything else that every biologist knows to be false. It is the merest truism that evolution proceeds within the constraints set by physical and chemical laws; hence it is the merest truism that these principles "drive" evolution. It is also well known that there are many other processes involved in evolution besides natural selection. That natural selection is a critical factor in evolution is not in question. But just what role it plays is always very much in question. There is no point in worshipping at a shrine that every biologist knows to be that of a false god.

JB *So, why not use adaptative thinking in order to guess what elements human nature might contain?*

NC How could anyone object? I certainly do not. Those who find such thinking helpful for their guesses should by all means proceed that way. But as is well known, when put this way, the task is both too easy and too hard. Too easy, because there are innumerable guesses that quickly come to mind in any interesting case—say, the evolution of human language. Too hard, because we know much too little to be able to evaluate the guesses. But, again, if adaptationist speculations lead to guesses that have some plausibility and that can be explored, who could possibly object? Certainly I never have.

JB *Of course, any such guess has to be checked independently, but isn't that a natural way to proceed? Again, isn't your opposition politically motivated, because the Darwinian view of human nature is not as optimistic as the one that your anarchist views lead you to believe?*

NC Since I have no opposition, it cannot be motivated, either politically or in any other way. There is no "Darwinian view of human nature" of any substance, so it cannot be optimistic or pessimistic. My "anarchist views" are no more or less Darwinian than views held by those who prefer to adopt pessimistic conclusions, for whatever reason—certainly not biologically or historically well-grounded reasons. There may well be political motivations behind the routine dismissal of Kropotkin's "Darwinian" speculations that mutual aid is a factor in evolution and that human evolution therefore tends naturally toward communist anarchism, or for preferring the view of many Marxists and others that there is "no human nature apart from history" (whatever that is supposed to mean), or that Thatcher's TINA is forced on us by human nature, and thus Adam Smith,

David Hume, and other founders of classical liberalism were completely wrong in fundamental ways. It may even be interesting to ask why such ideas become prevalent among intellectuals. I think it actually is interesting to ask why, and I have suggested some tentative answers. But I, at least, can think of and know of nothing serious to say at the level of generality at which the questions are posed.

PART III

LINGUISTIC THEORY AND LANGUAGE PROCESSES

5
THE VARYING AIMS OF LINGUISTIC THEORY

CEDRIC BOECKX AND NORBERT HORNSTEIN

The "generative" program for linguistic theory is now about fifty years old. During its short history, the aims and methods of the program have changed, as is to be expected of any scientific approach to natural phenomena. This essay outlines three periods within the generative enterprise. The phases can be (roughly) identified in terms of the different goals that generativists set for themselves, each bringing with it different standards of success and suggesting (somewhat) different research agendas. All three goals are still with us and animate related yet different kinds of linguistic investigation, so getting some clarity on these historical periods might serve to clarify current practice.

The three phases or periods that we wish to consider can be called the *combinatoric*, the *cognitive*, and the *minimalist*. Each offers conceptual parallelisms with (and looks to inspiration from) better-developed sciences. Thus, the combinatoric phase "connects" at some level with engineering, the cognitive phase with biology, and the minimalist phase with physics.

Each period is also associated with a central text (or two) by Noam Chomsky, and the texts give the research within each period its distinctive technical look and theoretical flavor. The core text of the combinatoric period is *Syntactic Structures* (Chomsky 1957). The cognitive era has an earlier and a later part. The central text of the former is *Aspects of a Theory of Syntax* (Chomsky 1965), and that of the latter is *Lectures on Government and Binding* (Chomsky 1981). The last phase reflects the spirit of *The Minimalist Program* (Chomsky 1995). Though these books have many overlapping themes, each also has a peculiar research emphasis and each broadly identifies different criteria of success.[1] As such, they will serve as guideposts in the discussion that follows.

The paper is structured as follows. In section 1, we discuss the combinatoric stage. In section 2, we first highlight the conceptual shift that the generative program underwent and how the cognitive stage was defined (section 2.1). We then outline the logic of the central poverty-of-stimulus argument (section 2.2) and finally show how a successful research agenda eventually emerged (section 2.3). Section 3 focuses on the minimalist program, the most recent stage of the generative enterprise. Here we attempt to clarify the various questions and methods currently being explored in linguistic theory.

1. The Combinatoric Stage

In the beginning, there was *Syntactic Structures*. This slim volume, no more than one hundred pages long, initiated the generative turn in the study of language. The book focuses on developing an explicit (hence the term "generative") formalism adequate for representing linguistic phenomena. Chomsky expressed the broad aim of linguistics as follows: "The fundamental aim of linguistic analysis of a language L is to separate the grammatical sequences which are the sentences of L from the ungrammatical sequences which are not sentences of L. The grammar of L will thus be a device that generates all of the grammatical sequences of L and none of the ungrammatical ones" (1957, 13).

This passage gives the flavor of the enterprise. The primary aim is computational or combinatoric.[2] The problem is framed by two observations. First, the set of well-formed sentences of a natural language is infinite. Second, natural-language sentences naturally partition into two sets: the well formed and the ill formed. Given this partition into two infinite sets, the grammarian's goal is characterize them by finding a set of rules (a grammar) that will generate all the well-formed sentences and not generate any ill-formed ones. If successful, such grammars would constitute comprehensive theories of language comparable to the kinds of theories that chemists and biologists construct in their respective areas (this sentiment is especially made explicit in Lees' 1957 review of *Syntactic Structures*).

Syntactic Structures contains two important arguments that reflect this broad research agenda. They are developed in the context of considering the virtues and vices of alternative formalizations of natural-language grammars. Let's consider some of these.

First, Chomsky argues against the adequacy of finite-state grammars as adequate models for natural language. Chapter 3 is dedicated to this end. Chomsky notes that finite-state, so-called Markovian processes are formally incapable of modeling "languages" that display certain nonlocal dependencies between expressions in a string, e.g., languages that have *n* occurrences of *a* followed by *n* occurrences of *b*. These are ubiquitous in natural language. As Chomsky (1957, 22) notes (the numbering in this quotation has been altered):

> Let S_1, S_2, S_3, . . . , be declarative sentences in English. Then we can have such sentences as:

(A) (1) (i) If S_1, then S_2
(ii) Either S_3, or S_4
(iii) The man who said S_5, is arriving today

> In (1i) we cannot have "or" in place of "then"; in (1ii) we cannot have "then" in place of "or"; in (1iii) we cannot have "are" instead of "is." In each of these cases, there is a dependency between words on opposite sides of the comma. But between the interdependent words, in each case we can insert a declarative sentence, and this declarative sentence may in fact be one of (1i–iii). Thus, if in (1i) we take S_1 as (1ii) and S_3 as (1iii), we will have the sentence:

(B) If, either (1iii) or S_4, then S_5

> and S_5 in (1iii) may again be one of the sentences in (1). It is clear, then, that in English we can find a sequence a+S_1+b where there is a dependency between *a* and *b* and we can select as s_1 another sentence containing c+S_2+d where there is a dependency between *c* and *d*, then select as S_2 another sequence of this form etc. A set of sentences that is constructed in this way . . . will have all of the . . . properties . . . which exclude them from the set of finite state languages.

Chomsky concludes that the presence of (for all practical purposes, an infinite number of) sentences like this in English (and any other natural language one might investigate) precludes any theory of linguistic structure based exclusively on finite-state processes. In short, precisely because such grammars cannot generate *all* the grammatical sequences of English, they are inadequate formal models of grammar.

The next kind of grammar Chomsky considers is a phrase-structure grammar. These do not fail in the same way as finite-state grammars. However, Chomsky argues that such grammars also fail to be *fully* adequate because the grammars constructed exclusively in phrase-structure terms will be "extremely complex, ad hoc, and 'unrevealing'" (1957, 34). To put this another way, grammars restricted to phrase-structure formats will be incapable of expressing obvious, significant generalizations displayed by natural languages. Observe, the claim here is *not* that these grammars cannot draw the line between grammatical and ungrammatical sentences. Rather, it is that they cannot do so in a way that cleaves to the generalizations that the language reveals; they do so clumsily and without, as it were, cutting the language neatly at the joints.

Chomsky illustrates this kind of failure in several examples, the most famous being his discussion of the English auxiliary system in chapter 7. Lasnik (2000) offers a detailed explication of this argument. We will limit ourselves to offering a taste for Chomsky's reasoning by considering his discussion of the relation between active and passive sentences (1957, 42–44). Sentences such as (1) and (2) are related as active to passive.

(1) John ate a bagel
(2) A bagel was eaten by John

Active/passive pairs such as (1)–(2) have several interesting properties. First, the restrictions imposed by *ate* in (1) on the subject and the object is identical to that imposed by *was eaten by* in (2) on its object and subject. Thus, in general if (3) is fine, then the passive (4) is fine. But if it isn't, then neither is (4). This is illustrated in (5) and (6).

(3) NP_1 V NP_2
(4) NP_2 be V+en by NP_1
(5) a. John drinks wine
b. Wine is drunk by John
c. Sincerity frightens John
d. John is frightened by sincerity
(6) a. *Wine drinks John
b. John is drunk by wine
c. *John frightens sincerity
d. *Sincerity is frightened by John

The generalization that appears to be lurking here is that the restrictions that a transitive active verb places on its arguments (its subject and object) are identical to those that a passive version of that verb places on its arguments (the NP in the *by*-phrase and the subject). Coding these facts in a phrase-structure grammar requires a lot of redundancy, as any restriction that the grammar encodes on a transitive predicate must be reencoded for the passive *be+en* forms of these predicates. As such, phrase-structure grammars cannot reveal the generalization expressed in (5) and (6), and this, Chomsky maintains, argues against exclusively relying on such grammars as the formal machinery behind natural languages.

Active-passive sentence pairs are related in other ways. Chomsky notes that there are "heavy restrictions" on the selection of *be+en* in a phrase-structure operation. For example, *be+en* "can be selected only if the following verb is transitive (e.g., *was+eaten* is permitted but not *was+occurred*)." Nor can it be selected if the verb is followed by a noun phrase, e.g., **lunch was eaten John*. Nor can we get active sentences like *John is eating by lunch*. All of these restrictions would have to be built into the phrase-structure rules, adding to the "ad hoc–ness" and inelegance of the system. Chomsky therefore concludes that natural language goes beyond phrase-structure grammars and contains rules that transform structures. For example, he proposes that actives and passives are related by a passive transformation like (7).

(7) If S1 is a grammatical sentence of the form NP1-Aux-V-NP2
Then the corresponding string of the form NP_2-Aux+be+en-V-by-NP_1
is also a grammatical sentence.

Exploiting rules like (7) dispenses with the "inelegant duplication" required in a phrase-structure system and allows for all the special restrictions on *be+en* noted above. In short, grammars with transformational rules can *elegantly* generate these sentences, the elegance of the transformational process stemming from directly relating actives and passives transformationally.

Much of the work done in this first phase of generative grammar investigated various combinations of phrase-structure rules and transformational operations whose aim was to generate all and only the grammatical sentences of a given language L (e.g., English) and reflect the felt relations between sentence types that native speakers intuited. Proposals were evaluated in terms of these criteria. Thus a common criticism of a given proposal could be that a certain proposed transformation failed to generate a grammatical sentence of English, generated an ungrammatical sentence, or failed to "capture a generalization" between sentences that native speakers recognized.

At the time, such standards of evaluation were familiar from other domains of inquiry. For example, at about this time, logicians and philosophers were occupied with axiomatizing various forms of inference. The goal was to find a set of axioms from which it was possible to derive all and only the valid inferences. A secondary aim was to derive all inferences intuited to be of the same kind in the same way. The generative program was quite clearly parallel to this. Two tacit empirical assumptions lay behind these projects:[3] first, that it is possible to bifurcate the set of sentences into the grammatical and ungrammatical (or the class of inferences into the valid and invalid), and second, that it was possible to see what sentences were directly related to which others. In other words, it was assumed that native speakers could directly intuit a sentence's *grammaticality* as well as certain general relations of a grammatical nature between sentences. In this period, then, it was tacitly assumed that speakers had direct insight into the grammatical structure of their language and so could evaluate candidate rule systems in terms of whether they adequately respected these speaker intuitions. The empirical adequacy of a grammar reflected the extent to which it met these conditions. As we see in the next section, this soon changed, as the focus shifted from finding the right axioms (as it were) to solving what came to be known as Plato's problem.

2. The Cognitive Era

The cognitive period has two parts: An early part (section 2.1), during which the generative program indeed became "biolinguistics," and a latter part (section 2.3), where the biolinguistic research agenda experienced broad empirical and theoretical successes. A discussion of the poverty-of-stimulus argument (section 2.2) acts as a bridge between the two subsections.

2.1. Methodological Preliminaries

The first chapter of *Aspects of a Theory of Syntax* (hereafter, *Aspects*) firmly places the study of language in a cognitive, and ultimately biological, setting, and it arguably remains to this day the clearest statement of the generative enterprise as a whole.[4] Here Chomsky argues that the central problem of linguistics is to account for how children are able to acquire their native languages. He describes two standards for evaluating grammatical proposals, known as levels of adequacy.

A *grammar* is *descriptively adequate* "to the extent that it correctly describes the intrinsic competence of the idealized native speaker" (1965, 24). The following quotation clarifies this statement: the "distinctions that [the grammar] makes between well-formed and deviant, and so on . . . correspond[s] to the linguistic intuition of the native speaker . . . in a substantial and significant class of crucial cases." A *theory of grammar* is descriptively adequate "if it makes a descriptively adequate grammar available for each natural language."

Several points are worth emphasizing here. First, descriptive adequacy applies both to particular grammars and theories of grammar. In the former case, the grammar correctly describes what a speaker knows in knowing a particular language. Note that grammars here are evaluated not in terms of generating all and only the grammatical sentences but in more abstract terms: whether they correctly describe a certain cognitive state, specifically, a native speaker's knowledge of his native language. Second, not only are grammars evaluated for each language, but the focus of research is set more abstractly still in that we want our descriptively adequate grammars to follow from descriptively adequate theories of grammar. We can call a grammar *explanatorily adequate* if it meets this second condition.

Chomsky notes that the notions of adequacy become clearer if we consider them against the "abstract problem" of constructing an "acquisition model"[5] for language (1965, 25). The problem facing the child looks as follows:

> A child who has learned a language has developed an internal representation of a system of rules. . . . He has done this on the basis of observations of what we may call *primary linguistic data*. On the basis of such data, the child constructs a grammar—that is, a theory of the language of which the well-formed sentences of the primary linguistic data constitute a small sample.

This conceives of the central question of linguistics as follows: how does the child go from primary linguistic data (PLD), i.e., well-formed, short sentences of the target language, to a grammar for that language, i.e., a procedure for generating an infinite number of linguistic objects? The problem facing the child looks quite formidable when considered from this perspective, as it quickly becomes evident that the linguistic evidence available to the child during the period of language acquisition is simply too impoverished to account for how he or she generalizes from this small sample of cases to a grammar that generates the infinite set of the well-formed sentences of the language. In light of this gap, the broadest aim of linguistic theory is to discover the "innate linguistic theory that provides the basis for language learning" (1965, 25). In other words, the aim of a theory of grammar is to outline the biologically given cognitive structures that enable human children to so reliably and effortlessly project grammars from PLD. Candidate grammars meet the condition of *explanatory adequacy* to "the extent that a linguistic theory succeeds in selecting a descriptively adequate grammar on the basis of primary linguistic data" (1965, 25).

The emphasis on descriptive and explanatory adequacy (especially the latter) provoked a change in the kinds of work that grammarians pursued. In the earlier *Syntactic Structures* period, the aim was to develop rule systems that had the appropriate combinatoric properties. In the *Aspects* era, the aim of the generative enterprise is understood in a much wider cognitive setting. Goals shifted from finding grammars that generate all and only the grammatical sequences to finding grammars that native speakers have actually cognitively internalized ("cognized").

In the *Aspects* era, grammars are empirically motivated in two ways: *internally*, in that they respect a speaker's intuitions about the grammar, and *externally*, by being acquirable by a child in the circumstances that characterize language acquisition.

The first internal motivation is a descendent of the earlier combinatoric goal, somewhat revised. Grammars must respect speaker intuitions broadly construed. There is no suggestion (as there was in the earlier period) that grammaticality is an observable quality of sentences, one that speakers can detect by inspection. Rather, the kinds of "data" against which internal evaluations are made might include the following: a native speaker's judgments about the acceptability of sentences (e.g., this one sounds funny, I wouldn't use it, it sounds archaic, etc.), its relative acceptability when compared with other sentences (this sentence sounds better than that one), its meaning (this one can be adequately paraphrased in two ways, while this one is hard to understand with such and such a meaning), or its acceptability with a given meaning (this sentence is acceptable with this interpretation but not that one). The primary descriptive notion is not "grammaticality" but "acceptability." The former is a theoretical term; the latter is observational. Sentences that are intuited to be acceptable should, by and large, be grammatical (i.e., generated by a descriptively adequate grammar). Sentences judged unacceptable should not be so generated. However, it is important to see that this is true *for the most part*. It is recognized that acceptable sentences might be ungrammatical and grammatical sentences might be unacceptable, and what native speakers can reliably intuit is acceptability (or acceptability under an interpretation), *not* grammaticality.

Let us put this point another way. To get the prior combinatoric program off the ground required reliably bifurcating linguistic objects into two groups, the grammatical and the ungrammatical. Once this is done, the goal of finding systems that generate all of the former and none of the latter can begin. However, if being (un)grammatical is not something that speakers can directly judge, then it is unclear how to get the program going. Thus, the combinatoric enterprise rests on the tacit assumption that speakers, by consulting their intuitions, can say whether or not a sentence is grammatical. Upon reflection, this assumption seems unfounded. Speakers do have intuitions about the linguistic objects in their native languages, but these intuitions are raw and unlabeled. They can reliably tell *how* a given "sentence" strikes them. However, they cannot reliably tell whether it is grammatical. Grammaticality is a theoretical assessment made by the linguist, not an eyewitness report that a speaker makes by introspective examination of his intuitions. This affects the central aims of grammatical theory. Why? Because what it means to generate all and only the grammatical sentences is actually to provide a description of whatever it is (presumably some kind of cognitive state or states) that underlies a native speaker's ability to consistently judge a sentence's acceptability. Against this larger backdrop, a speaker's grammatical knowledge results in his judgments about sentences but is not co-extensive with it. The immediate problem is not just combinatoric but a project first in cognitive psychology and ultimately in biology more generally.

Something else becomes evident at this stage as well. Finding a system that accounts for a speaker's judgments is just the first step in a wider problem, viz., finding out how the speaker acquired the knowledge that underlies these judgments. This brings us to the second concern, the external justification of grammar. Say that what native speakers know (in part) is some transformational grammar of their language. The question immediately arises as to how they came to possess this knowledge. External justification hinges on outlining explanatorily adequate grammatical theories, theories embedded in accounts of how the grammars postulated could have arisen. The main hurdle here is that the descriptive resources of transformational grammars are very broad. The set of *actual* human transformational grammars is a very small subset of the possible ones. Why then do children reliably and effortlessly converge on the ones that they do and not on the others? The answer cannot completely rest with the fact that children build grammars to conform to the language they hear around them (although this is, *of course*, part of the answer), for, when looked at closely, there are too many ways for children to generalize from the linguistic input available to them to rules consistent with these inputs. Nonetheless, most of these logically possible options are not taken. So what constrains the child?

In *Aspects*, this question is approached by looking for invariances across natural-language grammars that can be taken as *innately* constraining the class of possible grammars. In other words, there must be principles of universal grammar that restrict the candidates for a possible language's particular grammars. There are several ways of hunting for such invariances. One is to consider many languages and see what commonalities, if any, emerge. Call this the typological approach.[6] However, the way *Aspects* frames the problem suggests an alternative, more abstract method of investigating the invariant properties of the language faculty. The logic behind this approach is known as the poverty-of-stimulus argument (POS). Because the argument is central to the enterprise, and because it has been repeatedly misunderstood, we sketch it in the next subsection.

2.2. The Poverty-of-Stimulus Argument

In this subsection, we first set the stage for the POS by discussing a concrete example (question formation, section 2.2.1), then turn to the logic of the POS in relation to the example under discussion (section 2.2.2), and conclude that POS is virtually forced upon us. Finally, we reexamine the premises of the POS and show that they are equally compelling (section 2.2.3).

2.2.1. THE ACQUISITION PROBLEM

Consider how English forms yes/no questions (questions whose answer is "yes" or "no").

(8) a. Is Mary at home? (Answer: Yes, Mary is at home)
b. Can Bill sing? (Answer: Yes, Bill can sing)
c. Will Mary be at the party tomorrow? (Answer: Yes, Mary will be at the party tomorrow)

The questions seem to be related to their (affirmative) answers as follows.

(9) To form a Y[es]/N[o] question concerning some state of affairs described by a structure S, transform S as follows: Find the Auxiliary of S and put it at the front.

So, in (8a), the proposition of interest is described by the sentence "Mary is at home." The auxiliary in this sentence is *is*. The rule says that one moves this to the front to derive the Y/N question: *Is Mary at home?*[7]

The procedure in (9) works fine for these simple cases, but it fails for more complex sentences, like (10).

(10) Will Mary believe that Frank is here? (Yes, Mary will believe that Frank is here)

(10) is problematic because there is more than one auxiliary, so the injunction to move *the* auxiliary is inapposite. We must specify which of the two (/*n*) auxiliaries gets moved. To accommodate (10), we can modify (9) in several ways. Here are several options:

(11) a. Move the main clause Aux to the front
b. Move the leftmost Aux to the front
c. Move any Aux to the front

Each of these revisions of (9) suffices to generate (10). However, with the exception of (11a), they all lead to unacceptable sentences as well. Consider how (11c) applies to the affirmative answer of (10). It can form the Y/N question depicted. However, it also can form the Y/N question (12) if the rule chooses to move *is*. In other words, (11c) overgenerates.

(12) *Is Mary will believe that Frank here?

(12) is English word salad and will be judged highly unacceptable by virtually any native speaker. Thus we know that native speakers of English do not use a rule like (11c). We are also confident that they do not use rules like (11b), based on sentences like (13).

(13) The man who is tall will leave now.
The Y/N question that corresponds to (13) is (14a), not (14b). The latter is terrible.
(14) a. Will the man who is tall leave now?
b. *Is the man who tall will leave now?

(11b) predicts just the opposite pattern. Thus, (11b) both over- and under-generates.

However, (11a) runs into no similar difficulty. The main clause auxiliary is *will*. The auxiliary *is* resides in an embedded clause and so will not be moved by (11a). Thus it appears that we have evidence that the rule that native speakers of English have acquired is roughly that in (9) as modified in (11a).

Now the typical *Aspects* question is: how did adults come to internalize (11a)? There are two possible answers. First, the adults were once children and as children they surveyed the linguistic evidence and concluded that the right rule for forming Y/N questions is (11a). The other option is that humans are built so as to only consider rules like (11a) viable. The reason they converge on (11a) is not that they are led there by the linguistic data but because they never really consider any other option.

The second answer is generally considered the more exotic. Some continue to resist it. However, the logic that supports it is, we believe, impossible to resist. It also well illustrates the POS strategy, as we will now show.

2.2.2. THE LOGIC OF THE POS

Let us assume, for the sake of argument, that the correct rule, (11a), is learned. This means that children are driven to this rule on the basis of the available data, the PLD. A relevant question is: what does the PLD look like? In other words, what does the linguistic input that children use look like? What is the general character of the PLD? Here are some reasonable properties of the PLD: First, it is finite. Children can only use what they are exposed to and this will, by necessity, be finite. Second, the data that the children use will be well-formed bits of the target language, e.g., well-formed phrases, sentences. Note that this *excludes* ill-formed cases and the information that it is ill formed. E.g., (12) and (14b) above will *not* be part of the data that the child has access to in divining the Y/N rule; that is, they are not part of its PLD for this rule. Third, the child uses relatively simple sentences. These will be short simple things like the sentences in (8) by and large. *If* this is the correct characterization of the PLD available to the child, then we can conclude that some version of the more exotic conclusion above is correct. In other words, it is not that the child learned the rule in the sense of using data to exclude *all* relevant alternatives. Rather, most of the "wrong" alternatives were never really considered as admissible options in the first place.

How does one argue for this second conclusion? By arguing that the PLD is insufficient to guide the observed acquisition. Consider the case at hand. First, native speakers of English have in fact internalized a rule like (11a), as this rule correctly describes which Y/N questions they find acceptable and which they reject. Second, one presumably learns the rule for Y/N questions by being exposed to instances of Y/N questions rather than, for example, seeing objects fall off tables or being hugged by one's mother. Say that the PLD relevant for this are simple well-formed instances of Y/N questions, sentences analogous to the examples in (8). On the basis of such examples, the child must fix on the correct rule, roughly something like (11a). The question now is: does the data in (8) suffice to drive the child to that rule? We already know that the answer is no, as we have seen that the data in (8) is compatible with *any of the rules* in (11)! Given that there is only a single auxiliary in these cases, the issue of which of several to move never arises. What of data like (10)? These cases involve several auxiliaries, but once again all three options in (11) are compatible with both the data in (10) and the data in (8).

Is there any data that could decisively lead the child to (11a) (at least among the three alternatives)? There is. We noted that examples like (14a) argue against (11b) and that (14b) and (12) provide evidence against (11c). However, the child could not use these sorts of cases to converge on rule (11a) *if she only uses simple well-formed bits of the language as data*. In other words, if the PLD is roughly as described above, then sentences like (14b) and (12) are not part of the data available to the child. Examples (12) and (14b) are excluded from the PLD because they are unacceptable. If such "bad" sentences are rarely uttered, or, if uttered, are rarely corrected, or, if corrected, are not attended to by children, then they will not form part of the PLD the child uses in acquiring the Y/N question rule. Similarly, it is quite possible that examples like (14a), though well-formed, are too complex to be part of the PLD. If so, they too will not be of any help to the child. In short, though there *is* decisive linguistic evidence concerning what the correct rule is—i.e., it is (11a) and not (11b) or (11c)—there need not be such evidence in the PLD, the evidence available to the child. And this would then imply that the child does not arrive at the right rule *solely* on the basis of the linguistic input of the target language. But *if* it does not use the linguistic input (and what other sort would be relevant to the issue of what the specific rule of Y/N question looks like in English?) and all native speakers of English come to acquire the rule in (11a), it must be the case that this process is guided by some process internal to the language learners. In other words, this implies that the acquisition is guided by some biological feature of children rather than some property of the linguistic input. The conclusion, then, is that children have some biological endowment that allows them to converge on (11a) and not even consider (11b) or (11c) as viable options.

This is a brief example of the POS argument. The logic is tight. Granted the premises, the conclusion is ineluctable. What then of the premises? For example, is it the case that children only have access to acceptable forms of the language, i.e., not cases like (12) or (14b)? Is it true that children do not use complex examples? Before considering these questions, let us reiterate that if the premises are granted, then the conclusion seems airtight: if the acquisition does not track the contours of the linguistic environment, then the convergence to the correct rule requires a more endogenous, biological explanation. So, how good are the premises?

2.2.3. THE PREMISES OF THE POS REVISITED

For the PLD to be the main causal factor in choosing between the options in (11), we would, at the very least, expect the relevant data to be robust in the sense that *any* child might be expected to encounter sufficient examples of the decisive data. Recall that virtually all native speakers of English act as if (11a) is the correct rule. So the possibility that *some* children might be exposed to the decisive sentences is irrelevant given that *all* speakers converge on the same rule. Moreover, it must be robust in another sense. Not only must all speakers encounter the relevant data, but they must do so a sufficient number of times. Any learning system will have to be supple enough to ignore noise in the data. Learning cannot be a single-example affair. There must be a sufficient number of sentences like (12) and (14b) in the PLD if such sentences are to be of any relevance.

It is regularly observed that the PLD *does* contain examples like (14).[8] Again, however, this is not the relevant point. What is required is that there be enough of it. To determine this, we need to determine how much is enough. Legate and Yang (2002) and Yang (2002) address exactly this problem. Based on empirical findings in Yang (2002), they propose to "quantify" the POS argument. To do that, they situate the issue at hand in a comparative setting and propose "an independent yardstick to quantitatively relate the amount of relevant linguistic experience to the outcome of language acquisition" (Yang 2002, 111). The independent benchmark they propose is the well-studied use of null subjects in child language. They note that subject use reaches adult levels at around three years of age. This is comparable to the age of children Crain and Nakayama (1987) tested for yes/no questions (youngest group: 3.2 years). The core examples that inform children that all English (finite) sentences require phonologically overt subjects are sentences involving expletive subjects (e.g., *there is a man here*). Such sentences amount to 1.2 percent of the potential PLD (all sentences). Legate and Yang suggest, quite reasonably, that the PLD relevant to fixing the Y/N question rule should be of a roughly comparable proportion. To be generous, let's say that even 0.5 to 1 percent would suffice.

Pullum (1996) and Pullum and Scholz (2002) find in a sentence search of the *Wall Street Journal* that about 1 percent of the sentences have the shape of (14), putting it within our accepted range. However, as Legate and Yang note, the *Wall Street Journal* is not a good surrogate for what children are exposed to. A better search would be in something like the CHILDES database, a compendium of child-caretaker linguistic interactions. In a search of this database it appears that sentences like (14) amount to between .045 to .068 percent of the sentences, well over an order of magnitude *less* than is required. In fact, as Legate and Yang observe, this number is so low that it is likely to be negligible in the sense of not being reliably available to every child! Just as interesting, of roughly 67,000 adult sentences surveyed in CHILDES (the kind of data that would be ideal for the child to use), there is not a single example of a Y/N question like (14). If this survey of CHILDES is representative of the PLD available to the child (and there is no reason to think that it is not), then the fact that the *Wall Street Journal* contains sentences like (14) is irrelevant. Recall, however, that it is these sorts of sentences that would provide evidence for choosing (11a) over (11b). And if they are missing from the PLD, as seems to be the case, then it seems that the PLD is too poor to explain the facts concerning the acquisition of the Y/N question rule in English. In short, the conclusion of the POS argument outlined above follows.

We have here dwelled on this issue because it has been recently advanced (once again) as a refutation of the nativist conclusions of the POS argument. However, to be fair, we should observe that our discussion above is too generous to the opponents of the POS. The discussion has concentrated on whether examples like (14a) occur in the PLD. Even if they did, it would not undermine the argument presented above. The presence of sentences like (14a) would simply tell us that the PLD *can* distinguish (11a) from (11b). It does not yet address how to avoid generalizing to (11c). This option must also be removed, or the nativist assumptions are once again required. However, (14a) does not bear on (11c) at all. It is (12) and (14b) that are relevant here. Such data, often called "negative evidence," is what counts. Is negative evidence present in the PLD? If it is, how would it be manifest?

One way would be if adults made the relevant errors and corrected themselves somehow. However, nobody makes mistakes like (12) and (14b). Such sentences are even hard for native speakers to articulate. A second possibility would be that children make errors of the relevant sort and are corrected somehow. However, this too is virtually unattested. Children never make errors like those in (12) and (14b) even when strongly primed to do so (see Crain and Nakayama 1987). If they do not make the errors, then they cannot be corrected. Moreover, there is plenty of evidence that children are very resistant to correction (see McNeill 1966; Jackendoff 1994, 22ff.). Thus, even when mistakes occur, children seem to ignore the best-intentioned efforts to help them along grammatically. A third option is to build the negative-evidence option into the learning process itself. For example, we might say that children are very conservative learners and will not consider structures as possible that they have not observed instances of. This is often referred to as "indirect negative evidence." The problem with this, however, is that it is difficult to state the restriction in a way that won't be obviously wrong. Recall that children are exposed to at most a finite number of sentences and, therefore, to at most a finite number of sentence patterns. Recall that it seems that a negligible number of sentences like (14a) occur in the PLD, so if children were too conservative they would never form such questions. Moreover, mature native speakers can use and understand an unbounded number of sentences and sentence patterns. If children were conservative in the way hinted at above, they could never fully acquire language at all, as they would never be exposed to most of the patterns of the language. Thus, any simple-minded idea of conservatism will not do, and we are left with the conclusion that the assumption that children do not have access to negative data in the PLD is a reasonable one.

To get back to the main point, if what we have said above is correct, then why children do not opt for rules like (11c) is unaccounted for. Recall that only negative data tells against (11c), as the correct option, (11a), is simply a proper subcase of (11c). It would seem, then, that both the logic and the premises of the POS argument are sufficient to lead us to conclude that language acquisition is not explicable *solely* on the basis of the linguistic input. More is needed. In particular, we follow Chomsky in asserting the need for some biological, human-specific mechanism for language development.[9]

2.3. Taking Stock

The *Aspects* project of developing theories that had explanatory adequacy brought with it several abstract requirements that Chomsky enumerated (1965, 31).

We must require of such a linguistic theory that it provide for

(A) (i) an enumeration of the class $s_1, s_2 \ldots$ of possible sentences
(ii) an enumeration of the class $SD_1, SD_2 \ldots$ of possible structural descriptions
(iii) an enumeration of the class $G_1, G_2, \ldots$ of possible generative grammars
(iv) specification of a function f such that $SD_{f(I,j)}$ is the structural description assigned to sentence s_i by grammar G_j, for arbitrary i,j
(v) specification of a function m such that $m(i)$ is an integer associated with grammar G_i as its value (with, let us say, lower values indicated by higher numbers)

A device that met these requirements could utilize the PLD to form grammars adequate to the input. The fifth condition, the "evaluation metric," orders the biologically available grammars along an accessibility hierarchy. The language-acquisition device (i.e., the child) chooses the most highly valued grammar—the one with the lower integral value in (v)—compatible with assignment of structural descriptions for every sentence of the PLD. Thus, the evaluation metric in combination with the PLD selects a grammar. This is what language acquisition amounts to.

The empirical challenge is to specify the evaluation function in (v) and the class of possible generative grammars in (iii). Restricting the class of possible grammars proved to be quite successful. For example, POS arguments like the one above led to the conclusion that human grammars could only use "structure dependent" operations, ones that exploited hierarchical (rather than linear) structure. This would exclude rules like (11a), which exploited linear notions like "leftmost" from the class of possible grammatical operations. Similarly, research into the properties of transformations led to the discovery that certain grammatical configurations were immune to alterations of certain sorts, thereby forming so-called islands. Here the work of J. Ross (1967, 1986) deserves special mention. Interestingly, Ross argued based on examples like (15) that displacement of a grammatical category outside of a relative clause gives rise to unacceptability. (The original position of the displacement element, *who*, is marked by being indicated as ~~<*who*>~~.)

(15) *who did John meet [the woman [that met ~~<who>~~]]?

Note that the constraint on displacement just illustrated, known as the complex NP constraint, prevents the generation of examples like (11c), as it would involve movement out of a relative clause. In sum, many grammatical restrictions were discovered that served to cut down the class of possible operations and so restrict the space of admissible options. (Aside from Ross's seminal thesis, see especially Chomsky 1973 and Emonds 1970.)

There was, however, little progress on point (v) above. Stated from a cognitive perspective, the issue is this: The acquisition problem is bounded by two big facts. First, the POS nature of the acquisition process. Second, the fact that languages (and their grammars) differ. The problem facing the child is to choose a grammar that fits the PLD from the class of possible grammars. The evaluation measure arrays the class of possible grammars in a descending order of desirability. The task, then, is to take the PLD and find the "best" grammar (i.e., highest ranked) that fits it.

Although this characterization is abstractly correct, it proved to be hard to implement. In fact, it is fair to say that the abstract characterization in point (v) was only made usable empirically in the early 1980s with the introduction of the "principles and parameters model." Thus, though the problem was clearly identified in *Aspects* and the general form of a solution sketched out, a *workable and usable* proposal was not. Put bluntly, nobody quite knew how to specify the evaluation metric. A workable proposal emerged in *Lectures on Government and Binding* (hereafter *LGB*; Chomsky 1981), in the guise of a "principles and parameters" architecture, to which we now turn.

2.3. *Principles and Parameters*

Since *Aspects*, the central problem in linguistics has been identical to the one in the branch of biology known as "theoretical morphology" (see McGhee 1998). Those that Kauffman (1993) dubbed the "rationalist morphologists," such as Goethe, Cuvier, and St. Hilaire, had already recognized that extant organismal forms are only a subset of the range of theoretically possible morphologies. The primary question of theoretical morphology parallels the one within generative grammar:

> The goal is to explore the possible range of morphologic variability that nature could produce by constructing *n*-dimensional geometric hyperspaces (termed "theoretical morphospaces"), which can be produced by systematically varying the parameter values of a geometric model of form. . . . Once constructed, the range of existent variability in form may be examined in this hypothetical morphospace, both to quantify the range of existent form and to reveal nonexistent organic form. That is, to reveal morphologies that theoretically could exist . . . but that never have been produced in the process of organic evolution on the planet Earth. The ultimate goal of this area of research is to understand why existent form actually exists and why nonexistent form does not (McGhee 1998, 2).

Aspects essentially identified this "ultimate" goal, but no workable mechanism for generating the "theoretical morphospaces" was available until the 1980s.

In *LGB*, the issue was conceived as follows.[10] Children come equipped with a set of principles of grammar construction (i.e., universal grammar [UG]). The principles of UG have open parameters. Specific grammars arise once values for these open parameters are specified. Parameter values are determined on the basis of PLD. A language-specific grammar, then, is simply a specification of the values that the principles of UG leave open. This conceives of the acquisition process as sensitive to the details of the environmental input (as well as the level of development of other cognitive capacities), as it is the PLD that provides the parameter values. However, the shape of the knowledge attained (the structure of the acquired grammar) is not limited to information that can be gleaned from the PLD, since the latter exercises its influence against the rich principles that UG makes available. Much of the work since the middle of the 1970s, especially the countless studies inspired by Kayne (1975), can be seen, in retrospect, as demonstrating the viability of this conception. And viable it was judged to be! There was an explosion of comparative grammatical research that exploited this combination of fixed principles and varying parametric values, and this research showed that languages, despite apparent surface diversity, could be seen as patterns with a common fixed core. An example, based on Pollock (1989), should provide a flavor of this research.

Consider the placement of adverbs in English and French. In English, an adverb may not intervene between the verb and the direct object, in contrast with French.

(15) a. *John eats quickly an apple
b. Jean mange rapidement une pomme
c. John quickly eats an apple
d. *Jean rapidement mange une pomme

The paradigm in (15) appears to be the result of a parametric variation between the grammar of English and that of French. In both languages, the clause has a structure roughly as in (16).

(16) [$_S$ Subject [Inflection [Adverb[$_{VP}$ Verb Object]]]]

What makes a sentence finite are features in the inflection position. These must be added to the verb in both languages (call this the "inflection-attachment principle"). The languages differ, however, in how this happens (call this the "inflection-attachment parameter"). In English, inflection lowers onto the verb, whereas in French the verb raises to inflection. The difference is illustrated in (17).

(17) a. [$_S$ Subj [~~<Infl>~~ [Adverb[$_{VP}$ V+Infl Obj]]]]
b. [$_S$ Subj [Infl+V [Adverb[$_{VP}$ ~~<V>~~ Obj]]]]

Note that this one difference explains the data in (15). In English, since the verb doesn't raise, and the adverb is assumed to stay put, the adverb will be to the left but not the right of the finite verb (17a), whereas in French the opposite holds, due to V-movement across the adverb (17b). Thus one parametric difference accommodates the facts in (16). Note, by the way, that we keep the basic clausal structure the same in the two languages. Likewise, the demand that inflection be attached to the verb remains constant. What changes is how this attachment takes place.[11]

As should be clear, this sort of account can be multiplied to accommodate all sorts of differences between languages (see Baker 2001), and a good part of research in the 1980s involved exactly these sorts of analyses. It proved to be very insightful, and grammarians came to the conclusion that something like a principles-and-parameters account of the language faculty was essentially correct. Note that this does *not* say which of the many possible principles-and-parameters theories is the right one. It only says that the right theory should have this general architecture. This consensus opened the door to the most current shift in grammatical theory, the minimalist program.

3. The Minimalist Program

The brief history above has taken us through two periods of grammatical research. The first succeeded in developing adequate formal tools for the study of natural-language grammars. The second placed the enterprise firmly in a broader cognitive, ultimately biological, context and succeeded in framing a general kind of solution to the acquisition problem most broadly conceived. The principles-and-parameters proposal has three great virtues: (a) it accommodates the fact that what language a person ends up speaking is closely related to the one that he or she is exposed to, (b) it accommodates the fact that acquisition takes place despite a significant poverty of the linguistic stimulus by having the PLD act against a fixed backdrop of invariant principles, and (c) it is immediately applicable in day-to-day grammatical research. In particular, in contrast to the vagaries of the evaluation metric, the parameter-setting model has been widely used to account for grammatical variation. These three facts have led to a general consensus among linguists that the language faculty has a principles-and-parameters (P&P) architecture.

This consensus invites a new question: granted that the language faculty has a P&P character, which of the many possible P&P models is the right one? In other words, what other conditions on grammatical adequacy are there, and how can they be used to move the generative enterprise forward? Minimalism is an attempt to answer this question. However, because the legitimacy of a minimalist program for linguistic theory has been disputed (see Lappin, Levine, and Johnson 2000), we first want to note that the minimalist turn is fully in line with the research agenda initiated in *Aspects* (see Freidin and Vergnaud 2001 on this point) and pursues questions quite common in the well-developed sciences.[12]

Succinctly put, the minimalist program conjectures that the computational system ("syntax") central to human language is an "optimal" solution to the central task of language: relating sound and meaning. This thesis will be vindicated once the complexities apparent in earlier approaches (such as *LGB*) are eliminated or shown to be only apparent, following from deeper and simpler properties.

Stated thus, the minimalist conjecture is no different from the emphasis in theoretical morphology to "model existent form with a minimum of parameters and mathematical complexity" (McGhee 1998, 2). In fact, minimalism responds to a deep-seated urge characteristic of the sciences. As Feynman (1963, 1:26) puts it: "Now in the further advancement of science, we want more than just a formula. First we have an observation, then we have numbers that we measure, then we have a law which summarizes all the numbers. But the real *glory* of science is that *we can find a way of thinking* such that the law is evident." Or, in the words of Einstein, "[the purpose of physics is] not only to know how nature is and how her transactions are carried through, but also to reach as far as possible the Utopian and seemingly arrogant aim of knowing why nature is thus and not otherwise" (cited in Weinberg 2001, 127).

We suspect that this "seemingly arrogant" aim of the minimalist program is what many have found irksome. But as Feynman's statement reveals, once observational, descriptive, and explanatory levels of adequacy are reached, the desire to go "beyond explanatory adequacy" (Chomsky forthcoming) naturally emerges and makes sense in the context of a naturalistic approach to language (Chomsky 2000a).

Seen in this context, minimalism emerges from the success of the *LGB* program. Because the P&P approach "solves" Plato's problem, more methodological criteria of theory evaluation revolving around simplicity, elegance, and other notions that are hard to quantify but are omnipresent in science can become more prominent. Until *LGB*, solving the acquisition problem was the paramount measure of theoretical success. Once, however, this problem is taken as essentially understood, then the question is *not* how to solve it but how *best* to do so. By its nature, this question abstracts away from the POS problem and points toward other criteria of adequacy, that is, "beyond explanatory adequacy."

The successes of *LGB* are important in a second way within minimalism. The government-binding (GB) theory is a very well-developed P&P theory with wide empirical coverage and an interesting deductive structure. It thus provides a foil for methodological reflection, a starting point for explanatory refinement. This sort of "benchmarking" is well illustrated by the discussion of levels of representation in the first minimalist paper (Chomsky 1993).

GB is a theory that identifies four important kinds of grammatical information, associated with four distinctive grammatical "levels." These four levels are d(eep)-structure, s(urface)-structure, logical form (LF), and phonetic form (PF). The latter two name the two points (technically, levels of representation) where the grammar interfaces with other cognitive components. Thus, as has been observed since Aristotle, language pairs sounds and meanings. LF is that part of the grammar that feeds cognitive components that deal with intentions, beliefs, and other forms of conceptual knowledge. PF is what grammar contributes to the sound structure of language.

Note that this conceives of grammar as interacting with other parts of the mind/brain. Moreover, it assumes that this interaction is modular; not all parts of the language faculty interact with all parts of the other mental modules. Rather, they interact in specific ways and at specific points. Chomsky notes that virtually everyone who thinks about language has levels similar to LF and PF: points of interaction between the grammar and other cognitive domains. As such, having these two levels within GB is not surprising or unique to that framework. Virtually any reasonable account would have something analogous. Thus, these levels are motivated not on *narrow* empirical grounds (say because of a desire to represent quantifier scope or sandhi effects) but on very broad (almost conceptual) grounds. *Any* reasonable theory would have phonetic and semantic interfaces.

This is not so for DS and SS. These are theory-internal levels within the language module. If they exist, they are motivated for narrow empirical reasons (this is not a criticism, just an observation), not on broader conceptual grounds. Chomsky then makes the following methodological argument: it is better to have a two-level theory that only has LF and PF than a four-level theory that has DS and SS in addition. Methodologically (what is sometimes referred to as the "weak minimalist thesis"), Ockham's razor would support the conclusion that multiplying levels is conceptually costly unless it has strong empirical motivation. Conceptually (what Chomsky likes to refer to as the "strong minimalist thesis"), it would be surprisingly nice if language made use of only those levels that are necessary to relate sound and meaning.

Interestingly, Chomsky (1993) manages to show that the bulk of the evidence for DS and SS is less empirical than technological. By making slightly different technical assumptions, it is possible to cover the same empirical ground without requiring levels akin to DS and SS. If this is correct, then a better kind of theory, conceptually speaking—one without DS and SS levels—is no less empirically adequate than the standard one. This, then, is an example where descriptive and explanatory adequacies can go hand in hand.

Levels of representation are but one area of minimalist concern. And, as one might expect, there are various ways of carrying out the minimalist program. Not surprisingly, the research directions that linguists have pursued are similar to those in the more developed sciences. Two approaches were clearly identified by Dirac in 1968.[13] One method consists in removing the inconsistencies, "pinpointing the faults in [the theory] and then tr[ying] to remove them . . . without destroying the very great successes of the existing theory." The other method consists in unifying theories that were previously disjoint. Let us refer to the first method as the "vertical method" (digging out inconsistencies) and the second as the "horizontal method" (embracing disjoint sets of phenomena and laws). Together, these form the axes of scientific research, trying to deepen understanding, and both can be seen at work within the minimalist program for linguistic theory.

As an illustration, take the fact that GB theory is modular in the technical sense; that is, it is conceived as having independent interacting subsystems, e.g., the binding module, the X′ module, the control module, the movement module, the case module, etc. These modules have their own properties (principles or "laws"), concern themselves with different aspects of grammatical structure (e.g., anaphora versus case), and operate on different scales (e.g., governing categories versus subjacent domains). These modules have proven to be very important in explaining the basic facts about grammatical structure across a variety of languages. Nonetheless, from a methodological perspective, an account with fewer modules is superior to one with more. Thus, one kind of minimalist project is to reduce the number of grammar-internal modules as far as possible, preferably to one. Doing this in an empirically responsible manner requires showing that the generalizations that the different modules have coded can be accommodated in a theory with a less modular format. In effect, the aim is to unify the subcomponents and show that the various generalizations and restrictions that characterize them are really all aspects of the same underlying principles and "laws." This would be an instance of horizontal minimalism. The paradigm of this sort of enterprise is found in physics, which has long had the ambition of unifying all the fundamental forces. The enterprise in the case of linguistics is not as grand, but the ambition is similar. Is it possible to unify these various domains and show that they all reflect the same underlying principles and grammatical forces? This is not the place to outline specific proposals in tune with this reductive impulse. Suffice it to say that tentative proposals toward unification of the modules have been advanced, in particular in unifying the theories of movement, control, and binding with case and agreement (see Hornstein 2001, Boeckx 2003, and references therein).

There is a second, more vertical strand of minimalist inquiry. This strand focuses more on trying to rationalize grammatical properties in other terms, most typically in terms of some version of computational complexity. The aim here is to show that the properties that emerge are just those that an optimal computational device charged with linking the sound system and conceptual system would have. This project shares the ambitions of another kind of reductive style within physics, the one reducing thermodynamics to statistical mechanics.

The motivations of vertical minimalism are visible in proposals that argue that the grammars do the least work necessary to produce objects usable by the sound/meaning interfaces. So, for example, if something must move to meet some requirement, then the movement must be the shortest possible, or if some requirement must be satisfied, it must be so satisfied by the first available expression that can do so (Collins 1997; Kitahara 1997). If a search is required, then the system is designed to ensure that the search is optimal, that the relevant information is easy to get to, and that the relevant operations are easy to implement. Consider an example of this form of reasoning. English existential constructions like (18) have a property widely seen across languages. The verb (*is, are*) agrees in number with a subject not in its canonical subject position to the left of the verb (19).

(18) a. There was/*were a man in *the halls*
b. There were/*was men in *the hall*

(19) a. A man was in the halls
b. Men were in the hall

As indicated in (18), the agreement pattern manifested is tightly restricted. Note that in (18a) we require *was* and cannot get *were*, and the opposite is true in (18b).[14] In effect, the verb *must* agree with the nearest element, in this case the noun phrase immediately to its right.[15] Or put negatively, it cannot agree with the underlined noun phrases (note that we say *the halls were/*was and the hall was/*were*). The reason advanced for this state of affairs is that the grammar is optimally designed and so agreement must be with the *nearest* possible agreeing element. In (18) and (19), *man/men* is nearer than *hall/halls*, so agreement must be with the former and not the latter. In other words, the patterns we see fit with what a well-designed system would deliver.

This kind of explanation can be pressed into very general service. For instance, Chomsky (forthcoming) has argued that many of the properties of grammars, e.g., the fact that there are labeled phrases (VPs and NPs), is to facilitate search in the process of a derivation. Similar sorts of suggestions have been made concerning the binary branching property of grammars and the local nature of movement.

Note that the vertical and the horizontal methods are not mutually exclusive. As Dirac already observed, it is much harder to unify theories if they don't contain any inconsistencies. Thus, the vertical method can make the horizontal method easier. Likewise, unification will often reveal anomalies that the vertical method will seek to remove. This is important, as it emphasizes the fact, often noted by Chomsky, that there is not just one way to do minimalism. Typically, a research paper will invoke a blend of vertical and horizontal considerations. However, as a matter of practice, the two styles differ, and we hope to have given a feel of this to the reader. Despite the differences, however, both constitute attempts to move grammatical theory in new directions toward greater explanatory depth. Only time will tell whether these impulses lead in fruitful theoretical and empirical directions.

4. Conclusion

Language is part of the biological world. Once this fact is the focus of research, as it has been in generative grammar, it "makes sense to think of this level of inquiry as in principle similar to chemistry in the twentieth century: in principle that is, not in terms of the depth and richness of the 'bodies of doctrine' established" (Chomsky 2000b, 26).

Like any other scientific enterprise, linguistics has focused on different goals over the years and has resorted to different methods to achieve them. We have here distinguished three periods in which distinct goals and methods were emphasized: the combinatoric stage, the cognitive stage, and the minimalist stage. We hope to have conveyed the sense that the evolution of linguistic theory has not been erratic but instead has followed a coherent direction of inquiry, similar in spirit to what we find in more basic and successful sciences. The results obtained so far are promising, and their use has already extended beyond linguistic matters.[16] Indeed, if the minimalist conjecture about the optimal character of the language organ turns out to be tenable, one will be able to draw "conclusions of some significance, not only for the study of language itself" (Chomsky forthcoming, 25) but for the biological world at large. In many ways, these are exciting times for linguistics.

NOTES

1. As a matter of fact, the three phases we focus on are arguably present (albeit in embryonic form) in Chomsky's magnum opus *The Logical Structure of Linguistic Theory* (hereafter *LSLT*, Chomsky 1955/1975), on which *Syntactic Structures* was based. As Chomsky notes, "[*LSLT*] ha[s] just about everything that [he] has done since, at least in a rough form" (1988, 129). Lightfoot (2003) notes how easy the transition from the combinatoric stage to the cognitive stage was, despite the absence of explicit discussion of "cognitive" themes in *Syntactic Structures*. Likewise, Freidin and Vergnaud (2001) highlight the presence of economy and simplicity considerations, now central to the minimalist program, in Chomsky's earliest writings (1951, 1955).

2. The following, we believe, is less a description of what Chomsky intended than how he was understood. Chomsky himself was never particularly moved by the combinatoric perspective. However, we believe that many others in the field took him to be advocating what we outline here.

3. Although, see Chomsky (1955, chap. 5) for the seeds of the rejection of this assumption.

4. The biolinguistic program was clearly influenced by Eric Lenneberg (1967).

5. The term "acquisition" is inappropriate in a generative setting. Unfortunately, it appears to have fossilized in the literature, so we will stick to the term here, noting that "growth" or "development" are better.

6. The core text here would be Greenberg (1963). Chomsky (1965, 118) casts doubt on the typological approach thus: "Insofar as attention is restricted to surface structures, the most that can be expected is the discovery of statistical tendencies, such as those presented by Greenberg (1963)." More recently, in light of the work that extended Kayne's (1994) antisymmetry hypothesis, Chomsky has qualified his position, as expressed by the following (1998, 33): "There has also been very productive study of generalizations that are more directly observable: generalizations about the word orders we actually see, for example. The work of Joseph Greenberg has been particularly instructive and influential in this regard. These universals are probably descriptive generalizations that should be derived from principles of U[niversal] G[rammar]." To understand this last qualification, the reader is referred to section 2.3.

7. There are many additional bells and whistles that one can add to the rule to make it more complete. For example, not all sentences have overt auxiliaries. The details of the process were already discussed in detail in *Syntactic Structures* and the process of do-support was offered as a way of regularizing this process. However, for current concerns, these additional details are of no moment. So let's stick with simple cases like (8).

8. See, most recently, Pullum and Scholz (2002), Cowie (1998), and Sampson (1999).

9. Exactly which brain property corresponds to the mental property under discussion is far from clear. Hopefully, informed research in neurolinguistics will help us bridge this gap and unify the mind/brain.

10. For an introduction to the principles-and-parameters approach to language variation and language acquisition that develops this theme in detail, see Baker (2001).

11. Lasnik (2000) claims that the core difference between English and French is that Infl is affixal in the former (hence requiring the operation of affix hopping/lowering), whereas it is featural in the latter (forcing upward movement).

12. The only note of caution worth bearing in mind is that the minimalist program may be premature (Chomsky 2001, 1).

13. The mathematical procedure is another term for what Husserl dubbed the "Galilean Style of Science," characterized by Weinberg (1976) as "making abstract mathematical models of the universe to which at least the physicists give a higher degree of reality than they accord the ordinary world of sensation."

14. Note that we set aside semiformulaic, nonproductive instances like *there's two men in the room*.

15. Using terms like "to its left/right" is just a manner of speaking. Ultimately, closeness is defined in structural terms (depth of embedding), the details of which are immaterial to the present discussion.

16. For instance, Searls (2001) shows how the results of the Chomsky hierarchy (Markovian, finite state machines → phrase-structure grammars → transformational grammars) and the formalism of *Syntactic Structures* can be extended to model interactions inside the genetic code. For further ramifications, see Jenkins (2001; forthcoming).

REFERENCES

Baker, Mark. 2001. *The atoms of language: The mind's hidden rules of grammar*. New York: Basic Books.

Boeckx, Cedric. 2003. *Islands and chains*. Amsterdam: John Benjamins.

Chomsky, Noam. 1951. *Morphophonemics of modern Hebrew*. MA thesis, University of Pennsylvania. [Published, New York: Garland, 1979]

——. 1955. The logical structure of linguistic theory. Thesis, Harvard/MIT. [Published in part, New York: Plenum, 1975]

——. 1957. *Syntactic Structures*. The Hague: Mouton.

——. 1965. *Aspects of the theory of syntax*. Cambridge, Mass.: The MIT Press.

——. 1973. Conditions on transformations. In *A festschrift for Morris Halle*, ed. S. Anderson and P. Kiparsky, 232–286. New York: Holt, Rinehart, and Winston.

——. 1981. *Lectures on government and binding*. Dordrecht: Foris.

——. 1993. A minimalist program for linguistic theory. In *The view from Building 20*, ed. K. Hale and S. J. Keyser, 1–52. Cambridge, Mass.: The MIT Press.

——. 1995. *The minimalist program*. Cambridge, Mass.: The MIT Press.

——. 1998. Noam Chomsky's minimalist program and the philosophy of mind. An interview [with] C. J. Cela-Conde and G. Marty. *Syntax* 1: 19–36.

——. 2000a. *New horizons in the study of language and mind*. Cambridge: Cambridge University Press.

——. 2000b. Linguistics and brain science. In *Image, language, and brain*, ed. A. Marantz, Y. Miyashita, and W. O'Neil, 13–28. Cambridge, Mass.: The MIT Press.

——. 2001. Derivation by phase. In *Ken Hale: A life in language*, ed. M. Kenstowicz, 1–52. Cambridge, Mass.: The MIT Press.

——. Forthcoming. Beyond explanatory adequacy. In *Structures and beyond*, ed. A. Belletti. Oxford: Oxford University Press.

Collins, Chris. 1997. *Local economy*. Cambridge, Mass: The MIT Press.

Cowie, Fiona. 1998. *What's within? Nativism reconsidered*. Oxford: Oxford University Press.

Crain, Stephen, and Mineharu Nakayama. 1987. Structure dependence in grammar formation. *Language* 63: 522–543.

Dirac, Paul. 1968. Methods in theoretical physics. In *From a life in physics: Evening lectures at the International Center for Theoretical Physics*, Trieste, Italy. A special supplement of the International Atomic Energy Agency Bulletin, Austria. [Reprinted in *Unification of fundamental forces*, ed. A. Salam, 125–143. Cambridge: Cambridge University Press.]

Emonds, Joseph. 1970. Root and structure preserving transformations. Doctoral dissertation, MIT. [Published as *A transformational approach to English syntax*. New York: Academic Press, 1976]

Feynman, Richard. 1963. *The Feynman lectures in physics*. Reading, Mass.: Addison-Wesley.

Freidin, Robert, and Jean-Roger Vergnaud. 2001. Exquisite connections: Some remarks on the evolution of linguistic theory. *Lingua* 111: 639–666.

Greenberg, Joseph. 1963. Some universals of language with special reference to the order of meaningful elements. In *Universals of language*, ed. J. Greenberg, 73–113. Cambridge, Mass.: The MIT Press.

Hornstein, Norbert. 2001. *Move! A minimalist approach to construal*. Oxford: Blackwell.

Jackendoff, Ray. 1994. *Patterns in the mind*. New York: Basic Books.

Jenkins, Lyle. 2001. *Biolinguistics*. Cambridge: Cambridge University Press.

———, ed. Forthcoming. *Variations and universals of biolinguistics*. London: Elsevier.

Kauffman, Stuart. 1993. *The origins of order*. Oxford: Oxford University Press.

Kayne, Richard. 1975. *French syntax: The transformational cycle*. Cambridge, Mass.: The MIT Press.

——. 1994. *The antisymmetry of syntax*. Cambridge, Mass.: The MIT Press.

Kitahara, Hisatsugu. 1997. *Elementary operations and optimal derivations*. Cambridge, Mass.: The MIT Press.

Lappin, Shalom, Robert Levine, and David Johnson. 2000. The structure of unscientific revolutions. *Natural Language and Linguistic Theory* 18: 665–671.

Lasnik, Howard. 2000. Syntactic Structures *revisited*. Cambridge, Mass.: The MIT Press.

Lees, Robert E. 1957. Review of *Syntactic Structures*. *Language* 33: 375–407.

Legate, Julie, and Charles Yang. 2002. Empirical reassessment of stimulus poverty arguments. *The Linguistic Review* 19: 151–162.

Lenneberg, Eric. 1967. *Biological foundations of language*. New York: John Wiley.

Lightfoot, David. 2003. Introduction to *Syntactic Structures*, 2nd ed., v–xviii. Berlin: Mouton/de Gruyter.

McGhee, George. 1998. *Theoretical morphology*. New York: Columbia University Press.

McNeill, David. 1966. Developmental psycholinguistics. In *The Genesis of Language*, ed. F. Smith and G. Miller, 15–84. Cambridge, Mass.: The MIT Press.

Pollock, Jean-Yves. 1989. Verb movement, universal grammar, and the structure of IP. *Linguistic Inquiry* 20: 365–424.

Pullum, Geoffrey. 1996. Learnability, hyperlearning, and the poverty of the stimulus. Paper presented at the parasession on learnability, 22nd Annual Meeting of the Berkeley Linguistic Society.

Pullum, Geoffrey, and Barbara Scholz. 2002. Empirical assessment of stimulus poverty arguments. *The Linguistic Review* 19: 9–50.

Ross, John R. 1967. Constraints on variables in syntax. Doctoral dissertation, MIT. [Published as *Infinite Syntax!* Norwood, N.J.: ABLEX, 1986]

Sampson, Geoffrey. 1999. *Educating Eve: The language instinct debate.* London: Cassel Academic Publishers.

Searls, David B. 2002. The language of genes. *Nature* 420: 211–217.

Weinberg, Steven. 1976. The forces of nature. *Bulletin of the American Academy of Arts and Sciences* 29, no. 4: 13–29.

——. 2001. *Facing Up.* Cambridge, Mass.: Harvard University Press.

Yang, Charles. 2002. *Knowledge and learning in natural language.* Oxford: Oxford University Press.

6

LANGUAGE, THOUGHT, AND REALITY AFTER CHOMSKY

GENNARO CHIERCHIA

I. Introduction

In 1956, *Language, Thought, and Reality*, a famous book by B. L. Whorf, was published. It addressed some fundamental and very hard questions, which have remained at the heart of an intense debate since. Such questions include the following:

(1) a. What is the relationship of language to thought?
   i. Is there thinking without language? Does the structure of our language determine how we think?
   ii. How are our linguistic abilities related to our general intelligence?
b. What is the relation of language to reality? Is language a mirror of the world, or does language determine the way we view the world?
c. What is the relation of language to culture?
   i. Is language a cultural and historical artifact (on a par with other human institutions)?
   ii. Is the design of language wholly and optimally functional to our communicative needs?

I am formulating these questions in an informal and impressionistic way. In the second half of the past century, they have been made considerably sharper thanks to the interaction of and debate across the disciplines that study our various cognitive abilities (linguistics, psychology, neuroscience, philosophy, anthropology, etc.). In the present paper, I would like to show how semantics in generative grammar (the research paradigm initiated by Chomsky nearly fifty years ago) is helping put some such questions on firmer grounds. I will use an example taken from the grammar of nouns (the mass-count distinction).

Let me begin by fleshing out a bit more the questions in (1). We interpret language quickly and with hardly any awareness of what we are doing. Decoding our language is almost as immediate as seeing. In the thousands of ways we use language in our daily life, we constantly and effortlessly go from words to meaning, and, vice versa, we put what we intend to convey into words. In doing so, we rely on some internalized way of coding information in the expressions of our language. Language embodies a way of systematically relating symbols arranged in specific ways and meanings, which we take completely for granted in our own language, even though it looks mysterious and complex when we look at some other languages. How do expressions code meanings? Why are we so good at using expressions to mean in our native language but find it so hard when we try to do so in some other language?

There are no quick and easy answers to such questions. One traditional illustration of their difficulty is the problematic character of translation. The notion of "exact" translation is very elusive. Take, as an example, the merely modest successes of the attempts to translate mechanically, using machines. Much research (with substantive financial investment) has been put into the development of programs for automatic translation. Although there is progress, we are still far from having truly usable tools. The problematic character of translation is a source of endless discussion. Something that clearly comes out of even a superficial look at it is that differences among languages are not just phonetic (the shape of sounds used). It's not simply that things or concepts get different phonetic labels in different languages. Many words do not readily translate from a language to the next, and when it comes to translating whole sentences, things become pretty involved. Crosslinguistic differences seem to affect the whole system of how meanings are coded. The parallel with cultural differences (in habits, institutions, traditions, and history) comes readily to mind. And with this parallel naturally comes the idea that language (and more specifically the way in which language encodes information about our surroundings) embodies a worldview, a conceptual frame, a form of life. This would explain the difficulty we find in translating. We view things differently, depending on our histories. And no two forms of life, worldviews, or histories match perfectly. If languages are cultural manifestations on a par with political institutions, wedding rituals, monetary systems, and so on, no wonder we cannot translate any more readily than we can convert a currency into another or find the exact equivalent of the U.S. president among, say,

Italian institutional figures (the Italian president has quite different functions, etc.). Whorf, building on work by Sapir and others, formulated a particularly perspicuous version of these ideas, a position that has come to be known as "linguistic relativism." According to this view, language is mostly a cultural product and embodies and determines the way in which a community views reality.

Although "perfect" translation may well not exist, partial translations are possible and effective. There are two reasons for this, presumably. First, humans are made alike. They have the same physical structure, the same cognitive resources, and so on. Hence there will be some commonalities in the way in they process and code information. Second, there typically is enough common ground across cultures, worldviews, etc. so as to allow us, with some care, to interpret one another. The key to this is human intelligence, a complex and integrated set of cognitive resources. Such resources enable us to extract complex information from our environment and exchange it with one another. This enhances our capacity to cope with our environment. Since our experiences differ, we may categorize information in different ways. Nonetheless, we are remarkably successful at such tasks and we can convert our categorizations across our different experiences in a fairly successful manner.

If we were to sum up these commonplace observations in a form that addresses the concerns in (1), we might do it as follows:

(2) a. Men are endowed with intelligence, i.e., a uniform, plastic capacity that
  i. enables us to extract regularities from our environment and
  ii. endows us with advanced problem-solving skills.
 b. Language is our intelligence's reply to our need for communication.
 c. The specific forms that languages take reflect:
  i. how our intelligence copes with the external environment (language as mirror of reality)
  ii. how our intelligence gives shape to our forms of life (language as cultural artifact)

This summary sets the Sapir-Whorf hypothesis within a broader view of how thought and language may interact, a view that enables us to do justice to the idea that language may well not be wholly cultural. We may not know (and disagree on) how much is cultural and how much is natural/instinctual in human language. But at any rate, the roughly six to seven thousand languages we observe in the world are mainly the reaction of a highly developed, adaptable capacity located in the brain to external pressure from nature and to the culture we develop.

Even though this broad picture looks quite plausible, I think that it is to large extent wrong. The main thesis I will defend, in contrast with (2), is that language is a largely specialized system characterized by specific computational devices autonomous from other abilities constitutive of "human intelligence" (like statistical resources or problem-solving strategies). Chomsky often talks in this connection of a sort of "language organ" located in the brain. More specifically, when it comes to semantics, people often think of it in terms of world knowledge (be it perceptual, encyclopedic, or what have you) linked to linguistic representations in a largely conventional way. In modern generative linguistics, semantics has taken a rather different form. It is the investigation of logical forms and denotational structures (similar to those first exposed by logical semantics) linked to the syntax of particular languages by a set of universal mappings. Such structures systematically constrain the way we refer to things and reason about them.

In what follows, I discuss some relevant evidence, drawing from the mass-count distinction. Such a distinction has been discussed extensively and lends itself particularly well to giving us a relatively quick overview of what ongoing debates are about.

2. Masses and Individuals

One characteristic of the nominal system of many languages, including English, is that it differentiates two sorts of nouns: mass and count. Typical count nouns are *chair, table, man,* etc. Typical mass nouns are *water, blood, garbage,* etc. Count nouns seem to refer to classes of well-identified, discrete objects that appear to be readily accessible to us. Mass nouns seem to refer to substances or amounts not evidently made up of discrete parts. This difference manifests itself in a number of morphosyntactic phenomena that we will now review. To talk about single tables or chairs, we use the singular; to talk about groups we have to use the plural:

(3) a. That table is cheap
    b. Those chairs are cheap
    c. *Those table is cheap
    d. *That chairs are cheap

In (3a), a singular noun (*table, chair*) combines with a demonstrative determiner (*that*); the resulting noun phrase (abbreviated NP, a syntactic unit characterized by the presence of a noun) refers to or denotes a particular entity (for example, the table we are pointing at). The plural noun phrase in (3b), instead, refers to a group. Sentences (3c, d) are deviant, and this shows that combinations of demonstratives and nouns have to match in (or agree on) singularity/plurality. While plural marking is a very useful and general device, it is subject to certain constraints; in particular, with mass nouns, we cannot readily use plural morphology:

(4) a. This rice is tasty
b. *These rices are tasty
c. The water in this tank is dirty
d. *The waters in this tank are dirty

The deviance of sentences (4b, d) is not absolute. It is not the case that producing and understanding sentences such as (4b) or (4d) is totally impossible. But mass nouns in the plural have a very different "feel" from ordinary count nouns in the plural. For example, we can say things like *Mary broke her waters*. But this is an idiomatic expression; we cannot simply reconstruct its meaning from the meaning of *breaking* and the meaning of *water*. In order to understand it, we have to know that *breaking waters* refers to the onset of labor. We might also sometimes say *the waters of the sea divided in front of us*. But this sounds "poetic"; it evokes biblical images. If you drop a glass of water on the floor, it would sound very funny to say *you spilled waters all over*.

A further characteristic of mass nouns is that they do not combine with numerals. Consider the following example:

(5) a. I need three chairs
b. Those chairs are three
c. *I drank three waters
d. *I bought one hundred rices

If I ask: "how many loaves of bread did you buy?" you can answer "three." But not if I ask, "how much bread did you buy?" In that case, you must answer something like "three loaves" or "three kilos." In order to "count" bread, blood, rice, etc. we have to use either "measure phrases" (kilos, liters, and so on) or "classifier phrases" (i.e., things like *drops* of blood, *loaves* of bread). Classifier phrases are used to indicate countable objects typically associated with the relevant substances or "standard servings" in which such substances can be divided.

The behavior of articles is also telling. The definite article *the* combines with both count and mass nouns.

(6) a. The rice in this tank
b. The chair over there
c. The tables I inherited from aunt Robertina

But the indefinite article combines only with count nouns:

(7) a. an expensive chair
b. *a rice that I bought

The noun phrase in (7b) can only have the "type" interpretation (a type of rice I bought). This contrast is not limited to articles. It generalizes to many components of the noun phrase. The syntactic structure of a noun phrase can be schematized as follows:

(8) a. [Det N]
the man
b. Det: the, a, this, every, most, one, two, . . . , more than three, etc.

The tree in (8) tells us that a typical noun phrase is constituted minimally by a common noun and a determiner; determiners include articles, demonstratives, and things like *every, many, several*, etc. Such elements determine, roughly speaking, how many things (of the sort specified by the common noun) are under consideration. We will say that a common noun constitutes the *restriction* of a determiner in structures like (8). Some determiners are morphologically simple (i.e., single morphemes or words, like *the, every*, etc.); others may be complex (*more than three, a lot of*, etc.). It is easy to observe that determiners are picky as to what kinds of nouns they combine with. For example, *every chair* sounds fine, but *every blood* is weird (to the extent we can use it, it can only be understood as "every type of blood"). The determiner *all* is nearly synonymous with *every*, yet unlike the latter, *all* readily combines with mass nouns (cf. *all water*, or, in combination with the definite article, *all the water*). This "pickiness" of various determiners appears to permeate the whole system of determiners.

Consider, for example, determiners like *several* or *many*. They can readily be used with plural count nouns (e.g., *I saw several/many students*), but not with mass. It's strange to say *I bought several/many rices* (marginally acceptable on the "type" reading). On the other hand, the determiner *much* goes with mass (*I ate too much rice*) but not with count (*I saw too much student* is strange indeed). *Much* is quite clearly a mass counterpart of *many*. It is clear that the whole determiner system is affected by the mass-count distinction; several generalizations centered on this distinction emerge.

Overall, the mass-count distinction has a series of related morphosyntactic manifestations having to do with plural morphology, numerals, and the determiner system. It's worth summarizing them in schematic form:

(9) Count nouns:
(i) take readily plural morphology
(ii) combine with numerals (one, two, three, . . .)
(iii) do not need to use a classifier/measure phrase for counting
(iv) work with *a, many*, . . . , but not with *much, little*, . . . etc.
Mass nouns:

(i′) have no plural morphology
(ii′) do not combine with numerals
(iii′) must use a classifier/measure phrase for counting
(iv′) work with *much, little, . . .* , but not with *a, many, . . .* , etc.

As is evident even from these cursory remarks, the contrast between mass and count nouns appears to be well entrenched in the way we speak. There are a host of patterns in which this distinction manifests itself. The question is why. Where does this restriction come from? Why do so many languages make it? In what follows, we will consider a number of possible answers that have been given.

3. Possible Extralinguistic Roots of the Mass-Count Distinction

If we look at the world, on the one hand we find things like pebbles, dogs, chairs, etc.; on the other hand, we also find substances like water, air, or (if we are lucky) gold. Objects form discrete, readily countable units; substances do not. Substances tend to be scattered around, often mixed with other stuff. They do not have evident minimal parts and thus are not readily countable (even though they can be measured with appropriate devices). Perhaps language simply reflects this feature of the world. If a noun is used to refer to objects, it is a count noun; if it is used to refer to a substance, it is a mass noun. The morphosyntactic properties summarized in (9) are merely the linguistic manifestation of extralinguistic, semantic properties of the things nouns refer to. In this kind of approach to the mass-count distinction we see a concrete exemplification of the view of language as a mirror of the world (one we might apply to other grammatical distinctions). The guiding principle is that language is used to talk about reality. We attach names to things, much like we attach labels on medicines or on books in a library. Labeling is useful. If we label things systematically, we can identify, locate, and retrieve them as needed. Perhaps language is a somewhat complex, spontaneous form of labeling. Our capacity to mean really is our capacity to label. The mass-count contrast might well be good prima facie evidence in favor of such a view.

This general approach explains in very simple terms how language comes to carry information about the world. As we attach names (i.e., symbols) to things, we can, in virtue of this implicitly assumed association, use arrays of symbols to express how things are arranged in the world. For example, if *A* stands for John and *B* for Bill, we might use *AB* to represent that *B* follows *A* and the reverse order *BA* for *A* follows *B*. The different symbolic arrays *AB* and *BA* represent two different ways in which John and Bill are related. This is very simplistic. But you can see how the idea might be developed further and how, in fact, it can be used to code elaborate information. Perhaps language is just a rather complex code that functions according to these principles. Versions of this view of language have been put forth very authoritatively in the history of thought, from Aristotle to Wittgenstein. As will become apparent in what follows, there is a lot that is right about this view.

There is work in cognitive and developmental psychology that may provide further support for the view of the mass-count distinction that I have sketched. In particular, Spelke (1985) and Soja, Carey, and Spelke (1991), among others, have argued that babies a few months old (well before they speak) have an articulated theory of the world. They believe that solid objects have boundaries, are cohesive (i.e., their parts stick together), and move as a whole (without, e.g., splitting or merging) along continuous paths. In contrast with this, children believe (or, we should rather say, they *know*) that nonsolid substances like liquids or powders are not as cohesive. As they move and contact each other, they may not retain their boundedness; they may merge or split. How can we impute such an elaborate view of the world to babies? The experimental paradigm that has been used to demonstrate these claims is of the following type. Imagine an object, say a teddy bear, on a table and a screen lying flat in front of the object. The screen then slowly rotates upward, covering the teddy bear. In one condition, the screen rises all the way vertically and occludes the teddy bear from view. This is a "normal" state of affairs (the expected condition). In the second condition, some sort of "magic" happens (from the adult's point of view). The screen keeps rotating and, as it were, goes through the space occupied by the teddy bear (the unexpected condition—in fact, as the screen rotates upward, the teddy bear is removed by the experimenter without the observer being able to see the removal). It turns out that children tend to stare longer at events of this sort than at those of the normal, expected type. That is, they show surprise at "abnormal" events, which in turn suggests that they have the expectation that objects (like teddy bears or bottles) persist in their location and are solid. What is striking is that this surprise is manifest at three months of age, when the infants cannot possibly have elaborated a theory of solid objects from experience. Hauser (cf. 1996) has pushed this line of inquiry further, showing that rhesus monkeys are endowed with a similar theory of discrete objects versus substances. Going back to children, it is highly plausible that such knowledge, which children appear to be endowed with at birth, guides them as an identification criterion for novel objects versus substances they encounter; later on, such knowledge guides them in acquiring language. For example, upon encountering a class of solid objects, say bottle openers, the child identifies some key properties of the objects (say, shape and function) and then generalizes it to other objects of the same sort (forming the concept of a uniform class of objects, bottle openers in general). Upon encountering, instead, a new paste or powder, one again identifies some of its key properties (in this case it won't be shape but, say, texture and what one typically does with it) and then generalizes such properties to other portions of the same substance (see, e.g., Soja, Carey, and Spelke 1991). Knowing that things are set up in this way (i.e., that they are naturally sorted in substances and objects) makes identification and naming easier. When language comes in, common nouns will naturally fall into two categories accordingly.

The view we have developed so far can be summarized as follows: The world is structured in objects and substances (defined in terms of the way they behave as they move across space and interact with each other). Children (and, in fact, other primates) seem to have inborn knowledge that the world is so structured. It might be tempting to speculate on the kind of adaptive advantages that would descend from such knowledge. But be that as it may, the presence and pervasiveness of the object-substance distinction seems to be uncontroversially true. The mass-count distinction registers this fact.

4. The Grammatical Basis of Mass and Count

There are two main types of argument against the position outlined in section 3. The first is based on the existence of what we might regard as "funny" substances or objects. The second is based on crosslinguistic variation.

The first kind of problem has to do with the observation that the mass-count distinction is operative with nouns that do not readily map into objects versus substances as we have defined them. Consider, for example, the following pairs:

(10) a. generosity, virtue/virtues
b. knowledge, belief/beliefs
c. pride, prejudice/prejudices
d. fun, joy/joys
etc.

We are, with the examples in (10), in the realm of abstract nouns (often derived from verbs). Now notice that the first member of each pair in (10) is typically mass and the second typically count. For example, we can say "the generosity you displayed toward me . . . "; we cannot say "*a generosity you displayed toward me . . . ." Or we can say "I haven't encountered much generosity there" but not "*I haven't encountered many generosities there." On the other hand, *virtue* appears to be perfectly count: "I see many virtues in you, my young apprentice," says Obi Wan Kenobi to Anakin Skywalker. I'll let readers construct their own examples with the items in (10). Why are these examples problematic for the view presented in section 3? Clearly, it is hard to view, say, generosity as a "substance" and virtue as an "object" or knowledge as a substance and beliefs as objects—and certainly not in Spelke's sense, as she bases her distinction on, for example, how things move in space. Trying to extend the object-substance distinction to abstract objects clearly involves going beyond the conceptual system that was found in prelinguistic babies and other primates.

A related issue stems from the observation that there are things that are hard not to classify along with pasta, rice, and salad. And yet they are count: lentils, beans, sprouts, etc. There are, furthermore, things that clearly are discrete objects for which English uses a mass noun. Consider the chair I am sitting on. It is a comfortable chair (count). But it is also cheap furniture (mass). Furniture, footwear, cutlery, and drapery are made up of perfectly discrete objects (tables and chairs, knives and forks, curtains, etc.), yet grammatically they are mass nouns: they don't pluralize naturally (and if they do, they acquire the "type of" interpretation), they don't combine with numerals, and so on. These nouns have a collective flavor, in the sense that they are typically used to refer to a plurality of objects. But nonetheless they are mass (and contrast with count collective nouns like *bunch*, *pile*, *group*, etc.). Moreover, it is not absolutely obligatory to use nouns like *furniture* only for pluralities. I can point to a single couch and say "I inherited this furniture from aunt Robertina." If furniture is made up of countable objects, why can't we combine it directly with a numeral (i.e., why can't we say "*those three furnitures are all that is left after the fire")? It is as if the mass-count distinction ceases to be a "real" distinction and becomes a way we can categorize things.

This circumstance becomes particularly evident once we look at it from a crosslinguistic standpoint. What we find is that even closely related languages have somewhat diverging sets of mass and count nouns. For example, in English one says "I cut my hair." In Italian, one has to say "mi sono tagliato i capelli" (*I cut my hairs*). *Hair*, used to refer to what grows on our head, seems to be mass in English and count in Italian. Yet clearly we are referring to the same stuff. Your hair doesn't change as we change languages. And the prelinguistic conceptual system of infants doesn't change if they happen to be exposed to Italian or English, either. There are quite a few mismatches of this sort. Here is a small list:

(11)

Count	Mass
capello	hair
bagaglio	luggage
servant	*servitù*
mobile	furniture
calzatura	footwear
posata	cutlery
relative	*parentela*

In all these cases, the same entities seem to be classified as mass or count depending on the language.[1]

One final point. In a given language, one can find pairs of closely related nouns, one mass and the other count. This may give us an insight into what the crosslinguistic differences just discussed are really like:

(12) a. coin — change
b. shoe — footwear
c. *mobile* (a piece of furniture) — *mobilia* (furniture)

Perhaps the way in which the counterpart of a count noun like man "feels" in a classifier language (like, say, Chinese) is similar to the contrasts in (12). In the mass counterpart of a count noun (say, change versus coin), there is a sort of neutralization of the plural-singular distinction; a set of coins may qualify as change but sometimes a single coin can too. The most minimal pair I was able to find is (12c). Italian has a lexically simple word that means, essentially, "a piece of furniture" next to one that means "furniture."

The existence of pairs such as those in (12) is partly related to another observation. There are nouns that have both count and mass uses. Consider for example *wine*, *water*, *beer*. When we say "there is beer all over the floor," we are using it in a masslike way. When we say "we ordered three beers," we are using it as a count noun. In the latter case, *beer* comes to mean something like "standard serving of beer" (e.g., a pint glass, a bottle, etc.). In fact, to a certain degree, virtually every noun that has a predominant use can be pushed into the other use. We have already noted that most mass nouns can be shifted into a count mold and get the "type" reading. One might understand in this way utterances like "in this lab we store three bloods." We now see that shifts from mass to count can also be accomplished by conceptualizing the substance in terms of a notion of "standard serving/portion." Let us imagine, for example, a community of English-speaking vampires. In such a community, standard servings of blood might become common. In a bar, one might well say things like "we ordered three bloods." Similarly, count nouns can have mass uses. For example, *apple* is count, but it has mass uses, as in "there is apple in the salad." Here too you can imagine pushing the idea further. "After my son attacked it with his tools, there was car all over the floor." And so on. D. Lewis discusses this in connection with the "Universal Meatgrinder." You put anything through it and you get a substance. By the same token, as Pelletier and Schubert (1989) point out, we also seem to have some kind of mental "Universal Packager," which turns substances into discrete portions.

Summing up, we have made the following observations:

(13) a. The mass-count distinction doesn't coincide with the prelinguistic notion of object versus substance; it extends to abstract objects and eventualities and doesn't coincide with the prelinguistic notion of solid object.
b. There is crosslinguistic variation as to what is categorized as mass and what is categorized as count.

c. Even within the same language, there are nearly synonymous noun pairs, one mass the other count, and there can be shifts from one category to the other.

What conclusions can we draw from this? For the view of language as a mirror of the world, we can draw mostly negative ones. The world can well be made of substances and discrete entities. And the prelinguistic child may well know this (or, more accurately, the child may well be endowed with an articulated theory of objects versus substances). But the mass-count distinction is something else. For one thing, the two distinctions simply do not coincide. Moreover, languages appear to have some freedom in how they classify their nouns. We must conclude that the mass-count distinction does not appear to be readily and completely reducible to any known extralinguistic one. It obviously bears a resemblance to the object-substance distinction, but it then seems to acquire a life of its own. It becomes a formal, abstract marking (think of its application to abstract objects). This passage from a "substantive" to a "formal" distinction happens a lot in language: this may be the key to understanding how language works. Think, for example, of gender distinctions. In English, gender is simply based on natural gender (a female being is a *she*, a male one is a *he*), and stereotypes occasionally kick in (so that a car may be referred to as a *she*). But in many languages, this becomes a formal marking. For example, in Italian, *tavolo* (table) is masculine and *tavola* (table) is feminine, with hardly any semantic distinction. Or think of the fact that we typically tend to package information into a topic ("as for John") and a comment ("he is a real bastard"). This is clearly related to the subject-predicate structure we see in sentences ("John is a bastard"). But the two pairs (topic/comment versus subject/predicate) simply do not coincide.

Does this mean that once language takes over some extralinguistic concept it voids it of its original character? For example, is the mass-count distinction void of any systematic "cognitive" role (other than bearing a family resemblance to objects versus substance)? Probably not. Otherwise, why would so many languages make it? Why would grammar fastidiously insist on making numeral expressions and other determiners behave in the complex ways we have seen, if there was no communicative purpose behind it? Moreover, with all the crosslinguistic variation we have observed, there are also fairly evident universal tendencies in this domain. Nouns of liquids, for example, do tend to always have a mass lexical entry. Furthermore, while, as we saw above, it has been argued that in some languages every noun is mass, there seems to be no language in which every noun is count (a corollary of the observation that nouns of liquids are universally mass). Universal tendencies of this sort are hard to explain if the mass-count distinction is pure morphosyntax—pure form—with no semantic effects. There is a puzzle here worth understanding in more depth. On the one hand, the mass-count distinction appears to be formal, in the sense that we can apply it to just about anything (the same "thing," e.g., your hair, can be viewed as mass or count). On the other hand, such a distinction does not appear to be purely formal (a pure matter of morphosyntax), as its correlation with "real" objects and substances is not arbitrary. Addressing this puzzle involves understanding more about the relationship between semantics and cognition, a hard and controversial issue.

5. A Semantic Model

In what follows, I will first make a proposal as to what is behind the mass-count distinction, namely a universe of discourse structured in specific ways and linked to the nominal system by universal mapping principles (the "denotational structure" mentioned in section 1). Then I will discuss how an approach along such lines addresses the questions raised above.[2]

5.1. Universes of Discourse

(14)

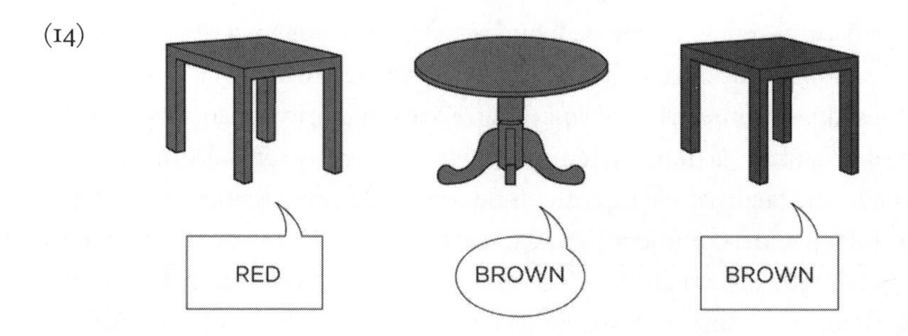

Suppose those tables in (14) are all the tables in the world. With the noun phrase *this table* (a singular definite noun phrase), I can refer to any of the tables in (14), i.e., table *a*, table *b*, or table *c*. With the noun phrase *these tables* (a plural definite noun phrase), I can refer to any group made up of the tables in (14). For example, I can point to table *a* and table *b* and say *these tables are cheap*. Or I can point to all of them and say *I like these tables*. The same holds if I use the definite article; *the brown tables* (as in "the brown tables are made in Hong Kong") refers to table *b* and table *c* taken together. *The red table* (as in "I inherited the red table from aunt Robertina") refers to table *a*, etc. So, in general, singular definites can refer to any individual table and plural definites to any group thereof. A related class of examples illustrating the same point is the following: I can point to table *b* in (14) and say *this is a (nice) table*. Or I can say of table *a* and *b*: *those are (solid) tables*. In this sort of sentence, the subject is a definite NP; the predicate is also an NP, albeit an *in*definite one: *a (nice) table/(nice) tables*. NPs in their predicative uses are used to classify things (rather than referring to them).

Where are we headed with these common-sense observations? The important point is that given a set of objects like the tables in (14), we can, using plurals, group them together any which way we want. It is as if we are thinking in terms of a structure with this form:

(15) {a, b, c}
{a, b} {b, c} {a, c}
a b c

Those at the bottom represent the individual tables. The curly brackets represent all the possible ways of grouping them (i.e., all the possible sets or groups we can construct out of *a*, *b*, and *c*; the curly-bracket notation is borrowed from set theory). Singular definites can refer to any of *a*, *b*, or *c* (i.e., to tables individually taken); plural definites can refer to any of the groups or sets in curly brackets. Let us call the individual entities "atoms." By that I do not mean that they are atoms in the physical sense. What I mean is that in talking about *the red table* we disregard or put in parentheses its parts (its legs, drawers, etc.) and consider it as a unit. We can generalize this way of thinking. Whenever we engage in a conversational exchange, we typically have a certain universe (or domain) of discourse in mind, i.e., a series of salient entities and groups thereof (the things we see in our environment, those that constitute the topic of our conversation, etc.). The universe of discourse will vary from context to context, depending on the circumstances of the conversation, the intentions and interests of the speakers, and so on. However, such universes will all have a structure similar to the one in (16), namely:

(16) {a, b, c, d} . . .
{a, b, c} {a, b, d} . . .
{a, b} {b, c} {a, c} {c, d} . . .
a b c d . . .

Our domains of discourse, in other words, are populated by entities that play the role of atoms or units (that we may refer to using singular definite noun phrases) and by groups or aggregates (that we may refer to using plural definite noun phrases). What do nouns do? Nouns like *table* partition the set of atoms dividing tables from nontables. In the example above, supposing that the tables are *a*, *b*, and *c*, we have something that may be schematized as follows:

(17) table → [a, b, c]

I am adopting here the convention of putting in square brackets the atoms of which *(is a) table* is true (applies truthfully). Other nouns (e.g., chair, car, etc.) would work in the same way. The role of plural morphology is that of turning something that applies only to atoms into something that applies to the corresponding groups.

(18)
$$\text{tables} \rightarrow \begin{bmatrix} \{a, b, c\} \\ \{a, b\}\ \{b, c\}\ \{a, c\} \end{bmatrix}$$

This is the starting point. When I combine a noun like *table* with a definite determiner such as *that* or *the*, I get to refer to *particular* atoms. For example, with the NP *that table* I may invite you to individuate from the class of tables the one I am pointing at. Similarly for plurals: if I say *those tables*, I am inviting you to consider from the totality of tables those I am indicating. And so on.

As a picture of how nouns and noun phrases work, this is very approximate. But it arguably constitutes a step in the right direction. The role of lexical nouns (without the determiner) is that of singling out a class of objects (which will have common traits: they may share a form, a function, or what have you). Plural morphology (*-s*, in English) turns nouns that apply to singularities into nouns that apply to groups. Simple as this may be, it has the right consequences. Suppose, for example, that *a* and *b* are tables. Suppose that *c* is also a table. It then follows inexorably that *a*, *b*, and *c* taken together are also tables and also, for instance, that *b* and *c* taken together are tables. This is something we know a priori, and it follows from our way of looking at plurality. Once we individuate through a singular noun a set of atoms, all of the groups (of various size) made up of those atoms will be something the corresponding plural is true of.

A further important consequence of this way of thinking about plurality is the following. It is immaterial whether we are talking of concrete or abstract objects. If *a* is honesty and *b* is perseverance, then *a* is a virtue and so is *b*, and *a* and *b* together (in our notation, {a, b}) will be virtues (e.g., John's chief virtues). Our schema applies equally well to concrete and abstract entities.

Where do mass nouns fit in? We seem to have created a world of neatly identified objects that seems prima facie only suitable to count nouns. Let us see. Imagine that the table and chairs in (19), a very modest dining room set, is all is left in the world after a huge fire.

(19)

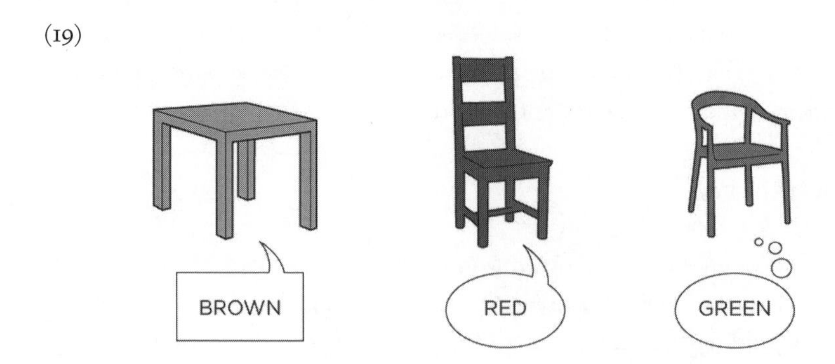

In this case, we know that *table* would apply only to *a* and *chair* to *b* and *c*. What would *furniture* apply to? The most plausible answer, it seems to me, is that it would apply indifferently to *a*, *b*, *c*, and any aggregate thereof. In other words, here is the structure of what the mass noun *furniture* applies to:

(20)

$$\text{furniture} \rightarrow \left[\begin{array}{ccc} & \{a, b, c\} & \\ \{a, b\} & \{b, c\} & \{a, c\} \\ a & b & c \end{array}\right]$$

This would be consistent with the observation that we can say that the table and the green chair are furniture, just like the chairs are, etc. Even the table in (19) by itself might be claimed to be (expensive) furniture, as opposed to the chairs that are cheap, and so on.

Note that mass nouns share the same part-whole structure as plurals. In particular, if two aggregates belong to the same category, their sum or fusion also does (if *a* and *b* are furniture, the aggregate that has *a* and *b* as parts, i.e., {a, b}, is also furniture). This common structure arguably explains, as we will see, their commonalities.

According to this view, in any given universe of discourse, a count noun individuates a set of atoms. This means that in order to use it, we must have a criterion to somehow identify singularities. A mass noun does not. A mass noun applies to aggregates of any size. This means that in order to use it, we must have a way to somehow recognize aggregates (no need to be able to single out atoms).

Brief aside: It is not the case that to use a noun we must be able to decide for each entity out there whether it belongs to the relevant kind/class/category of things or not. It is sufficient to have a general idea of how typical members of the class might look, what function they might play, etc. That seems sufficient for the noun to perform its communicative work (and to determine the grammatical status of the noun).

Back to the main point. You see that this way of thinking extends to liquids, powders, etc. Think, for example, of *a*, *b*, and *c* in (19) as grains of sand. (To make it more realistic, you have to imagine larger amounts). Quite clearly, *a*, *b*, *c*, and any aggregate thereof would be sand. Similarly, for water: *a*, *b*, *c*, and . . . might be water molecules (or whatever larger liquid aggregate of H_2O we need to have water). The important thing is this: a mass noun must be associated with a criterion to identify aggregates of something, not atoms—not, that is, minimal elements of the relevant stuff.

As was the case with count nouns, this way of conceptualizing things does not apply merely to concrete entities. In principle, it applies equally well to abstract things. Let for instance *a* be a piece of information (say, that the earth is round) and *b* be another piece of information (say, that the earth revolves around the sun). Then *a* is common knowledge (where *knowledge* is an abstract mass noun), and so are *a* and *b* taken together. Even though we may be hard pressed in giving the exact criteria for what may constitute a "minimal" piece of knowledge, we know enough about what constitutes knowledge in general to be able to use the corresponding abstract mass noun in a sensible manner.

Summarizing, our possible domains of discourse have a structure like the one in (16). Some entities play the role of atoms (concrete or abstract), others of aggregates thereof. Anything that can be the reference of a singular definite noun phrase is being regarded as an atom; anything we refer to through a plural definite noun phrase is an aggregate. Basic lexical items split our universe into classes (or kinds, or categories) of objects; they categorize our domains of discourse. Lexical nouns come in two sorts. They may identify a set of atoms. In this case, they are count. Or they may identify sets of aggregates (without caring to sort out atoms from nonatoms). In this case, they are mass. The denotation of a mass noun is "closed under sum"; i.e., if *a* is furniture and *b* is furniture, then their sum {a, b} is furniture. Singular count nouns are not; if *a* is a cat and *b* is a cat, then their sum is not a cat. Plurals, like mass nouns, are closed under sum. If *a* and *b* are cats, and *c* and *d* are cats, then their sum (*a*, *b*, *c*, and *d*) are also cats. Simple enough. Some readers may find this picture congenial; others will find it puzzling. The point is, however, that if you assume something of this sort, the properties of the mass-count distinction fall arguably right into place. Let us see how.

5.2. Deriving the Properties of the Mass-Count Distinction

Why can't we pluralize mass nouns? The relevant observation, in this connection, is that they are, in some sense, already plural. The purpose of the plural morpheme is to make it clear that we intend to talk about aggregates, but if a noun is mass, this will be already known, if our hypothesis is correct. Pluralizing a mass noun would be like trying to pluralize a plural: tree/tree-s/[tree-s]-s. Not possible.

Why can we directly count tables but not furniture? Here is the idea: A set of atoms is neatly divided up in discrete, nonoverlapping entities. A set of aggregates is not. They overlap, are made up of other aggregates of the same sort, etc. In order to count using language, we seem to need a domain of discrete nonoverlapping things. Count nouns provide domains with these characteristics; mass nouns do not.

What about plurals, then? Plurals too refer to aggregates, just like mass nouns. But we count with plurals; in fact, that's how we typically count: three chair*s*, four donkey*s*, seven sin*s*, etc. Still, there is crucial difference here. Plurals are derived from singulars. We first must understand what a table is; then we immediately understand what tables are. For mass nouns, it's very different. To understand a mass noun, we must have some idea of what a typical aggregate of that stuff is like, without necessarily having much of a clue as to its minimal parts. That's why a mass noun is not good enough for counting. From plurals, instead, we know we can retrieve atoms.

Consider next the case of articles. Why can't we say "a water"? Typically, the indefinite article is either identical to the first numeral (as in Italian *uno*, "one") or it historically derives from it (often through morphophonological "simplification," like "a" in English). In either case, the indefinite article presumably means roughly the same thing as the first numeral *one*. Hence, we can't say *a water* for the same reasons why we can't say *one water*; counting (or, in the case of *one*, singling out an atom) requires an atomic noun.

The case of the definite article *the* is more complex and interesting. Consider first how it works with plurals. Imagine we are discussing the behavior of a class in a high school and say:

(21) Today the girls are quiet; the boys, instead are very rowdy. It must be because of the soccer match.

What is the role of *the* here? It invites us to consider the *largest* group to which the accompanying name applies. For example, in (20) we are attributing rowdiness to the totality of boys (in the relevant domain, i.e., the classroom). Something very similar seems to be going on with mass nouns. If we say "the water in this region is polluted," again we seem to refer to the totality of the water (in the place where we are); i.e., we refer to the largest aggregate of water. So far so good. With singular nouns, however, something else seems to be going on. To see it, go back to our example (14) and imagine that that's our domain of discourse (i.e., the set of salient objects we are talking about). In such a situation, we could easily say, for example, "the round table is broken," but we couldn't say "the square table is broken too." The reason seems to be intuitively clear. There is only one round table, but there are two square ones. Saying "the square table" doesn't determine which of the two we mean, and hence we cannot readily interpret the sentence. We would have to say something like "the square table on the right," for there is only one such table. It looks like *the* requires a noun specific enough to apply to just one object (in the relevant domain). So to use *the* felicitously with singulars, we must choose a domain of discourse and a noun that guarantee that the noun is true of just one thing in that domain. This "uniqueness condition" on the use of singular definites was noted by Bertrand Russell and much discussed since. The upshot of these simple observations is that with mass nouns and plurals, *the* is a kind of "maximizer": it invites us to consider a totality of objects that fall under a certain category (boys, tables, etc.). With singular, it is used to refer to a single entity; the noun it combines with is accordingly required to single out such an entity. Now, these two functions are natural enough. They serve useful communicative purposes, each in its own right. But why do they hang together, so to speak? Why is there one single word that covers both functions, if they are distinct? Is it conceivable that there could a language in which two different words are used, one word for the singular referential *the* and one for the maximization function? As matter of fact, this doesn't seem to happen. If a language has the definite article, it typically has a maximizing function with plural and mass nouns and is subject to a uniqueness condition with singular count nouns. But why?

Our simple model provides us with an answer. Look back at (16), the hypothesized structure that a domain of discourse has. Such a domain is structured in singularities and groups or aggregates. Nouns divide such a domain into kinds of objects that, in some sense, belong together (they have a common trait or function). Suppose, now, that *the* applies to a category of objects and selects out of it the largest, if there is one (and otherwise, the result is unintepretable). Here "largest" does not mean heaviest or fattest or tallest. It means the object that contains all the others as a part (i.e., the largest with respect to the part-whole relation naturally associated with domains of discourse). Consider now the structure of the different noun types, repeated here:

(22)

$$\text{table} \rightarrow [\text{a, b, c}]$$

$$\text{tables} \rightarrow \begin{bmatrix} & \{\mathbf{a, b, c}\} & \\ \{\text{a, b}\} & \{\text{b, c}\} & \{\text{a, c}\} \end{bmatrix}$$

$$\text{furniture} \rightarrow \begin{bmatrix} & \{\mathbf{a, b, c}\} & \\ \{\text{a, b}\} & \{\text{b, c}\} & \{\text{a, c}\} \\ \text{a} & \text{b} & \text{c} \end{bmatrix}$$

If we apply *the* to the plural *tables*, it will find the largest object that contains as parts all the others (the one in boldface). This works similarly for mass nouns like *furniture*. But if we apply it to the singular *table* in a situation like the one in (23), we get stuck, for no table has any of the others as a part. Suppose on the other hand that we are in a situation in which there is only one table:

(23) table → [b]

Each object is (trivially) part of itself. So in such a case, *the* will find the largest object that contains all the others as part, namely *b*. Conclusions: *the* is always a maximizer (more precisely, a supremum operator on the structure in [16]); it always selects the largest object out of a category of objects. In the case of plurals and mass nouns, this will be the largest aggregate of the relevant type. In the case of a singular noun, *the* will be able to perform its function only if the noun applies to just one entity in the domain of discourse we are selecting. *The* has a uniform meaning, which, in interaction with the way singular, plural, and mass nouns are structured, determines how we use it.

Let us now turn to the issue of variation. Our idea is that the basic way to shift from mass to count and vice versa is the following:

(24)

count	mass
[{a, b, c}	[{a, b, c}
{a, b} {b, c} {a, c}]	{a, b} {b, c} {a, c}]
[a b c]	a b c

For concreteness sake, imagine that *a*, *b*, and *c* are tables and that they are all the furniture there is in our domain of discourse. The left column represents the organization of the noun *table*, the right column the one of the noun *furniture*. The same entities are structured in slightly different ways. Or if you think of *a*, *b*, and *c* as water molecules, then the left column represents the organization of the compound count noun *water molecule* and the right column the mass noun *water*. And so on. The different structuring exemplified in (24) is what is responsible for the different behavior of mass versus count nouns, along the lines discussed above. According to this model, the components of the distinction are the same; the structuring may vary. Whence the possibility that languages may choose somewhat different ways to organize their lexicon. Let us look at this more carefully.

The mechanism in (24) is the basic one. But it interacts in interesting ways with other mechanisms we have discussed. Think of a and b as apples (count). Now divide each of them into halves and attach a label to each half. Each apple is now viewed as the sum (or aggregate) of its halves:

(25)

The half apples may constitute new atoms, and they can enter in our general schema in (24). For example, we can use two half apples to bake a cake and then say, "there is apple in this cake," a typical mass use of a count noun. In this case, we are viewing apples as having the structure of mass nouns (i.e., the structure on the right side of [24], with the relevant apple parts as "atoms").

Note it is not necessary to *physically* split things up to have a mass use of a count nouns. You can divide them into the relevant parts in your head, so to speak. In this way, you can view just about anything as a substance. It's D. Lewis's universal meatgrinder. The way to turn a count noun into a mass noun is to create a structure such as the one in (25). This may respect the units of the original count noun (as when we go from *piece of furniture* to *furniture*), or such units are first divided into parts and then the masslike structure is constructed out of those (as when we go from count to mass uses of *apple*).

A little warning is called for: compound nouns like *apple part* are perfectly count, which may be very confusing. Yet grammar leaves no doubt: *apple part* pluralizes, combines with numerals, etc. In terms of our approach, *apple* (in its basic count sense) and *apple part* both individuate atoms. They do so in different ways: what counts as an apple part may vary greatly from one occasion of use to another, while our notion of apple is more stable. However, both *apple* and *apple part* pass all the tests for countability. Our mass uses of *apple*, therefore, (i.e., "there is apple in the cake") is not simply to be equated with "apple part." If anything, it should be equated to "apple part or apple parts." Such a mass use comes from conceptualizing apple parts as having the structure of mass nouns, i.e., the one schematized in the right column in (24), with *a*, *b*, and *c* viewed as the relevant "minimal" portions. To repeat, *apple* in its mass use applies to any aggregate of apple parts (down to the smallest ones).

It may be useful to step back and get a more general picture of what goes on, according to our model. On each context (or occasion of use), our domain of discourse has the same and by now familiar structure. What changes from one occasion to the next are the atoms. The change in atomic structure of our domain is not necessarily a physical change in the structure of the world (though sometimes it may well be). Often enough, the change is in our head; i.e., we modify what we regard as atoms (the apples or their parts). The change concerns, in other terms, how we map out the structure of our domain of discourse into the world. In this process, we are guided by the nouns in our lexicon, which categorize the domain of discourse, dividing it in masses, singularities, and groups. The categorization we do via the nouns thus constrain how we set up our domain. In particular, an object *a* will have to have certain characteristics to count as an apple; it will have to have different ones to count as an apple part. Apples are physically separated from each other (and typically have physically undetached parts). Apple parts may be physically separated or unseparated from each other.

With this in mind, we can now understand what is behind crosslinguistic variation. Essentially the mechanism is the one in (24). English *hair* is to Italian *capello* what *furniture* is to *piece of furniture* (or to Italian *mobile*). Except that what counts as "hair unit" is somewhat flimsier than what counts as "furniture unit." (Take a single threadlike growth on the head and cut it in two. In English you still have simply hair; in Italian you've got to decide whether you have got two hairs or two halves of a single hair. The decision is perhaps not momentous, but grammar requires it.)

At a more macroscopic level, one can perhaps see now why no language has only count nouns and why nouns for liquids are generally crosslinguistically mass. If the picture above is correct, through language we can categorize things in two ways: by singling out the smallest units (atoms) and by simply characterizing aggregates of arbitrary size. If the minimal components of a substance are not readily accessible to our perception, it would be difficult and awkward to use them, given the option. We would have to go for either a vague classification (like *water part*) or for an excessively specific one (like *water molecule*). Going for aggregates of arbitrary size is clearly simpler in this case. What these considerations suggest is that a language in which every noun is forced to be atomic would strain our perceptual apparatus without bringing any practical advantage.

We have conjectured that our domain of discourse, while varying from context to context, holds constant a certain structure, represented in (16). I have characterized such a structure in a naïve way, drawing pictures, describing in words how things work, etc. In fact, such a structure is known in algebra as "free atomic join semilattice." In terms of this structure, we have made a conjecture as to how mass and count nouns differ. Count nouns denote sets of atoms; mass nouns denote sum-closed sublattices of the whole domain. The phenomenology of the mass-count distinction seems to follow naturally from this assumption. In particular, our proposed account of the relation between the "formal/grammatical" and "substantive" aspect of count versus mass has the following form. Our semantics constrains how we refer to things (the denotational structure); these constraints interact with nonlanguage-specific aspects of our conceptual system (our knowledge of objects and substances) to yield tangible "real world" effects (no noun of liquid is mass).

6. Semantic Modeling and the Study of Cognition

Where does the mass-count distinction come from? Here is one way of restating our main findings: In our dealings with the environment, we recognize entities, classes of entities, substances, etc. Some such things are perceived as being constituted of discrete components; others are not. In particular, it has emerged in the work of cognitive psychologists such as Spelke and investigators of animal cognition such as Hauser that human infants (as well as other primates) have a robust notion of solid object versus nonsolid substance. It is clear that the mass-count distinction has some structural properties in common with these prelinguistic notions. But it is also clear that it is not at all the same thing. It simply does not coincide with it. So, at present, there does not seem to be a reliable extralinguistic (or prelinguistic) contrast to which the mass-count distinction can be wholly reduced. One might speculate that the mass-count distinction is a philogenetic development of the prelinguistic distinction between substances and objects that we share with many other species. But for now this is sheer speculation.

Be that as it may, the main feature of the mass-count distinction, what sets it aside from the contrast between solid, discrete objects and substances is its *formal* character. "Formal" here is opposed to the "material" makeup of things. Even though mass and count maps naturally into the distinction between solid object and nonsolid substance, it also cuts across the latter (in the sense that solid objects can be grammaticized as mass, and the other way around). In a way, the distinction ceases to be anchored to a specific domain (that of everyday substances and discrete objects) and becomes applicable to any domain we can talk about, whatever its material makeup.

One might be tempted to say that the mass-count distinction is, nonetheless, rooted in our prelinguistic knowledge of objects and substances extended to other domains *by analogy*, in order to optimize communication by riding on a very pervasive and basic distinction. The problem I see with this is that there are many ways in which the analogy can proceed. And there is no one best way to optimize communication. Suppose we do indeed start out with our knowledge of substances and objects. A novel domain comes along, to which notions like "moving as a whole through continuous paths" do not apply (say, the domain of abstract entities). Why should we extend *both* modes to such a new domain? Wouldn't it be more economical to extend just one of the two available modes? Choosing both modes seems to force us to apply the original criteria in a substantially arbitrary way. (It seems, in fact, inappropriate to wonder whether knowledge, belief, fun, joy, etc. come in discrete units or not.) Moreover, there are other available resources we could reasonably exploit in a highly economical and efficient manner. For example, there are proper names (i.e., items like *Bill*, *London*, *Mars*, *July*). They do not display a mass-count contrast and have a morphosyntactic behavior different from that of common nouns (for example, they barely combine with determiners, with some combinations much worse than others, cf. things like **all the John*, **all the London* versus *all the water*, *all the knowledge*, etc.). In dealing with abstract entities, we might well resort to proper names, i.e., pick a way of referring to them with the same syntax as items like *John* or *London*. Yet this doesn't happen in any language. Languages, in referring to abstract nouns, do insist in picking either a mass or count syntax. So, of the many ways in which things can be analogized and communication optimized, only *particular* ones appear to be attested. This is left unexplained by the appeal to analogy or optimization of communicative needs.

We seem to be left with one possibility: the mass-count distinction is specific to human language and to its grammar. It is a feature proper of the architecture of human language. But this immediately entails that language is not merely a manifestation of intelligence or culture (for some of its features are autonomous of them).

Before addressing this big question more extensively, I would like to point out a further piece of relevant evidence. Such evidence comes from the study of language deficits. It is well known that there are specialized areas of the brain in the left hemisphere to which the language capacity appears to be linked. One is known as Broca's area and is located at the rear of the lower portion of the frontal area; the other, known as Wernicke's area, is located more posterior, near the primary auditory cortex. Subjects that suffer from brain damage involving these areas may lose in toto or in part the capacity to speak while sometimes retaining intact other cognitive capacities (including articulated problem-solving skills). The condition that ensues is the complex syndrome known as *aphasia*, which may be temporary or permanent. The study of language deficits has proven to be a formidable source of evidence concerning language and the way it is organized in the brain. By studying the various patterns of aphasic speech, one can put to the test linguistic hypotheses.

The relevance of this to the point under discussion (the mass-count distinction) is that deficits specifically affecting such a distinction have been reported in the literature. In particular, Semenza, Mondini, and Cappelletti (1997) describe the case of F. A., a seventy-three-year-old housewife and a speaker of a standard variety of Italian who suffered from a vascular lesion in the left temporal region of her brain. Before her illness, she was working in her husband's business (partly in accounting, partly in public relations) and she had no record of linguistic abnormalities. After the onset of her condition, she suffered from aphasia, from which she had since mostly recovered, retaining only what looked like a mild type of anomia (a difficulty in retrieving and using nouns). Upon closer scrutiny in tasks involving various noun types, it emerged that she had a specific impairment in the use of mass nouns. One of the tasks on which she was tested, for example, involved detecting grammatical errors with mass versus count nouns. She was asked to correct sentences like "there is much desk in the classroom" and "there is a sand on the beach." She had no problem in detecting the problem with sentences of the first type and in correcting it appropriately ("there are many desks in the classroom"). But she had severe difficulties in detecting and correcting the problem in the second kind of sentence, involving mass nouns. This difficulty persisted across other types of tasks. For example, she was asked to produce sentences using pairs like ship/sea versus roll/butter. With the first pair, she would produce perfectly grammatical sentences ("in the sea there are many ships"); with pairs of the second type, she had a remarkably high error rate ("I spread *a* butter on a roll"). The problem affected the whole range of mass nouns, from the most canonical and frequent, like *water*, to the less frequent ones of the collective sort, like *furniture*. All in all, F. A. seemed to have a serious impairment localized to the grammar of mass nouns, while displaying normal performance in other linguistic and nonlinguistic tasks. In our terms, F. A. forced every noun into an atomic mold. For her, every noun seemed to denote classes of atoms.[3] Semenza et al. also report finding the opposite dissociation (severely agrammatic patients who perform like normals on tasks involving mass nouns like those used in studying F. A.).

What does this tell us about the nature of the mass-count distinction? If such a distinction is the manifestation of some capacity *not* specific to language, the case of F. A. is very hard to comprehend, for one would expect that such a patient should display some kind of independent cognitive deficit also detectable in nonspecifically linguistic tasks. For example, there should be a difficulty in distinguishing, say, liquids and substances from discrete objects in general. Or there should be some other problem in the communicative abilities of the patient. But according to Semenza et al., this was not the case. If, on the other hand, the mass-count distinction is a specific feature of the architecture of grammar, which is realized in our brains, then it is conceivable that such a feature may be somehow damaged in isolation while leaving relatively intact the rest of our capacities.

7. Concluding Remarks

What does it mean to say that the conceptual distinction between mass and count is a feature specific to grammar? How does this claim mesh with the way in which language fits with the rest of our cognitive abilities? In section 1, we sketched, somewhat impressionistically, the following rough picture of such a relation. Intelligence is a generalized capacity to detect regularities in our environment and to solve problems that arise as we negotiate our survival with the environment (and with our kin). Language is the response of this generalized capacity to one of the problems that comes up in this process, that of communicating.

Evidence considered in this chapter seems to point in a different direction. Such a direction can be characterized (again, in just an impressionistic and approximate way) along the following lines: Our mind is constituted by a series of relatively autonomous systems, each using a specific vocabulary and mode of operating. One such system is language, which should be viewed as a sort of "mental organ"; grammar constitutes our inborn knowledge of language. Particular languages arise through the creation of particular lexica and the use of the combinatorial apparatus constitutive of grammar. Such an apparatus allows for a limited number of options, which can be set differently across languages (these options are called the "parameters"). This view of language is associated with the research program that has come to be known as generative grammar, which was originated by Noam Chomsky some fifty years ago.

The view of intelligence as arising from the interaction of differentiated autonomous components has come to be known as the "modular view."

Our hypothesis on the nature of the mass-count distinction has two components: an assumption on the structure of the domain of discourse and an assumption on how such a structure maps onto the morphosyntax of nouns. Thanks to this, we have argued, we see how different morphosyntactic combinations (e.g., the combination of plural morphology with a lexical noun) come to have their communicative effect (their meaning) and why some combinations are impossible or require special subsidiary interpretive procedures. I have used throughout without much discussion notions like reference, denotation, and truth, all of which have become central in the study of linguistic meaning. Such notions are steeped in the tradition of logic and philosophy of language. It is out of the encounter of such a tradition with generative grammar that modern linguistic semantics has emerged. The analysis of the mass-count distinction informally sketched here is a first, rough exemplification of how semantics works. The main idea that has emerged is that the way in which we refer to things (the way in which we categorize through nouns, etc.) is partly constrained by grammar. In the specific case under discussion, there are two ways in which the reference of nouns may be set (classes of atoms versus classes of aggregates closed under sum). One such way (the mass one) is obviously better suited to conceptualize liquids than the other. But this makes sense only because grammatical modes of reference exist autonomously.

The structure we have hypothesized is a good candidate for being universal; it is a way in which all humans structure domains of discourse. This exposes one of the key methodological assumptions of the research paradigm exemplified here. Even though languages seem to vary indefinitely, there are limits to their variation, mostly concerning certain aspects of the lexicon. When it comes to structure, a considerable uniformity emerges, often in unsuspected ways. A metaphor used by Chomsky in this connection is that if a Martian were to land on Earth, it would think that humans all speak the same language (apart from minor lexical differences). Accordingly, it is useful to conjecture that if a structure is present in a language, it is present in every language (albeit in different forms). We may go wrong, but we typically learn a lot in the attempt of pursuing it.

Consider the way in which this universalism applies to the case at hand. How is it compatible with the variation we observe across languages? The English word *hair* differs from its Italian translation *capello*. This difference is, if my proposal is on the right track, a semantic one. It is the same difference that manifests itself within a single language between, say, *coin* and *change* and perhaps, on a larger scale, throughout the lexicon of classifier languages. Yet this obviously does not entail a linguistically triggered difference in worldviews, a linguistic relativism a la Sapir/Whorf. It is, rather, a locally differentiated use of universal grammatical resources.

This last conclusion meshes well with recent psychological investigations of the Sapir-Whorf hypothesis. For example, a fascinating conjecture directly inspired by Sapir and Whorf was put forth by the psychologist A. Bloom (1981), regarding Chinese. His idea was that since Chinese lacks a distinct way to mark counterfactual hypotheses, Chinese speakers would have difficulty in reasoning involving false premises. However, the psychologist Au (1983) has shown that this is not so. Chinese speakers have no trouble with counterfactuality (if the tasks are presented in properly idiomatic Chinese). He comments: "how can something so fundamental and pervasive in human thinking [i.e., reasoning about counterfactuals] be difficult in any human language?" Similarly, on the view adopted here, there is certainly no expectation that Chinese speakers have a different worldview regarding how substances and kinds of objects are categorized. At the same time, it is not unconceivable that more "local" linguistic differences (like hair/*capello*) may have some subtle effect on nonlinguistic tasks. This, in turn, would leave some room for arguments aimed at showing that different vocabularies on basic color terms have an effect on how colors are categorized (Lucy 1992) or that differences in spatial terms may affect tasks of spatial orientation (Levinson 1997).

The overall outcome of our discussion can be summarized with the following claims:

(26) a. The mass-count distinction appears to be a grammar-specific property of the way in which nouns are mapped onto their reference.
b. Languages may vary in the way meanings are carved out.
c. The variation consists in (slightly) different ways of using the same universal structure.

If Chomsky is right in thinking that universal grammar is a innate schema (a sort of mental organ located in our brains), then part of such a schema is the tendency to refer to things in specific ways.

NOTES

1. An extreme case of this sort is constituted by so-called classifier languages such as Chinese. In such languages, numerals cannot be directly combined with any nouns. One cannot say things like *two men* or *three tables*. One has to use a classifier:

a. liang li mi
two CL rice "two grains of rice"

b. liang zhang zhuozi
two CL table "two pieces of table"

For discussion of the relevant issues see Chierchia (1998a, 1998b). For a different view, see Cheng and Sybesma (1999).

2. There is extensive work on the semantics of mass nouns and plurals, from which the proposal presented here draws. For a general overview and bibliographical references of both the philosophical and the linguistic approaches to the mass-count distinction, see Pelletier (1979)

and Pelletier and Schubert (1989). The proposal presented here is articulated more extensively in Chierchia (1998a, 1998b); its closest antecedents are Link (1983) and Landman (1991). The account of the definite article sketched below is originally from Sharvy (1980). On the acquisition of the mass-count distinction by children, see Gathercole (1982) and Gordon (1982). I should add that I think the main point of these papers would remain valid for any current formal semantic analysis of the mass-count distinction.

3. So F. A. seems to be doing something that no known language does: have only count nouns. This makes her case particularly interesting. But we cannot speculate as to what might be going on in her grammar within the limits of the present work. The point crucial to our concerns here is that the grammar of mass nouns can be selectively impaired.

REFERENCES

Au, T. K. 1983. Chinese and English counterfactuals: The Sapir-Whorf hypothesis revisited. *Cognition* 15: 155–187.

Bloom, A. H. 1981. *The linguistic shaping of thought*. Hillsdale, N.J.: Lawrence Erlbaum Associates.

Cheng, L., and R. Sybesma. 1999. Bare and not-so-bare nouns and the structure of NP. *Linguistic Inquiry* 30, no. 4: 509–542.

Chierchia, G. 1998a. Plurality of mass nouns and the notion of semantic parameter. In *Events and Grammar*, ed. S. Rothstein. Dordrecht: Kluwer.

——. 1998b. Reference to kinds across languages. *Natural Language Semantics* 6: 339–405.

Gathercole, V. 1986. Evaluating competing linguistic theories with child language data: The case of the mass-count distinction. *Linguistics and Philosophy* 9, no. 2: 151–190.

Gordon, P. 1982. *The acquisition of syntactic categories: The case of the mass/count distinction*. Ph. D. diss., Massachusetts Institute of Technology.

Hauser, M. 1996. *The evolution of communication*. Cambridge, Mass.: The MIT Press.

Landman, F. 1991. *Structures for semantics*. Dordrecht: Kluwer.

Link, G. 1983. The logical analysis of plural and mass terms: A lattice theoretic approach. In *Meaning, use, and interpretation of language*, ed. R. Bauerle, C. Schwartze, and A. von Stechow. Berlin: de Gruyter.

Levinson, S. 1997. From outer to inner space: Linguistic categories and nonlinguistic thinking. In *The relationship between linguistic and conceptual representation*, ed. J. Nuyts and E. Pederson. Cambridge: Cambridge University Press.

Lucy, J. A. 1992. *Language diversity and thought: A reformulation of the linguistic relativity hypothesis*. Cambridge: Cambridge University Press.

Pelletier, J., ed. 1979. *Mass terms: Some philosophical problems*. Edmonton: University of Alberta Press.

Pelletier, J., and L. Schubert. 1989. Mass expressions. In *Handbook of philosophical logic*, vol. 4, ed. D. Gabbay and F. Guenthner. Dordrecht: Kluwer.

Semenza, C., S. Mondini, and M. Cappelletti. 1997. The grammatical properties of mass nouns: An aphasia study. *Neuropsychologia* 35: 669–675.

Sharvy, R. 1980. A more general theory of definite descriptions. *The Philosophical Review* 89: 607–624.

Soja, N., S. Carey, and L. E. Spelke. 1991. Ontological categories guide young children's induction of word meaning: Object and substance terms. *Cognition* 38: 178–211.

Spelke, E. 1985. Perception of unity, persistence, and identity: Thoughts on infants' conception of objects. In *Neonate cognition: Beyond the blooming and buzzing confusion*, ed. J. Mehler and R. Fox. Hillsdale, N.J.: Lawrence Erlbaum Associates.

Whorf, B. L. 1956. *Language, thought, and reality*. Cambridge, Mass.: The MIT Press.

7

GENERATIVE SYNTAX IN THE BRAIN

YOSEF GRODZINSKY

Since its inception, generative grammar has positioned itself as the branch of human psychology that focuses on the language faculty. Thus an important programmatic article by Chomsky and Miller (1963) formulated three related questions as those that the language sciences must answer:

1. What is knowledge of language?
2. How does it arise in the individual?
3. How is it put to use?

The work of a linguist who tries to discover principles of linguistic knowledge is on this view closely related to that of a developmental psycholinguist who studies patterns of language acquisition (and their underlying mechanisms) or to the work of an experimentalist who investigates processes of speech and language generation and recognition. Psycholinguistics and developmental linguistics are first cousins, if not sisters, of theoretical linguistics.

Yet when considered from a broader perspective, language and its psychological basis are biological phenomena and in fact should be studied as such. That is, what Chomsky and his followers propose is a biological research program. Indeed, to many, the standard stuff from which linguistic facts are carved—intuitions about grammaticality, ambiguity, synonymy, coreference, and meaning—are bona fide biological phenomena: responses to grammaticality quizzes count as behaviors, we humans are organisms, biology is (among other things) about the behaviors of organisms, hence linguistics is biology. This in a nutshell seems to be the logic behind Chomsky's oft-cited saying that "linguistics is part of psychology, ultimately biology."

That the above inquiry is closely related to the study of human biology was clear at the outset. Still, neither Chomsky and Miller nor others went as far as stating what might have been obvious; they stopped short of formulating a fourth question regarding the biological underpinnings of grammatical systems (although see Lenneberg 1967, for an early attempt). And these, one could assert even then, relate most centrally to our brain and to our genetic makeup. The choice to dodge biological issues may have been reasonable then: few studies existed that focused on the way grammar is represented in the brain, let alone in the human genome. With the exception of Roman Jakobson's thoughts on aphasia (Jakobson 1941; Jakobson and Halle 1956), and some sparse preliminary results from linguistically oriented experiments on grammatical deficits subsequent to focal brain damage (e.g., Goodglass and Berko 1960; Goodglass and Hunt 1958), there were no neurolinguistic results for Chomsky and Miller to hang on to or even consider as a clue for a biology of language.

Yet times have changed: there is now a rich literature available on the biology of language, which, in the context of grammar, leads to a fourth question, Chomsky and Miller style:

4. How is knowledge of language represented in a human brain?

We now have at our disposal some preliminary answers to this question. Empirical evidence amassed over the past thirty years or so by psychologists and neurologists provides important hints that language is a truly biological entity, hints that come from varied and sometimes unexpected empirical angles. New approaches, novel experimental methods, and advanced scientific instruments have yielded new sources of empirical evidence that will, one hopes, not only play a central role in future linguistic theorizing but also help bridge the gap between disciplines. In this chapter, I would like to go over a couple of such approaches. I will tell you in brief what we now know about these and how we came to know it.

1. Modularity in Anatomy and Linguistics

Linguists often talk about the modular structure of linguistic theory. Jerry Fodor's monograph (1983) emphasized modularity as an issue in cognitive science as well. Less known is the long history of this concept in anatomy. Paul Broca—a founding father of aphasiology—was already engaged in a fierce debate on this matter in the 1860s. His (rather compelling) case for a distinct anatomical region for language met a fierce opponent, Pierre Flourens, who put forth the argument that the brain makes neither functional nor structural distinctions (and who also beat Victor Hugo in a contest for a seat in the Académie Française). The notion they debated was quite similar to present-day modularity, and, like now, the debate was empirical in nature—both Broca and Flourens referred to pathological data, discussing the relation between selectively impaired behaviors of brain-damaged patients and lesion sites as revealed by postmortem operations. Broca observed a selective loss of expressive language capacity subsequent to focal damage to the left frontal lobe (or more accurately, the foot of the third gyral convolution thereof), to which Flourens retorted with counterevidence (if somewhat dubious) to show that damage to other regions had similar consequences (cf. Zeki 1993, chap. 2, for an insightful and amusing review). Later, John Hughlings-Jackson (1878), Sigmund Freud (1895), and others from the French school (see Kandel et al. 2000, chap. 1, for a review) joined in, further promoting a "holistic" view that permitted no distinctions among types of mental capacities and denying the claim that brain areas specialize. For them, language was part of a general symbolic ability that resided virtually everywhere in the brain.

This debate on modularity marked the first round in a series to be continued throughout the years to come. Just before the turn of the twentieth century, the modularist position returned, this time from a microanatomical, rather than clinical, angle. Up to that time, it was believed that the brain was not composed of cells but was, rather, an undifferentiated part of the organism (at least at the microscopic level). Golgi devised new methods for the staining of neural tissue, which Ramón y Cajal used to produce the first microscopic evidence for the existence of basic building blocks—neurons. Without the neuron theory, a clearly modular notion, it is difficult to see how any physiological justification could be given to a modular theory of cognition. Brodmann subsequently carried out a microscopic morphological analysis of the brain, and his findings further supported a modularist position. He succeeded in distinguishing among cortical regions on the basis of differences in the spatial arrangement, density, shape, and connectivity of cells within these regions. His work culminated in a famous partitioning and numbering of the human cortex into cytoarchitectonically defined regions, as in figure 7.1 (see Amunts and Zilles, 2001, for a recent review).

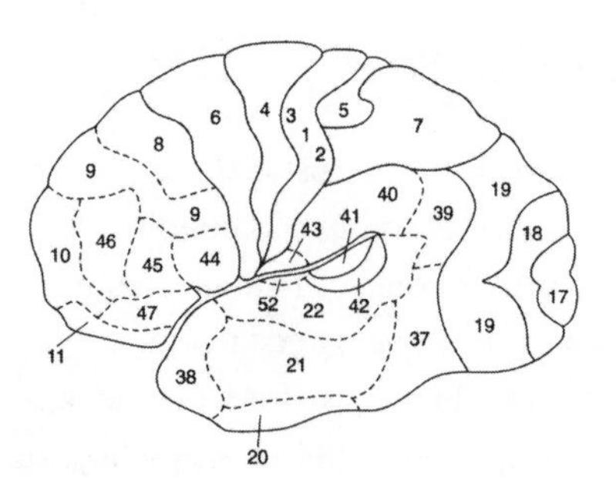

FIGURE 7.1
Brodmann's division of the left cortical surface into numbered areas (Broca's region is BA 44, 45).

Golgi, Ramón y Cajal, and Brodmann are credited with much of the solid foundation on which current functional neuroanatomy is now based. Their ideas, methods, and speculations on functional divisions in the cortex and their views on structure/function relations paved the way to current thinking. Still, this was not enough. The compelling empirical evidence these anatomists presented did not suppress the opposing view, championed by Karl Lashley (1951). On the contrary, the antimodular position, that language was nothing special—a mere reflection of a general sequencing skill—prevailed for most of the twentieth century in both psychology and neurology (and hints of it can still be found in current textbooks).

Yet since the 1960s, it was becoming increasingly clear that Broca was right: language (or at least important parts thereof) lives in the left hemisphere (see Geschwind 1965). The main source of evidence was lesion data—mostly aphasic deficits. Geschwind, reiterating Broca, Wernicke, and Lichtheim (cf. Lichtheim 1885, for a famous presentation), claimed that language centers exist. As clinicians, he and his followers emphasized communicative skills, viewing language as a collection of activities practiced in the service of communication: speaking, listening, reading, writing, naming, repetition, etc. The characterization of the language centers derived from this intuitive theory, with one cerebral center per activity posited. The resulting theory of localization uses these activities as building blocks, taking them to be the essence of human linguistic capacity (for clinical perspectives, cf. Geschwind 1979; Damasio 1992).

2. From Neurology to Neurolinguistics

The birth of a subdiscipline is not just an intellectual but also a social event, and it should be reported as such. It is difficult to grasp the evolution of neurolinguistics outside its historical context, and it is for this reason that we should approach current thinking as an interim ending of an ongoing historical tale. The appearance of modern linguistic theory and, in particular, generative syntax, did not pass unnoticed. Inspired by the zeitgeist, psycholinguists in the 1970s began using theoretical and experimental tools borrowed from linguistics and psycholinguistics (e.g., Blumstein 1972; Zurif and Caramazza 1976). Not denying the relevance of activities that humans engage in during communication, they began devising experiments that focused on linguistic distinctions that emanate from the new theoretical apparatus. Language was now seen as a piece of knowledge *about* structure, one that is organized in distinct components. The new experiments were fashioned accordingly: they aimed to test aphasics' abilities in phonology, morphology, syntax, and semantics. Soon thereafter, new findings were obtained indicating that the brain distinguishes between types of linguistic information. These results could not be couched in the activity-based view, and the centers had to be "redefined" (Zurif 1980). Each anatomical center was now said to contain devices used for the analysis and synthesis of linguistic objects. Roughly, Broca's region (Brodmann's area, BA 44, 45, 47; see fig. 7.1) was said to house syntax (for both receptive and productive language), while semantics resided in Wernicke's area (BA 22, 42, 39). This shift marked a new kind of debate: while earlier ones were about modularism versus holism, at issue now was the proper unit of analysis of brain/language relations.

Yet as findings accumulated—from different tasks, languages, stimulus types, and laboratories—new contradictions began to surface: in some cases, Wernicke's aphasics showed syntactic disturbances; Broca's patients, on the other hand, while failing certain tasks that probed syntactic abilities, succeeded in others. Serious doubts were cast on the new model, in which Broca's (but not Wernicke's) area supported receptive syntactic mechanisms. Finer linguistic concepts had to be introduced. At long last, linguistics and neuroscience were beginning to come closer, and harbingers of a new discipline—neurolinguistics—were appearing, with the goal of mapping linguistic abilities onto the cerebral cortex in a manner that takes linguistic theory as a central tool to describe behavior. At the same time, new methods, techniques, and concepts to study the brain were put to use. The next section briefly reviews the main sources of neurological evidence that are currently available.

3. Two Types of Experimental Neurolinguistic Arguments

3.1. Lesion/Aphasia Studies Through the Measurement and Analysis of Errors

Focal brain damage may selectively impair the linguistic system, and the resulting pattern of impairment and sparing may be gleaned from linguistically guided investigations into the patients' aberrant language behavior. Impairment to a particular rule system is isolated and linked to the damaged brain region. This is the "new localization" of grammar in the brain. Conversely, the lesion-based method provides an insight to the way the brain carves out linguistic ability, thereby providing critical neurobiological information as to the internal structure of linguistic-rule systems.

Since Roman Jakobson's famous study (1941) of language deficits in childhood and aphasia and their implications for phonological theory (see Grodzinsky 1990, chap. 6, for a reconsideration), a number of linguistic claims have been made based on aphasia data (e.g., Blumstein 1972, in phonology; Badeker and Caramazza 1998, in morphology; Avrutin 2000; 2001; Friedmann 2001; Friedmann and Grodzinsky 1997; Grodzinsky 1984; 1990; 2000a; Grodzinsky et al. 1993; Lonzi and Luzzatti 1993, in syntax). These works have also benefited from studies of the (deviant) time course of sentence comprehension in aphasia (for prominent examples, see Shapiro et al. 1993; Zurif 1995).

3.2. Blood Oxygenation Level Dependent (BOLD) Signal Monitored in PET/fMRI

Brain regions engaged in cognitive activity demand higher oxygen levels, increasing the blood flow into these regions (albeit with some delay). Change in blood flow is detected by advanced instruments: fMRI measures minute changes in the magnetic field that increased blood flow causes; when molecules with unstable isotopes are injected into the circulation, PET can detect their radioactive mark in cerebral areas in which their concentration increases due to oxygen demands. These techniques provide good spatial (but poor temporal) resolution, and the experimenter can carry out comparisons between stimulus types in terms of the location and intensity of the signal they evoke. Syntactic stimuli of various sorts can be thus investigated (Just et al. 1996; Stromswold et al. 1996; Embik et al. 2000; Ben Shachar et al. 2002, in press; Röder et al. 2002).

Below, I present two examples of these arguments in some detail and attempt to seek convergence among seemingly disparate sets of findings. I argue that the two methodologies produce highly consistent results, all leading to the conclusion that the left inferior frontal gyrus (LIFG, also known as Broca's region) is critically involved in the computation of phrasal movement in receptive language.

4. Damage to LIFG Results in a Receptive Deficit to Phrasal Movement

4.1. Trace Deletion: The Basics

Left Broca's region is topographically the triangular and opercular parts of the left inferior frontal gyrus (LIFG), or Brodmann's areas (BA) 44, 45, respectively. Focal insult to the vicinity of this region impairs linguistic ability in highly specific ways. The etiology of this condition may be stroke, hemorrhage, protrusion wound, tumor, or excision of tissue. Work carried out in many laboratories, using varied experimental methods and on several languages, has indicated that the receptive abilities of Broca's aphasic patients in the syntactic domain are compromised in that they are unable to link traces to (phrasal) antecedents, whereas other syntactic abilities remain intact. The various formulations of this claim have become known as the trace-deletion hypothesis, or TDH (Grodzinsky 1984, 1986, 1995, 2000a). The basic observation has been that Broca's aphasics suffer comprehension difficulties only when phrasal movement takes place: they fail to determine the grammatical status of sentences once traces are critically necessary, yet they are quite successful in both comprehension and judgment of other constructions. When asked to match sentences to pictures in a binary-choice experimental design (i.e., on tasks that require the matching of two arguments in a sentence to two actors in a scene, which amounts to a θ-role assignment task), their performance on the structures in (1) is at chance level (as measured by tests that consist of ten to twenty trials per construction type), yet they are well above chance in comprehending the controls in (2), whose representations contain no movement (for the moment, I ignore movement of the subject from [Spec,VP] to [Spec,IP] and other string-vacuous movement; see below). Further, their ability to link reflexives to their antecedents is intact; they also correctly rule out the binding of a pronoun to a local antecedent (3) (except when the antecedent is an R-expression, hence subject, perhaps, to constraints that go beyond the binding conditions as well as beyond the scope of this paper; see Grodzinsky et al. 1993), indicating that their deficit is restricted to movement (and is not a result of a generic "working memory" deficit or an overall linking failure; see Caplan and Waters 1999). Also, upon being asked to judge the grammaticality of sentences as in (4), they vacillate. Thus, they fail to consistently accept the (i) cases and, conversely, to detect the ungrammaticality of the (ii) cases. By contrast, they are quite agile when requested to determine the grammatical status of constructions whose status does not depend on traces of phrasal XP-movement (5). Finally, they successfully determine the status of constructions that involve head movement (6) (Grodzinsky and Finkel 1998; see Lima and Novaes 2000, for a replication in Brazilian Portuguese).

Chance Comprehension

(1) a. The man that the woman is chasing is tall
 b. Show me the man who the woman is chasing
 c. It is the man that the woman is chasing
 d. The man is chased by the woman

Above-Chance Comprehension

(2) a. The woman who is chasing the man is tall
 b. Show me the woman who is chasing the man
 c. It is the woman that is chasing the man
 d. The woman is chasing the man

(3) a. This is Goldilocks. This is Mama Bear. Is Mama Bear touching herself?
 b. This is Goldilocks. This is Mama Bear. Is every Bear touching her?

Failed Determination of Grammatical Status

(4) a. (i) John seems likely to win
(ii) *John seems that it is likely to win
b. (i) Which woman did David think saw John?
(ii) *Which woman did David think that saw John?
c. (i) I don't know who saw what
(ii) *I don't know what who saw

Successful Determination of Grammatical Status

(5) a. (i) Who did John see? (ii) *Who did John see Joe?
b. (i) The children sang (ii) *The children sang the ball over the fence
c. *The children threw (ii) *The children threw the ball over the fence
(6) a. (i) Could they have left town? (ii) *Have they could leave town?
b. (i) John did not sit (ii) *John sat not

This pattern of performance is rather intricate. The TDH claims that traces of phrasal movement are deleted from the patients' representations (I suppress considerations pertaining to the copy theory of movement [Chomsky 1995; Fox 2002]). A movement failure is discerned, cutting across construction types, tasks, laboratories, and languages.

4.2. *Mapping Representations Onto Performance*

Assuming trace deletion, albeit limited to XP-traces, predicts the pattern of success and failure that the patients exhibit in grammaticality judgment: they fail when an ability to represent traces of XP-movement is critical (4) and succeed in detecting other violations (5)–(6). Yet, in the domain of comprehension, trace deletion does not predict the data. Such results cannot follow from linguistic distinctions in the usual way, because a syntactic distinction cannot in itself map onto performance level in an experiment that is aimed to detect errors. Thus, an explicit mapping from structural deficiency to behavior (= error rate) is necessary, in the form of a set of premises from which actual performances can be deduced. Mere trace deletion, then, does not elucidate the chance performance on the constructions in (1). Moreover, as already noted, the patients' success in constructions that involve the movement of the VP-internal subject to [Spec,IP] is also left unaccounted for. Hence additional tools are needed, from which this intricate pattern would follow. As θ-assignment is the main task used in the above experiments, the solution derives the patients' aberrant performance from abnormal thematic representations they are said to possess. The main idea is to say that chance performance in a binary-choice θ-selection task actually follows from a θ-conflict. That is, the patient is receiving thematic information that indicates that both NPs in the sentence have the same θ-role and hence can each be matched to any of the actors in the sentence. The task of the theoretician, then, is to specify the conditions that would bring about this representation, that is, to create a situation in which the patient has to decide on agent and patient while, in his or her mind, both candidate NPs are linked to the same θ-role. This situation should lead to a θ-conflict, which would lead to guessing on the task at issue. We might want to characterize the desirable scenario a bit more generally: guessing behavior might follow when any two (potentially different) θ-roles are assigned to two NPs in a sentence, as long as both θ-roles are on a par on some universal thematic hierarchy. What follows is how this result is obtained.

The interpretation of moved constituents depends crucially on traces; without traces, the semantic role of a moved constituent cannot be determined. Under the TDH, moved constituents (*italicized* below) are uninterpretable. Assume that in such a situation, a nonlinguistic, linear order–based cognitive strategy is invoked, in an attempt to salvage those uninterpreted NPs. The strategy links θ-roles to (θ-ess) serial positions thus: <NP_1 = agent; NP_2 = theme>. In English, the strategy will force moved constituents in a clause-initial position to be agents: they are moved, hence linked to a trace; trace deletion hinders θ-assignment, and these NPs fall under the scope of the strategy. The idea is to view aphasic sentence interpretation as a composite process—an interaction between an incomplete (traceless) syntactic representation that may lead to a partial thematic representation and a compensatory cognitive strategy. In certain cases, for example, subject questions (7) or subject relative clauses (8), the default strategy should compensate correctly for the deficit:

(7) a. Which man *t* touched Mary? *Above chance*
b. Which man did Mary touch *t*? *Chance*

(8) a. The man who *t* is touching Mary is tall *Above chance*
b. The man who Mary is touching *t* is tall *Chance*

In the subject relative (8a), the object of the relative clause (*Mary*) is assigned the theme role without the mediation of a trace. The head of the relative (*the man*) is moved and receives its semantic interpretation (or thematic role) via the trace. A deleted trace renders this process impossible in Broca's aphasia. Thus only the object ends up with a grammatically assigned role; to save the uninterpretable subject of the relative (*the man*), the strategy is invoked, assigning it the agent role. This interaction between (deficient) grammar and (nongrammatical) strategy yields the correct semantics: NP_1(*the man*) = agent by strategy and NP_2(*Mary*) = theme by the remaining grammar. The same considerations hold of *which* subject questions (7a). By contrast, the TDH system predicts error in the object question (7b) and relative (8b): in these cases, an agent role is assigned to the subject of the relative or question (*Mary*), and another agent role is assigned by the strategy to the moved object (acting subsequent to trace deletion). Now the interaction between grammar and strategy gives rise to a misleading representation: NP_1 (*the man*) = agent by strategy, and NP_2 (*Mary*) = agent by the grammar. The result is a semantic representation with two potential agents for the depicted action, which predicts guessing behavior.

These assumptions lead to predictions that are confirmed once experiments are set up correctly so that they satisfy discourse requirements (although some of the data have been ignored here, for the sake of simplicity and focus; cf. Hickok and Avrutin 1995; Grodzinsky 1989). The TDH thus captures the selective nature of the comprehension deficit in Broca's aphasia.

We have ignored movement of subjects from [Spec,VP] to [Spec,IP], that is, movement that does not change the position of the subject relative to the predicate-object complex (at least in the cases for the data reviewed). Consider now a schematic solution to this problem: as the subject does not change its linear position relative to other major constituents in the string, the strategy compensates correctly, and above-chance performance in active sentences is expected, as is indeed the case. A broad range of experimental results is derived, and with it a conclusion: Broca's region is critically involved in the representation of traces of movement (or in linking them to their antecedents).

4.3. Cross-Linguistic Variation

It may be important to show that the comprehension problem in Broca's aphasia is not tied to a particular construction type (object relative, object question, passive, etc.) but is rather best accounted for by trace deletion cum default, i.e., the TDH. (1) We show that despite their success in comprehension tasks with subject relatives (due to their use of the default strategy), there are tasks in which their problems surface, namely such tasks in which the strategy is of no use or cannot be invoked. This is observed in real-time processing tasks. (2) We show that aphasics have comprehension problems in subject relatives in languages whose phrasal geometry is different from English.

Consider real-time processing in the healthy brain. It has long been known that neurologically intact subjects access the antecedents to traces at the gap position in real time. This is demonstrated by cross-modal priming tests, in which subjects listen to sentences such as (9a) while watching a screen onto which a visual probe of the types in (b–d) may be projected at points 1, 2, or 3 in the sentence. Their task is to make a lexical decision on the visually presented item:

(9) a. The passenger smiled at the baby[1] that the woman[2] in the pink jacket fed[3] __ at the train station
b. Diaper (related)
c. Horse (unrelated)
d. Strile (nonword)

At position (1)—immediately after the prime—access to the related target (9b) is obviously facilitated and reaction times are shorter. At position (2), there is a decay to this effect. And surprisingly, at (3) there appears to be facilitation—the prime gets reactivated at the gap position (Love and Swinney 1996).

When Broca's aphasics perform this task, they do not show normal priming at the gap (Zurif et al. 1993). This is in line with the TDH: if traces are deleted, they should not facilitate access to antecedents at the trace position. With that in mind, we can look at subject-gap relatives. If comprehension of such structures is intact, traces in subject position should be reactivated and the normal reaction-time patterns should follow; otherwise, in this case, too, performance should be aberrant. Indeed, Broca's aphasics demonstrate abnormal performance in this case, indicating that their impairment is not construction specific.

With this in mind, we can now move on and consider languages whose structural properties differ from English in ways that interact with the deficit in Broca's aphasia. Results obtained from a variety of language types lend support to the TDH in a surprising way. Consider Chinese, an otherwise SVO language, where heads of relative clauses (annotated by the subscript *h*) follow the relative (10a, 11a), unlike English, in which they precede it (10b, 11b):

(10)	a. [*t* zhuei gou] de mau_h hen da chased dog that cat very big	*Chance*
	b. the cat_h that [*t* chased the dog] was very big	*Above chance*
(11)	a. [mau zhuei *t*] de gou_h hen xiao cat chased that dog very small	*Above chance*
	b. the dog_h that [the cat chased *t*] was very big	*Chance*

This structural contrast leads to a remarkable prediction regarding performance in Broca's aphasia: Opposite English/Chinese performance patterns are expected. In English subject relatives, repeated as (10b), the head of the relative (*cat*) moves to the front (for concreteness, I assume a head-internal analysis of relative clauses, yet the analysis could be recast in other terms as well), lacks a role by the TDH, and is assigned as agent by the strategy, which leads to a correct representation, in which the cat indeed chases the dog. In Chinese (10a), the head (*mau*) also moves, but to sentence-final position, and the linear strategy assigns it the theme role. This representation has now two themes (*dog* and *cat*), and guessing follows. Similar considerations hold in object relatives (11a–b) and are left to the reader. This prediction is confirmed: the results in

Chinese are a mirror image of the English ones (Grodzinsky 1989; Su 2000; Law 2000). The mirror-image results correlate with a relevant syntactic contrast between the two languages—the position of the relative head. The θ-conflict now becomes a generalization, deriving chance performance from an agent/agent conflict in English relatives and from theme/theme conflict in Chinese.

Further intriguing cross-linguistic contrasts exist as well. Japanese scrambling, for example, results in two configurations:

(12) a. *Taro-ga Hanako-o nagutta* *Above chance*
Taro hit Hanako
Subject Object Verb
b. *Hanako-o Taro-ga t nagutta* *Chance*
Object Subject *t* Verb

As expected, Broca's aphasics are above chance in comprehending (12a) and at chance level on (12b), in keeping with the TDH (Hagiwara and Kaplan 1990). This result is robust and supported by a host of replications: it has also been obtained in Hebrew (Friedmann 2000), Spanish and Korean (Beretta et al. 2001), and German (Burchert et al. 2001). These results are important: they indicate that scrambling and cases of XP-movement form a neurological natural class (see Grodzinsky 2000b for further discussion).

4.4. A Brief Comment on Individual Variation

The cross-linguistic data coverage and the variety of constructions handled by the TDH do not resolve a nagging problem—that of individual variation. It has been the perception in aphasiology that replication of comprehension-test results is difficult. It has been pointed out that the evidential-basis claim to support the TDH is shaky, because most experimental results cited above are nonreplicable: performances on the passive have been found to vary greatly among patients in different studies (Berndt et al. 1996). These observations have been used in support of the claim that Broca's aphasia does not characterize a homogeneous group and should not be studied as such (Caramazza et al. 2001).

These claims are important. If the data are as dispersed as they appear, we should be very worried: that is, if behavioral aberrations among patients with the same lesion location vary arbitrarily, then there must be something fundamental that we are failing to understand about brain/behavior relations. Space limitations prevent further expansion here, so I will just say that in a series of studies (Grodzinsky et al. 1999; Drai and Grodzinsky 1999, forthcoming) we have demonstrated that while variation in the data does exist indeed, it cannot obfuscate the movement effects, which are extremely robust and leave no doubt about the critical role of Broca's region in this grammatical computation.

4.5. Summary

Three ideas pertain to the mapping from deficient syntactic representation onto performance rates in experiments that measure errors and collude to yield a uniform account of the receptive patterns in Broca's aphasia: (1) traces are deleted, (2) θ-assignment to moved NPs is augmented by a linear strategy, and (3) deviations from the expected pattern follow from the nature of chance-level performance: grouping patients' performances together yields a (corrected) Gaussian whose unimodal nature confirms (in fact for the first time) group uniformity; that is, it provides a formal argument in favor of the syndrome-based approach to Broca's aphasia. Next, when this observation is translated into a localizing claim, the conclusion is clear: Broca's region is critically involved the computation of XP-movement in receptive language. Having established that, we can proceed to view Broca's region from the angle of the healthy brain, as evidenced through functional imaging of movement. The TDH seems to have a fairly straightforward translation from the pathological to the normal functioning of left Broca's region: if the ability to compute movement is wiped out by damage to this area, then in the healthy brain, LIFG should be activated by syntactic movement.

5. Syntactic Operations in the Healthy Brain: LIFG and Movement

5.1. Step I: Imaging "Sentence Complexity"

Beginnings are always difficult, and thus the first studies of sentence comprehension were less syntactically detailed than one would have wished. Early efforts looked at the putative "processing difficulty" of different sentence types (Stromswold et al. 1996; Just et al. 1996), claiming that signal intensity in left and right Broca's and Wernicke's regions increases with "difficulty"; later, it turned out that anatomical overlap among studies was fairly poor (for critical appraisals, see Grodzinsky 2000a; Ben Shachar et al. 2002).

Experience with the linguistic interpretation of lesion data leaves one with a gnawing sense that the anatomically blurred picture may well be due to the fact that linguistic complexity (and subsequent "difficulty") is not a well-defined notion, and its varying interpretations may influence the way experimental materials are selected and thus may affect experimental results. Building on the aphasia data, we looked for stronger links between movement and the primary language areas of the brain.

5.2. Step II: Movement Activates Broca's Region in fMRI

When you tease movement apart from (whatever you construe as) complexity and test it in fMRI, a fairly clear picture emerges: Michal Ben-Shachar has led a series of tightly controlled fMRI experiments that our group conducted, which probe various aspects of movement in Hebrew, pitted against a fairly simple notion of complexity. Based on the aphasia data, we expected Broca's region to be activated when the movement contrast is tested. First, relative clauses (13a) were compared to sentences with CP complements (13b). The idea was to construct minimal pairs in which several straightforward complexity measures are kept constant (i.e., number of words, propositions, embeddings, verbs, ratio of functional to lexical categories, and more) and contrast object relative clauses (13a) with sentences that have main verbs that take CP complements (13b), so that the resulting minimal contrast would be movement (Ben Shachar et al. 2002):

(13) a. ʿazarti la-yalda [Se-Rina pagSa *t* ba-gina]
helped-I to-the-girl that-Rina met *t* in-the-garden
I helped the girl [that Rina met t in the garden]
b. ʿamarti le-Rina [Se-ha-yalda yaSna ba-gina]
told-I to-Rina that-the-girl slept in-the-garden
I told Rina [that the girl slept in the garden]

Complexity, in fact, to the extent that it entered into play, was pitted against our expectations. That is, the only potential difference in complexity between the conditions was the number of arguments of the predicate pairs in each sentence type. When this figure is calculated, the sentence type that does not involve movement (13b) has more argument slots. If anything, it should cause a conservative bias, increasing activation in the −movement case.

Subjects made grammaticality judgments. Each sentence had an ungrammatical counterpart created by the switching of the verbs in the embedded sentences in (13)—*meet* for *sleep*. We analyzed the grammatical sentences separately. We found a movement effect: (13a) produced a higher BOLD signal than (13b) in left Broca's region (LIFG, or BA 44, 45) and to a lesser extent in Heschl's gyri (BA 22; see fig. 7.1) of both hemispheres. Thus the core computational resource for movement structures is in areas 44, 45 of the left cerebral hemisphere, in keeping with the lesion data as described by the TDH. Auxiliary computations occur in temporal areas bilaterally. LIFG (seen in figure 7.2) was not only the region with the highest signal intensity but also the only region that exhibited left/right asymmetry.

An intriguing finding noted above—that scrambling patterns with movement in aphasia—actually converges on results from an experiment by Röder et al. (2001). This group has conducted an fMRI experiment in German, which

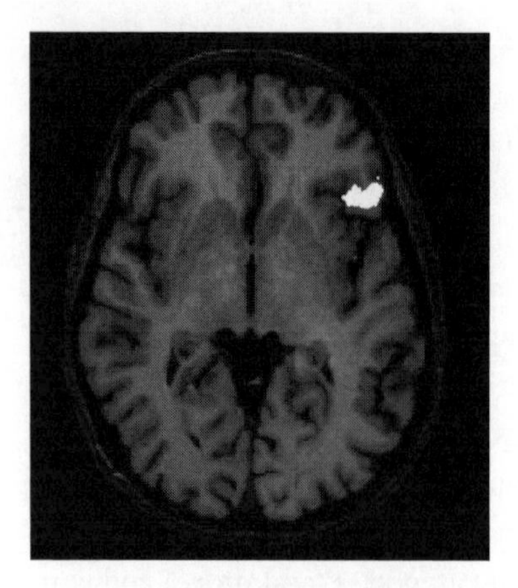

FIGURE 7.2
A statistical map associated with movement. Left IFG is the most activated region (Ben-Shachar et al., 2002). Note that left and right are reversed.

looked at different types of embedded clauses with double objects. The comparison they report gives a coherent (if partial) picture: when sentences in which both objects of an embedded double-object verb are scrambled with the embedded subject (14b) and compared to their nonscrambled counterparts (14a), activation is detected in the very same areas for which we found activations in the Hebrew fMRI experiments, that is, mostly in LFG, with some bilateral temporal activation.

(14) a. Jetzt wird der Astronaut dem Forscher den Mond beschrieben
Now will the astronaut [to] the scientist the moon describe
b. Jetzt wird dem Forscher den Mond der Astronaut *t t* beschrieben

Finally, additional studies in Hebrew (Ben Shachar et al. forthcoming) have indicated that this effect is not only localized but also very robust, generalizing over tasks (comprehension and grammaticality judgment) and over two additional contrasts in Hebrew: (1) embedded wh-questions versus yes/no questions and (2) object topicalized versus nontopicalized main clauses, e.g., (15a–b) versus (15c–d). All contrasts activate overlapping regions. These findings provide an imaging perspective that converges on the TDH. The same brain regions that implicate disorders in syntactic movement analysis are the most activated ones in the healthy brain, when syntactic movement operations are called for.

5.3. Step III: Separating Components: Double Objects

The experiments above provide an imaging perspective on the localization of movement operations. Despite that this research program is just at its beginning stages, we have attempted to take this perspective a step further. That is, we tried to see whether it is possible to use the location and intensity of the fMRI signal as a tool for the examination of specific linguistic hypotheses.

Ben-Shachar and Grodzinsky (2002) report a study of Hebrew double objects aimed at getting an imaging perspective on the linguistic analysis of this construction. As a first pass, we have focused on two main questions regarding this construction (e.g., Larson 1988; Aoun and Li 1988):

(I) What type of movement (if any) is involved?
(II) Which complement order (dative or double object) is base generated and which is derived?

We tried to answer these questions by constructing an activation map. As the materials in (15) crossed dative shift with topicalization, we compared datives and double objects (15c–d), on the one hand, and their topicalized counterparts (15a–b), on the other hand.

(15) a. ʻet ha-sefer ha-ʼadom Dani natan la-professor me-Oxford
Acc the-book the-red Dani gave to-the-professor from-Oxford
b. la-professor me-Oxford Dani natan ʻet ha-sefer ha-ʼadom
To-the-professor from-Oxford Dani gave Acc the-book the-red
c. Dani natan ʻet ha-sefer ha-ʼadom la-professor me-Oxford
Dani gave Acc the-book the-red to-the-professor from-Oxford
d. Dani natan la-professor me-Oxford ʻet ha-sefer ha-ʼadom
Dani gave to-the-professor from-Oxford Acc the-book the-red

Regarding issue (I), an activation-by-region interaction between the dative-shift contrast (15c–d) and the topicalization contrast (15a–b) would imply two distinct operations, while anatomical overlap in activation would suggest that a similar process is invoked in both cases. As to issue (II), following the same logic as in our previous experiments, the relative intensity of the signal in (15c–d) should indicate which is the derived order. Our study thus utilized two types of empirical argument: the anatomical locus of the fMRI signal as reflecting uniformity or distinctness of operations (topicalization versus dative shift) and the relative intensity of the fMRI signal within an anatomical region as reflecting more mental computation (double object versus dative).

The dative-shift contrast yielded two important results: First, it indicated a spatial pattern quite different from that for the relative/complement clause comparison and the topicalization contrast (to which it was compared directly). Specifically, the comparison between (15c) and (15d) activated two frontal regions in the *right cerebral hemisphere* and not any of the topicalization-related regions. This difference not only suggests that a different type of operation is involved but also, and quite surprisingly, it provides a preliminary indication against the commonly held belief that syntax is exclusively in the left hemisphere and is not represented on the right side of the brain (see Grodzinsky 2000a, for a review of this literature). Second, when the relative intensity of the BOLD signal is measured in the two right frontal regions that are sensitive to the dative-shift contrast, it is significantly higher for double objects (15d) than for datives (15c), suggesting that Hebrew double objects are more demanding than datives, providing an indication of their derived nature.

6. An Afterthought

The brain seems to be making finely grained syntactic distinctions that follow from current syntactic theory. In particular, left Broca's region handles movement of phrasal constituents. The studies reviewed demonstrate the direct neurological implications of the generative enterprise. Cross-methodological research programs such as the one just presented, which combine behavioral lesion studies with neuroimaging of healthy language users point to the great promise of neurolinguistics: they allow for the refined testing of modular linguistic as well as anatomical hypotheses.

REFERENCES

Aoun, J., and Y. A. Li. 1989. Scope and constituency. *Linguistic Inquiry* 20: 141–172.

Amunts, K., A. Schleicher, U. Bürgel, H. Mohlberg, H. B. M. Uylings, and K. Zilles. 1999. Broca's region revisited: Cytoarchitecture and intersubject variability. *Journal of Comparative Neurology* 412: 319–341.

Amunts, K., and K. Zilles. 2001. Cytoarchitectonic mapping of the human cerebral cortex. In *Neuroimaging clinics of North America on functional MR imaging*, ed. T. Naidich et al. New York: Harcourt.

Avrutin, S. 2000. Comprehension of Wh-questions by children and Broca's aphasics. In *Language and the brain: Representation and processing*, ed. Y. Grodzinsky, L. P. Shapiro, and D. A. Swinney, 295–312. San Diego, Calif.: Academic Press.

——. 2001. Linguistics and agrammatism. *GLOT International* 5: 3–11.

Badecker, W., and A. Caramazza. 1998. Morphology in aphasia. In *Handbook of Morphology*, ed. A. Zwicky and A. Spencer. Oxford: Blackwell.

Ben-Shachar, M., T. Hendler, I. Kahn, D. Ben-Bashat, and Y. Grodzinsky. 2002. The neural reality of syntactic transformations: Evidence from fMRI. *Psychological Science* 14, no. 5: 433–440.

Ben Shachar, M., and Y. Grodzinsky. 2002. On the derivation of Hebrew double objects: A functional imaging investigation. Paper presented at NELS 33, MIT, Cambridge, Mass.

Ben Shachar, M., D. Palti, and Y. Grodzinsky. Forthcoming. Neural correlates of syntactic movement: Converging evidence from two fMRI experiments. *NeuroImage*.

Beretta, A., C. Schmitt, J. Halliwell, A. Munn, F. Cuetos, and S. Kim. 2001. The effects of scrambling on Spanish and Korean agrammatic interpretation: Why linear models fail and structural models survive. *Brain and Language* 79: 407–425.

Berndt, R. S., C. C. Mitchum, and A. N. Haedinges. 1996. Comprehension of reversible sentences in "agrammatism": A meta-analysis. *Cognition* 58: 289–308.

Blumstein, S. 1972. *A phonological investigation of aphasia*. The Hague: Mouton.

Broca, P. 1861. Remarks on the seat of the faculty of articulate language followed by an observation of aphemia. In *Some papers on the cerebral cortex*, trans. G. von Bonin. Springfield, Ill.: Thomas, 1960.

Brodmann, K. 1909. Vergleichende Lokalisationslehre der Grosshirnrinde in ihren Prinzipien dargestellt auf Grund des Zellenbaus. Leipzig: Verlag von Johann Ambrosius Barth. Published as *Brodmann's Localisation in the Cerebral Cortex*, translated and edited by L. J. Garey. New York: Springer, 1999.

Burchert F., R. de Bleser, and K. Sonntag. 2001. Does case make the difference? *Cortex* 37: 700–703.

Caplan, D., and G. Waters. 1999. Verbal working memory and sentence comprehension. *Behavioral and Brain Sciences* 22: 77–126.

Caramazza, A., E. Capitani, A. Rey, and R. S. Berndt. 2001. Agrammatic Broca's aphasia is not associated with a single pattern of comprehension performance. *Brain and Language* 76: 158–184.

Chomsky, N. 1995. *The minimalist program*. Cambridge, Mass.: The MIT Press.

Damasio, A. 1992. Aphasia. *New England Journal of Medicine* 326: 531–539.

Drai, D., and Y. Grodzinsky. 1999. Syntactic regularity in Broca's aphasia: There's more of it than you ever imagined. *Brain and Language* 70: 139–143.

——. Forthcoming. Stability and variation in Broca's aphasia. McGill University and the Weizmann Institute.

Embick, D., A. Marantz, Y. Miyashita, W. O'Neil, and K. Sakai. 2000. Syntactic specialization for Broca's area. *Proceedings of the National Academy of Sciences* 97, no. 11: 6150–6154.

Fodor, Jerry. 1983. *The modularity of mind*. Cambridge, Mass.: The MIT Press.

Fox, D. 2002. Antecedent contained deletion and the copy theory of movement. *Linguistic Inquiry* 33: 63–96.

Freud, S. 1891. *On aphasia: A critical study*. Trans. E. Stengel. New York: International Universities Press, 1953.

Friederici, A., and P. Graetz. 1987. Processing passive sentences in aphasia: Deficits and strategies. *Brain and Language* 30: 93–105.

Friedmann, N. 2000. Agrammatic comprehension of OVS and OSV structures in Hebrew. *Behavioral and Brain Sciences* 23: 33–34.

Friedmann, N., and Y. Grodzinsky. 1997. Tense and agreement in agrammatic production: Pruning the syntactic tree. *Brain and Language* 56: 397–425.

Geschwind, N. 1965. Disconnexion syndromes in animals and man. *Brain* 88: 237–294, 585–644.

——. 1979. Specializations of the human brain. *Scientific American* (September).

Grodzinsky, Y. 1984. *Language deficits and linguistic theory*. Doctoral diss., Brandeis University.

——. 1990. *Theoretical perspectives on language deficits*. Cambridge, Mass.: The MIT Press.

——. 1995. A restrictive theory of agrammatic comprehension. *Brain and Language* 51: 26–51.

——. 2000a. The neurology of syntax: Language use without Broca's area. *Behavioral and Brain Sciences* 23, no. 1: 1–71.

——. 2000b. Anatomical variation and grammatical variation: A comparative approach to movement operations in the brain. Paper presented at the 31th meeting of the North Eastern Linguistic Society's special session on neurolinguistics, Washington, D.C.

——. 2002. Neurolinguistics and neuroimaging: Forward to the future, or is it back? *Psychological Science* 11: 188–193.

Grodzinsky, Y., A. Pierce, and S. Marakovitz. 1991. Neuropsychological reasons for a transformational analysis of verbal passive. *Natural Language & Linguistic Theory* 9: 431–453.

Grodzinsky, Y., and L. Finkel. 1998. The neurology of empty categories. *Journal of Cognitive Neuroscience* 10, no. 2: 281–292.

Grodzinsky, Y., M. Piñango, E. Zurif, and D. Drai. 1999. The critical role of group studies in neuropsychology: Comprehension regularities in Broca's aphasia. *Brain and Language* 67: 134–147.

Grodzinsky, Y., K. Wexler, Y.-C. Chien, S. Marakovitz, and J. Solomon. 1993. The breakdown of binding relations. *Brain and Language* 45, no. 3: 396–422.

Hagiwara, H., and P. Caplan. 1990. Syntactic comprehension in Japanese aphasics: Effects of category and thematic role order. *Brain and Language* 38: 159–170.

Hickok, G., and S. Avrutin. 1995. Comprehension of wh-questions by two agrammatic Broca's aphasics. *Brain and Language* 51: 10–26.

Hughlings-Jackson, John. 1878. *Selected writings of John Hughlings Jackson*. Vol. 2: *Evolution and dissolution of the nervous system and speech*. Ed. James Taylor. London: Hodder and Stoughton, 1931–1932.

Jakobson, R. 1941. *Kindersprache, Aphasie und algemeine Lautgesätze*. Translated as *Child language, aphasia, and phonological universals*. The Hague: Mouton, 1968.

Just, M. A., P. A. Carpenter, T. A. Keller, W. F. Eddy, and K. R. Thulborn. 1996. Brain activation modulated by sentence comprehension. *Science* 274: 114–116.

Kandell, T. J., and J. Schwartz. 2000. *Principles of neural science*. New York: Appleton & Lange.

Kluender, R., and M. Kutas. 1993. Bridging the gap: Evidence from ERPs on the processing of unbounded dependencies. *Journal of Cognitive Neuroscience* 5, no. 2: 196–214.

Larson, R. K. 1988. On the double object construction. *Linguistic Inquiry* 19: 335–391.

Lashley, K. S. 1951. The problem of serial order in behavior. In *Cerebral mechanisms in behavior*, 112–136. New York: Wiley.

Law, S.-P. 2000. Structural prominence hypothesis and Chinese aphasic sentence comprehension. *Brain and Language* 74: 260–268.

Lenneberg, E. 1967. *Biological foundations of language*. New York: Wiley.

Lichtheim, L. 1885. *Über Aphasie*. Leipzig: Deutsches Archiv für klinische Medicin, Leipzig, 36:204–268.

Lima R., and C. Novaes. 2000. Grammaticality judgments by agrammatic aphasics: Data from Brazilian-Portuguese. *Brain and Language* 74: 515–551.

Lonzi, L., and C. Luzzatti. 1993. Relevance of adverb distribution for the analysis of sentence representation in agrammatic patients. *Brain and Language* 45: 306–317.

Love, T., and D. Swinney. 1996. Coreference processing and levels of analysis in object relative constructions; Demonstration of antecedent reactivation with the cross modal paradigm. *Journal of Psycholinguistic Research* 25: 5–24.

Röder, B., O. Stock, H. Neville, S. Bien, and F. Rösler. 2001. Brain activation modulated by the comprehension of normal and pseudoword sentences of different processing demands: A functional magnetic resonance imaging study. *NeuroImage* 15: 1003–1014.

Stromswold, K., D. Caplan, N. Alpert, and S. Rauch. 1996. Localization of syntactic comprehension by positron emission tomography. *Brain and Language* 52: 452–473.

Su, Y.-C. 2000. Asyntactic thematic role assignment: Implications from Chinese aphasics. Paper presented at the LSA Meeting, Chicago.

Talairach, J., and P. Tournoux. 1988. *Coplanar stereotaxic atlas of the human brain*. Stuttgart: Thieme.

Tomaiuolo, F., J. D. MacDonald, Z. Caramanos, G. Posner, M. Chiavaras, A. C. Evans, and M. Petrides. 1999. Morphology, morphometry, and probability mapping of the pars opercularis of the inferior frontal gyrus: An *in vivo* MRI analysis. *European Journal of Neuroscience* 11: 3033–3046.

Zeki, S. 1993. *A vision of the brain*. Boston: Blackwell.

Zurif, E. B. 1980. Language mechanisms: a neuropsychological perspective. *American Scientist* (May).

——. 1995. Brain regions of relevance to syntactic processing. In *An invitation to cognitive science*, vol. 1, ed. L. Gleitman and M. Liberman. 2nd ed. Cambridge, Mass.: The MIT Press.

Zurif, E. B., and A. Caramazza. 1976. Linguistic structures in aphasia: Studies in syntax and semantics. In *Studies in neurolinguistics*, vol. 2. Ed. H. Whitaker and H. H. Whitaker. New York: Academic Press.

Zurif, E. B., D. Swinney, P. Prather, J. Solomon, and C. Bushell. 1993. An on-line analysis of syntactic processing in Broca's and Wernicke's aphasia. *Brain and Language* 45: 448–464.

PART IV

COGNITIVE SCIENCE AND THE PHILOSOPHY OF MIND

8

LEARNING ORGANS

CHARLES R. GALLISTEL

Harvey (1628) revolutionized physiological thinking when he showed that the heart circulates the blood and that its structure suits it to perform this function. Before Harvey, the modern conception of an organ as something whose particular structure enables it to perform a particular function did not exist. Physiological thinking centered not on organs but on humors. Humors had properties and effects, and pathological conditions were thought to arise from an excess or deficiency in one or more of them. The humors themselves did not have specific and limited functions. Much less did they have a structure that enabled them to perform a specified function. Organs, by contrast, have specific and distinct functions and a structure that enables them to perform them.

Chomsky reconceptualized learning in his *Reflections on Language* (1975). His reconceptualization is as radical in its implications for psychology and neuroscience as Harvey's work was for physiology. People generally conceive of learning as mediated by a small number of learning processes, none of them tailored to the learning of a particular kind of material. Examples are habituation, sensitization, and associative learning. (See any textbook on learning: for example, Domjan 1998, esp. 17–20, where he explicitly states and defends the general-process view, which is taken for granted in associative theories of learning.) These learning processes have properties and behavioral effects, but they do not have specific computational functions, nor do they have structures that enable them to perform those computations.

In the associative view of learning, the brain is plastic. It rewires itself to adapt to experience. There are no problem-specific learning organs that compute representations of different aspects of the world from different aspects of the animal's experience. Chomsky, by contrast, suggested that learning is mediated by distinct learning organs, each with a structure that enables it to learn a particular kind of contingent fact about the world. The noncontingent facts, the universal truths, are not learned; they are implicit in the structure of the learning organs, which is what makes it possible for each such organ to learn the contingent facts proper to it.

Chomsky's suggestion grew out of his recognition that learning was a computational problem—a view foreign to the associative conception of learning (Hawkins and Kandel 1984, Hull 1952, Skinner 1950) and to most neurobiological conceptions of learning. In Chomsky's view, the learner must compute from data a representation of the system that generated the data. The example Chomsky had foremost in mind was the learning of a language. He conceived of language learning as the computation from the limited number of often fragmented and often agrammatical sentences a learner hears the grammar of the system that generates all and only the good (well-formed) sentences in the language. The computational challenge this poses is so formidable that there is no hope of surmounting it without a task-specific learning organ, a computational organ with a structure tailored to the demands of this particular domain.

Whether this organ resides in a highly localized part of the brain or arises from a language-specific interconnection of diverse data-processing modules in the brain is irrelevant to whether it constitutes a distinct organ or not. Some organs are localized (for example, the kidney); others ramify everywhere (for example, the circulatory system). The essential feature of an organ is that it has a function distinct from the function of other organs and a structure suited to that function—a structure that makes it possible for it to do its job.

Although Chomsky (1975) had language foremost in mind, he clearly understood that his computational conception of learning implied that other forms of learning must likewise be mediated by problem-specific learning organs. From a computational point of view, the notion of a general-purpose learning process (for example, associative learning) makes no more sense than the notion of a general-purpose sensing organ—a bump in the middle of the forehead whose function is to sense things. There is no such bump, because picking up information from different kinds of stimuli—light, sound, chemical, mechanical, and so on—requires organs with structures shaped by the specific properties of the stimuli they process. The structure of an eye—including the neural circuitry in the retina and beyond—reflects in exquisite detail the laws of optics and the exigencies of extracting information about the world from reflected light. The same is true for the ear, where the exigencies of extracting information from emitted sounds dictates the many distinctive features of auditory organs. We see with eyes and hear with ears—rather than sensing through a general-purpose sense organ—because sensing requires organs with modality-specific structures.

Chomsky realized that the same must be true for learning. Learning different things about the world from different kinds of experience requires computations tailored both to what is to be learned and to the kind of experience from which it is to be learned. Therefore, there must be task-specific learning organs, with structures tailored both to what they are supposed to extract from experience and to the kind of experience from which they are to extract it. For computational reasons, learning organs may be expected to differ between species of animals, just as do sensory organs. Pit vipers sense infrared radiation, whereas we do not, because they have a sensory organ that we lack. We learn languages, whereas chimpanzees do not, because we have a language-learning organ, which they lack.

While Chomsky argued that there must be other learning organs, he did not specify what the other domains might be wherein they operated. We are now in a position to do this. In doing so, we make clear the computational nature of learning and why this leads to learning organs: mechanisms in the brain that carry out a particular kind of computation on a particular kind of data.

The Path-Integration Organ

Learning mechanisms enable us to acquire knowledge from experience. Among the most important kinds of acquired knowledge is knowledge of where we are. A fundamental computation by means of which animals of many kinds maintain moment-to-moment knowledge of where they are is path integration.

When a sailor sets out from her home port, she keeps a log of her position relative to her home port, based on her estimates of her speed and direction of movement. Her log might look something like table 8.1.

At the conclusion of each hour, she has recorded her estimated average speed during the past hour and her estimated direction of movement. She has then broken these estimates down into an estimate of how far she has progressed in a north/south direction and how far she has progressed in an east/west direction. For the first five hours, she sailed due west, so there was no change in her north/south position. For the first 4 hours her speed was 4 knots, putting her 16 nautical miles west of her port at 9:00. In the next hour, her speed picks up. She covers 6 miles in that hour, putting her 22 miles west. A change in the wind at 10:00 forces her to change her course to the northwest. Now, for every mile that she goes in the northwest direction, she moves 0.707 miles north and 0.707 miles west. Her speed picks up to 8, so after an hour of this, she is 0.707*8=5.7 miles north of her home port and a further

TABLE 8.1

Time	Est. speed (prev. hr.)	Est. direction	Northing (knts./hr.)	Easting (knts./hr.)	Change in position (nm)	
					N/S	E/W
5:00	leaving port					
6:00	4	W	0	−4	0	4.0 W
7:00	4	W	0	−4	0	8.0 W
8:00	4	W	0	−4	0	12.0 W
9:00	4	W	0	−4	0	16.0 W
10:00	6	W	0	−6	0	22.0 W
11:00	8	NW	5.7	−5.7	5.7 N	27.7 W
12:00	5	N	5.0	0	10.7 N	27.7 W
13:00	5	N	5.0	0	15.7 N	27.7 W
14:00	6	ESE	−3.0	5.2	12.7 N	22.5 W
15:00	7	ESE	−3.5	6.1	9.2 N	16.4 W
16:00	5	ESE	−2.5	4.3	6.7 N	12.1 W
17:00	4	ESE	−2.0	3.5	4.7 N	8.6 W
18:00	4	ESE	−2.0	3.5	2.7 N	5.1 W
19:00	4	ESE	−2.0	3.5	0.7 N	1.7 W

5.7 miles west (making a total of 27.7 miles west). Now the wind forces her to sail due north for 10 miles, at which point she decides to head for home. Looking at her log, she sees that she is 15.7 miles north of her port and 27.7 miles west. To head for home she has to sail more east than south, so she sets a course to 120°, that is, approximately to the east-southeast. For every mile she covers on this course, she progresses half a mile to the south and .87 miles to the east. After 6 hours on this course at varying speeds, she calculates that she is about 0.7 miles north of her home port and 1.7 miles west.

The computation laid out in this table is the path-integration computation. It is a computation performed on pairs of numbers. The pairs of numbers specifying for each hour the northing and easting (how fast and in what direction she has moved along a north/south axis and how fast and in what direction she has moved along an east/west axis) are her (estimated) velocities. It takes trigonometry to compute this number pair from her speed and course, which form the pair with which the computation begins. The number pairs in successive rows of the last two columns, which indicate her net change in positions, are obtained simply by adding up the hour-by-hour movements along each of the two directional axes.

It is easy to build physical devices that implement this computation. They do not, of course, work with numbers qua marks on paper; they work with physical quantities that could themselves be represented by such numbers, such as, for example, the bit patterns in the memory bank of a digital computer or the voltages in an analog computer. In the nervous system, such a computational mechanism works with neural activity proportional to the animal's speed and neural activity indicative of its direction. These neural signals take the place of the marks on paper (numbers) with which the human navigator works in computing her position. It is also easy to make the output of such a computation control the direction of a rocket or a robot, so no homunculus in the brain has to "see" and "interpret" these computations in order for their results to be manifest in the control of action. That the brain's behaviorally relevant activity must be understood in computational terms is the computational theory of mind, which is at the core of contemporary cognitive science. Chomsky's work did much to establish this conceptual framework, in which the brain's activity is conceived of in computational terms.

The brains of most animals contain neural machinery that implements the computation shown in the above table. That machinery—wherever it is found, however dispersed it may or may not be, and however it may be realized—constitutes a learning organ. It has a particular structure. It takes sensory signals indicating speed and sensory signals indicating direction and combines them to get signals indicating velocity, and then it adds up those velocities to get the net change in position. This structure has implicit in it a noncontingent purely mathematical truth about the world: position is the integral of velocity. That truth determines the structure of this learning organ in the same way that the truths of optics determine the structure of an eye. The function of this organ is to compute the animal's location from the sensory inputs it has generated in the process of moving to that location. Its structure suits it to perform that function and only that function. No one would suppose that an organ with this structure would be of any use in the learning of a language. Applying this learning organ to that learning problem would be like trying to hear with an eye or breathe with the liver.

Parameter Setting in the Organ That Learns the Solar Ephemeris

The path-integration computation requires a signal that indicates the compass direction in which the animal is progressing. Compass direction is direction relative to the axis of the earth's rotation, the north-south axis. Many animals use the sun's position for this purpose, which is remarkable, because the sun does not have a fixed compass direction. Its compass direction varies continuously during the day. Worse yet, the manner in which it does so depends on the season of the year and how far the observer is to the north or south of the equator. In short, the compass direction of the sun at a given time of day is a contingent fact about the world. If, however, the observer can overcome this problem, then the sun has advantages as a directional referent. It can be seen from almost any vantage point, and, most importantly, it is so far away that movements of an extent realizable in one day have very little effect on its direction. This is not true of earthly landmarks. The landmarks one can see are rarely farther away than the distance one can traverse in a small fraction of a day, which means that they change their compass direction as one moves. That makes them poor indicators of compass direction.

Human navigators also routinely use the sun as a compass referent; they have understood how to do so explicitly for centuries—and implicitly, probably for eons. The trick is to know the time of day and to learn the compass direction of the sun as a function of the time of day—that is, to learn the solar ephemeris.

Animals, including humans, have a built-in clock, which solves the first part of the problem, knowing the time of day. The circadian-clock mechanism is another example of the innumerable ways in which the enduring structure of the world they inhabit is reflected in the innate structure and functioning of the animal mechanism.

The second part of the problem is to know the compass direction of the sun for any given time of day. Bees and ants and birds (and probably many other animals as well) have an organ that learns this. Remarkably, bees, at least, learn the complete function from only a few observations, that is, without having seen the sun at most moments of the day. The learning of the solar ephemeris in bees illustrates the force and broad application of Chomsky's famous poverty-of-stimulus argument. Chomsky pointed out that the conclusions that humans draw about the grammatical and phonological structure of the languages they learn go far beyond what is justified by the limited samples from which they learn. Similarly, as we shall see, the conclusions that bees draw from limited observations of the sun's trajectory go beyond what is justified by those observations.

The structure of the organ that learns the solar ephemeris also illustrates a key idea in contemporary theories of language learning—the idea of parameter setting. As in Chomsky's thinking, the concept of a built-in function with settable parameters explains the ability of a learning organ to know more than it has observed, that is, to overcome the poverty of the stimulus.

The key experiment here was done by Dyer and Dickinson (1994), following earlier related work of a similar nature by Lindauer (1957, 1959). Dyer and Dickinson used the famous dance of the returning bee forager (Frisch 1967). The just-returned forager dances on the vertical surface of the honeycomb, inside the hive, out of sight of the sun. The dance is in the form of a figure 8. The bee circles first one way, then the other. In the middle stretch, where the loops of the 8 converge, the dancing bee waggles as it runs (flicks its abdomen from side to side). The direction of the waggle run relative to the vertical indicates the direction of the food source relative to the sun. If the bee runs

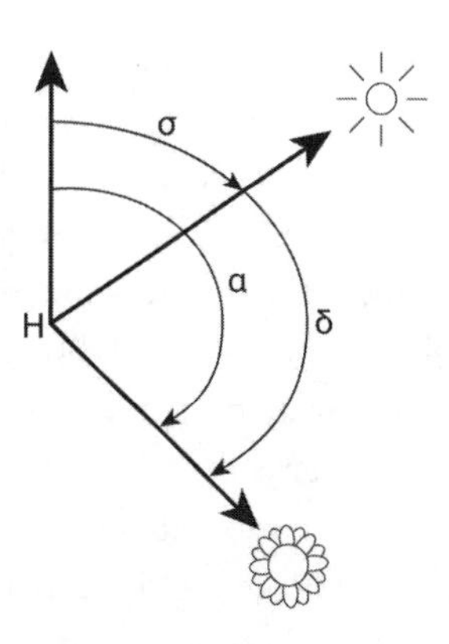

FIGURE 8.1.
Diagram of the angular (directional) relations. H = hive; N = north. σ = the compass direction of the sun. This direction is locally constant; it is the same regardless of where the bee is within its foraging range. α = the compass direction of the flower from the hive. δ = the solar bearing of the flower from the hive, which is the direction that the bee must fly relative to the sun in order to get from the hive to the flower.

straight up while waggling, it indicates that one must fly toward the sun; if it runs horizontally to the right, it indicates that one must fly with the sun on one's left; if it runs straight down, it indicates that one must fly away from the sun; and so on. The number of waggles during each waggle run indicates the distance: the more waggles, the farther the food.

The compass direction of the source from the hive (α in figure 8.1) equals the compass direction of the sun (σ in figure 8.1) plus the solar bearing of the source (δ in figure 8.1). Thus, when the experimenter, who observes the dance through a window on the hive constructed for that purpose, observes the direction of the waggle run, he can infer from it and from the compass direction of the source the direction in which the dancer believes the sun to be.

In Dyer and Dickinson's experiment, they raised bees in an incubator without a view of the sun. Then they allowed them to forage around their artificial hive, but only in the late afternoon, when the sun was sinking in the west. The bees learned to fly back and forth between the hive and a feeding station, which was located some tens of meters away to (for example) the west of the hive. When these foragers arrived back in the hive, they did a dance in which the waggle run was directed more or less straight up, indicating that, to get to the food, one should fly toward the sun.

Dyer and Dickinson allowed their bees to forage only in the late afternoon, until a morning came when the sun was completely hidden by a heavy overcast. Then, for the first time, Dyer and Dickinson allowed their bees to forage in the morning.

One might suppose that bees do not dance when cloud cover prevents their seeing the sun. How can you tell another bee how to fly relative to the sun if neither you nor the other bee can see it? But bees do dance when they cannot see the sun, other bees do attend to the dance, and they do fly in the direction indicated by the dance, even when they cannot see the sun as they fly. How can this be?

The key thing to realize is that once you have formed a compass-oriented map of the terrain within which you are navigating, you do not need to see the sun in order to set a course by it, provided you know the time of day and the sun's solar ephemeris, that is, its compass direction at different times of day. If I tell you to fly away from the sun at nine o'clock in the morning, a time at which you know the sun lies more or less due east, then I am in effect telling you to fly west. If you have a map of the terrain and you know the where the sun must be relative to that terrain at that time of day, then you can fly in the indicated direction by reference to appropriate terrain features. You fly toward the landmarks that lie to the west of your hive on your map, that is, landmarks that lie in the direction opposite the direction in which you know the sun must currently be.

The key to this scheme is the mechanism that learns the solar ephemeris, which, to repeat, is the compass direction of the sun as a function of the time of day. Learning the solar ephemeris—learning where the sun is at different times of day in relation to the terrain around the hive—is what makes the bee's map of that terrain compass oriented.

Returning now to the Dyer and Dickinson experiment: They allowed the bees to forage in the morning only when the sky was overcast, so that they could not see the sun. These bees, who had only seen the sun above the terrain lying to the west of the hive, flew to the food, and when they returned, they did the waggle dance, telling the other bees how to fly relative to the (unseen) sun. The astonishing thing is that the waggle run was now directed straight down rather than straight up. They were telling the other bees to fly away from the sun. This implies that they believed that the sun in the morning was positioned above the terrain in the direction opposite the food, even though, in their experience, it had always been positioned above the terrain in the direction of the food. Their experience of the sun, which was confined to the late afternoon, gave no grounds for their belief about where it is in the early morning (the poverty of the stimulus). Nonetheless, they had an approximately correct belief. They believed it to lie over the terrain to the east of the hive. For those who are uncomfortable with the attribution of beliefs to bees, the above can be reworded in terms of the compass direction of the sun specified by nerve signals in the brains of the dancers. These signals specified an eastward solar direction in the morning, even though the experienced sun had always been westward.

These fascinating results suggest the following model for the mechanism that learns the solar ephemeris (cf. Dickinson and Dyer 1996). Built into the nervous system of the bee is a genetically specified neural circuit that takes as its input signals from the bee's circadian clock and gives as its output a signal specifying the compass direction of the sun. If we imagine this signal to be the firing rates of two pairs of compass neurons in the brain, a N/S pair and an E/W pair, and if we imagine that different firing rates correspond to different compass directions, then the function in question is the one that determines the firing rates of the compass neurons at different phases of the bee's circadian clock.

The model further assumes that the general form of the function relating the output of the bee's circadian clock to the firing rates of its compass neurons is also genetically specified. It has the property that the firing rates for phases of the circadian clock twelve hours apart indicate opposing compass directions. One can imagine a genetically specified dynamic biochemical process in the brain closely analogous to the one that implements the bee's circadian clock. This mechanism leaves two things to be specified by the bee's experience: (1) the terrain views associated with a given set of firing rates (this anchors the compass signal to the local terrain) and (2) the kinetics of the change in the firing rates, that is, how they change as the circadian-clock mechanism runs through its cycle. The terrain views are analogous to the words in a lexicon. The kinetics of the ephemeris function are analogous to the grammar of a language. Observing the sun at a few different times of day suffices to determine the values of the free parameters in the innately specified ephemeris function. Fixing the values of those parameters adapts the universal function to the locally appropriate variant.

The interest of this model for present purposes is that it is an example of learning by parameter setting. The genes specify the general form that the relation between compass direction and time of day must take; in mathematical terms, they specify the function that is to be fitted to the data from experience. What is left for experience to specify are the values of a few parameters in the function (that is, the levels of a few physical variables in the mechanism that specifies where the sun is at different stages of the circadian cycle). What this learning mechanism does is allow the bee to get quickly to an approximately correct representation of the local solar ephemeris. It solves the poverty-of-stimulus problem. It enables the bee to estimate where the sun will be at times when it has never seen the sun, just as knowing the grammar of French enables one to speak and understand French sentences one has neither heard nor spoken before.

Modern theories of language learning in the Chomsky tradition are similar in spirit. The basic, high-level form of the grammar of any conceivable human language is assumed to be given by genetically specified language-learning machinery. This genetically specified general structure for any human language is what linguists call the universal grammar. Much research in linguistics is devoted to trying to specify what that general form is. The general form has in it a number of parameters whose value must be specified by experience. These parameters are generally thought to be binary parameters (the language must either do it this way or it must do it that way). Learning the language reduces to learning the parameter settings. When the child has got all the parameters set, it knows the language. It can produce and understand sentences it has never heard for the same reason that the bee can judge where the sun is at times of day when the bee has never seen the sun. As in the bee, the built-in structure of the learning organ, which is what makes learning possible, has as a consequence that the inferences the learner has drawn from limited experience go beyond what is justified by that experience.

This is a radically different conception of the learning process from the one that has dominated thinking in Western empiricist philosophy for centuries and in psychology and neuroscience more recently. It remains deeply controversial. Indeed, neuroscientists have not even begun to contemplate its implications. The idea that there may be no such thing as associative learning is unthinkable for most neuroscientists. Nonetheless, this conception is now very much, so to speak, on the scientific table. Chomsky has done more than anyone else to put it there.

REFERENCES

Chomsky, N. 1975. *Reflections on language*. New York: Pantheon.

Dickinson, J., and F. Dyer. 1996. How insects learn about the sun's course: Alternative modeling approaches. In *From animals to animats*, ed. S. Wilson et al. Cambridge, Mass.: The MIT Press.

Domjan, M. 1998. *The principles of learning and behavior*. Pacific Grove: Brooks/Cole.

Dyer, F. C., and J. A. Dickinson. 1994. Development of sun compensation by honeybees: How partially experienced bees estimate the sun's course. *Proceedings of the National Academy of Sciences, USA* 91: 4471–4474.

Frisch, K. v. 1967. *The dance-language and orientation of bees*. Cambridge, Mass.: Harvard University Press.

Harvey, W. 1628. *Exercitatio anatomica de motis cordis et sanguinis in animalibus*. Trans. A. Bowie. Frankfurt am Main.

Hawkins, R. D., and E. R. Kandel. 1984. Is there a cell-biological alphabet for simple forms of learning? *Psych. Rev.* 91: 375–391.

Hull, C. L. 1952. *A behavior system*. New Haven, Conn.: Yale University Press.

Lindauer, M. 1957. Sonnenorientierung der Bienen unter der Aequatorsonne und zur Nachtzeit. *Naturwissenschaften* 44: 1–6.

——. 1959. Angeborene und erlente Komponenten in der Sonnesorientierung der Bienen. *Zeitschrift für vergleichende Physiologie* 42: 43–63.

Skinner, B. F. 1950. Are theories of learning necessary? *Psych. Rev.* 57: 193–216.

9
INNATENESS, CHOICE, AND LANGUAGE

ELIZABETH SPELKE

Throughout his writings, Chomsky raises questions about human knowledge and freedom: What do we know, and how does our knowledge arise? Why do we act as we do, and how do we choose our actions? He offers strongly contrasting views of the progress of science in answering these questions. Although linguists and other cognitive scientists have made headway in understanding human knowledge, especially knowledge of language (e.g., Chomsky 1975), we have not even begun to understand how humans choose their actions (e.g., Chomsky 1988). In Chomsky's writings, knowledge of language and freedom of action are treated as distinct and separate problems. I believe, however, that there is a thread that connects them.

There is a tension in our conceptions of human action. On the one hand, we see ourselves and others as capable of choosing our actions and therefore as accountable for the choices we make. On the other hand, we see people's actions as shaped by the past, both by evolution and by past experience. This tension raises a question: if all our actions are shaped by our individual and collective history, how can people freely choose any course of action?

Studies of human cognitive development focused on the origins and growth of human cognitive capacities seem to suggest that true choice is not possible. These studies, I hope, will help tease apart what is constant versus what is changing over human development, what is universal versus what is variable across human cultures, and what is unique to humans versus what is shared by other animals. It looks, however, as if the answers psychologists give to these questions may obscure rather than clarify our human capacity for freely chosen action.

In the early years of experimental psychology, the dominant theory of human development rooted virtually all human cognitive achievements in a single capacity to associate co-occurring experiences (see Gallistel's contribution to this volume). This general learning capacity, together with our sensory apparatus, was thought to be innate; almost everything else was thought to be learned. According to associative-learning theory, people learned everything in the same way, from elementary perceptions of depth to concepts of number and ideas of virtue. Our learning processes, moreover, were the same as those of other animals, although humans learned more. The paradigm experiments in this tradition were conducted on animals, for example Pavlov's dogs, who learned to salivate upon hearing a bell that had previously been paired with food, or Thorndike's cats, who learned to press a lever that opened a lock and released them from a cage. Experimental psychologists showed that humans sometimes exhibit the same effects, salivating to a photograph of Oreos or pushing levers that deliver candy and other rewards.

Followers of this perspective are apt to conclude that the capacity for free choice doesn't exist, because people's actions depend entirely on their learning history. We would be mistaken to think that a criminal is morally responsible for her crime, from the perspective of associative-learning theory, because her criminal actions, like all her actions, are determined by her past experience. If anyone were to blame for her crime, it would be her teachers and parents—and they didn't choose their actions either. If human thoughts, values, and actions are wholly determined by learning history, then notions of freely chosen action and personal responsibility are illusions.

Over the last fifty years, the tide has turned against this view of human development, thanks in large part to the revolution that Chomsky's work initiated. It has been supplanted, for many psychologists, by the view that perception, thought, values, and actions depend on domain-specific cognitive systems, both in humans and in other animals. Each system has its own innate foundations and evolutionary history, and each system functions to treat specific kinds of information for specific purposes. The best-known example is the language faculty: an innate, special-purpose system with a distinctive internal mode of operation and a distinctive developmental course.

The language faculty is unique to humans, but many other faculties are shared by humans and other animals. A paradigm case of a special-purpose cognitive system was discovered by John Garcia in research with rats (e.g., Garcia and Koelling 1966). Garcia allowed rats to drink a liquid with a particular flavor (say, anise), which was delivered by a flashing, buzzing tube. Drugs or x-rays, unseen by the rats, then were used to make the rats nauseated. From one such experience, the rats learned to avoid the drink with the anise flavor, but they never learned to avoid the flashing, buzzing tube.

In contrasting experiments, other rats were given a shock after drinking the same liquid from the same tube. These rats learned to avoid the tube but not to avoid the anise-flavored drink. These findings suggest that rats have one system for learning about the edible properties of substances and a second system for learning about the mechanical properties of objects. Because these systems are separate, rats cannot learn to relate a food's taste to shock or an object's behavior to nausea. Psychologists are apt to give an evolutionary explanation for these special-purpose learning systems. In the environments in which rats evolved, for example, nausea tends to be caused by substances with characteristic odors and tastes, not by objects with characteristic lights and sounds. Rats' domain-specific food learning mechanism likely was shaped by this constraint.

One interesting discovery of the last few decades is that many of the special-purpose cognitive systems found in animals exist in humans; the "Garcia effect" provides an example (Rozin, Haidt, and McCauley 2000). If you get sick after eating creamed corn in a blue dish with Dennis, who is about to succumb to an intestinal flu, Dennis is the most likely cause of your sickness. Nevertheless, you may come to feel queasy, in the future, at the thought of corn, but you are unlikely to feel queasy at the thought of Dennis or the blue dish. Like rats, humans are predisposed to learn about certain environmental relationships over others.

More recent research provides evidence for other special-purpose cognitive systems both in animals and in humans (Spelke 2003a). One system serves to represent material objects and their mechanical interactions (e.g., Scholl 2001). A second system serves to represent persons as sentient beings whose actions are directed to goals (e.g., Woodward 1998). A third system serves to represent number and the operations of arithmetic (e.g., Dehaene 1997). A fourth system, to which I will return, serves to represent spatial layout for purposes of navigation (Jeffery 2004).

These findings accord with the view that human actions depend on a collection of innate, domain-specific, core knowledge systems: systems for reasoning and learning about food, systems for perceiving and learning about objects, systems for learning and using language, and the like. Many of these systems evolved before humans did, although language clearly is an exception.

Does this view shed light on human freedom and responsibility? Apparently not. Modern evolutionary psychology seems to leave as little room for freely chosen action as associative-learning theory did. If our evolutionary history determines our actions, we would seem to be predestined to repeat the actions of our ancestors. This sense of predestination may explain why many people resist evolutionary psychology or evolutionary theory in general. If natural selection has shaped men to dominate women, shaped groups of people to be aggressive to members of other groups, or shaped step-parents to harm step-children, people would appear to have no choice but to act in these ways.

In brief, psychology and cognitive science offer two notions—general-purpose learning and innate, domain-specific faculties—that both seem to undermine notions of human freedom and responsibility. But do these notions bear that weight? To introduce my own approach to this issue, let's return to the Garcia effect and consider a gap in the preceding account.

Rats, I suggested, learn to avoid a particular flavor but not a particular feeding tube after becoming ill because they have a core cognitive system that lets them think *this food made me sick* but not *this tube made me sick*. Humans, I suggested, share this core system. It is patently *not* true, however, that humans can't think the thought *this tube* [or *person*, or *plate*] *made me sick*. Possibly in contrast to rats, we can think all these thoughts. Although the sight of a particular person may never make us feel nauseated, the thought that the person has a contagious disease may lead us to choose to keep our distance.

If human cognition depends on a collection of core knowledge systems, how can one understand this capacity for freely assembling thoughts into new combinations? Linda Hermer-Vazquez, Anna Shusterman, and I have tried to approach this question by studying the development of this capacity in children, focusing on some learning tasks that are logically similar to Garcia's, although they differ in detail. I consider one of these tasks here.

We have studied adults and children as they perform tasks developed by Ken Cheng, C. R. Gallistel, Richard Morris, and other comparative psychologists for studies of navigation in rats and other animals. Their studies have revealed two distinct core systems subserving navigation and spatial memory in nonhuman animals. First, animals can learn to locate a hidden object by searching directly at a movable, visible landmark. If food consistently is hidden next to a white cylinder that is moved around a closed chamber from trial to trial, for example, rats learn to search for the food at the cylinder (e.g., Biegler and Morris 1996). Second, animals can learn to reorient themselves and locate a hidden object by recording the position of the object within a purely geometric representation of the environment. For example, if a hungry rat is placed in a rectangular, closed chamber, sees food hidden just to the left of a long wall, and then is disoriented, the rat will reorient himself by the shape of the enclosure and search for food to the left of one of the two long walls (Cheng 1986).

Rats, however, show no ability to combine information from these two systems. If one of the two long walls in the above experiment is painted white, for example, rats will ignore this lightness cue and search equally to the left of the long black and white walls (Cheng 1986). Rats evidently can represent the location of food as "at the white cylinder" or as "left of the long wall" but not as "left of the white wall." The last thought is unthinkable, Cheng and Gallistel suggest, because it requires a combination of information stored in separate cognitive systems.

To explore the capacities that allow humans but not rats to form such combinations, we have adapted Cheng's task for children and adults. We have built human-sized rooms similar to those used with rats, in which we hide a toy in one of various locations and then disorient the child or adult by slowly turning him or her with eyes closed. When children are then asked to find the toy, those ranging in age from eighteen months to four years perform very similarly to rats (Gouteux and Spelke 2001; Hermer and Spelke 1996; Wang, Hermer, and Spelke 1999). In a rectangular environment, for example, children readily find a toy hidden directly at a blue wall or to the left of a long wall, but they show no ability to combine geometric and color information to find a toy hidden left of a blue wall. Further studies have revealed that children's performance in this task shows all the signature limits of the performance of rats. These findings suggest that homologous cognitive systems underlie reorientation in rats and human children (Wang and Spelke 2002).

Human adults, in contrast, solve all of our reorientation tasks with ease: they can quickly find an object hidden left of a blue wall or in any other relation to any arbitrary landmarks (Hermer and Spelke 1996). Developmental studies provide evidence that this ability emerges in synchrony with the development of spatial language: especially the mastery of spatial expressions involving the terms "left" and "right," which occurs at about six years of age (Hermer-Vazquez, Moffett, and Munkholm 2001).

Further studies provide more evidence that flexible navigation is linked to natural language. In one set of studies, adults were tested under conditions of verbal interference (Hermer-Vazquez, Spelke, and Katsnelson 1999). While they participated in Cheng's reorientation task, they listened to a tape-recorded prose passage and repeated it continuously, word for word. While performing this "verbal shadowing" task, adults appeared to lose the ability to combine geometric and nongeometric information, and they performed instead like rats and young children. Their performance contrasted with that of a separate group of adults, who performed a nonverbal "rhythm shadowing" task by listening to a tape-recorded percussion sequence and repeating the sequence by clapping. During rhythm shadowing, adults continued to combine geometric with nongeometric information. These findings suggest that natural language serves not only in the construction of new spatial concepts but in their active use.

Finally, training studies are beginning to suggest that the performance of four-year-old children can be enhanced by teaching them the appropriate spatial language (Shusterman and Spelke 2004; Shusterman, Lee, and Spelke forthcoming). Although children of this age typically reorient only in accord with the geometry of the surrounding layout, those who have been trained to use and understand spatial expressions such as "raise your right hand" and "give me the car on your left" are more apt to perform like adults. They combine geometric and nongeometric information to guide their search for objects.

All these findings suggest that humans come to combine together information that is represented separately in the minds of human infants and nonhuman animals and that the acquisition and use of a natural language underlies this accomplishment. Because a natural language is first and foremost a combinatorial system, it may serve as a medium for constructing new concepts once words and expressions are linked to representations from multiple core systems (Spelke 2000, 2003b). These combinations, in turn, make available a new range of potential actions.

Findings such as these suggest a different picture of human cognition and action. Indeed, humans are endowed with a set of core knowledge systems, each with its own evolutionary history. These systems may give us the building-block concepts—like *food, plate, person, long, left,* and *blue*—that we assemble into thoughts. But humans also have a further cognitive system that is unique to our species: the language faculty. As children learn a natural language, this system may allow them freely to combine their existing concepts and form new ones.

Although the concepts and thoughts provided by core knowledge systems may be limited and fixed, the concepts humans construct with our combinatorial system are more numerous and flexible, and they are not directly constrained by our evolutionary history. Concepts like *food in a blue dish* or *left of a blue wall* may have been useless to ancestral humans, and we do not appear to have evolved any domain-specific cognitive system that expresses them. Our combinatorial capacity nevertheless makes these concepts, and indefinitely many other concepts, available to us. Humans therefore can act by choosing among the options made available by the combinatorics of natural language; we are not limited either to the options expressible within one or another core knowledge system or to the combinations that we can build through slow, piecemeal associative learning. Untrained nonhuman animals and prelinguistic infants may not be able to choose to avoid food in a particular container or to find an object left of a white wall, but older children and adults can and do make these choices.

The existence of a combinatorial, natural-language faculty that serves to combine previously encapsulated representations does not explain how people choose their actions. As Chomsky has argued, free will remains a mystery (e.g., Chomsky 1988). It is even possible that the combinatorial operations of natural language provide inherently inappropriate tools for understanding our capacity to act freely (McGinn 1993). Nevertheless, the combinatorial power of natural language leaves more room for choices that are unconstrained by our history (either in evolutionary or ontogenetic time) than either associationist or evolutionary psychology would seem to grant. Even if some actions depend on automatic processes of associative learning, as traditional learning theorists proposed, people's thoughts and choices are not limited by associative processes. A child with a language faculty and a finite stock of words can produce and understand an indefinitely large set of expressions that she has never learned, by association or otherwise. And even if humans have evolved cognitive systems that predispose us to act in certain ways, as evolutionary psychologists propose, no person of normal competence is compelled to act in accord with these predispositions. Instead, people can formulate an indefinitely large number of possible courses of action, most of which are not favored by any evolved, special-purpose systems, and they can choose which course to pursue.

Of course, mature human choices are subject to bias, because our learning history and evolutionary history will tend to predispose us to certain actions over others. For example, the finding that verbal interference leads adults to navigate like children and rats suggests that the building-block concepts and behavior patterns provided by core knowledge systems are automatic and resilient, whereas the new concepts that we construct by combining core representations are tenuous, effortful, and dependent on language use. Humans nevertheless can exercise our combinatorial capacity and move beyond both our evolutionary and our learning history.

With this ability come moral obligations. The capacity of normal, mature humans to combine information in new ways that go beyond the limits both of our evolved, special-purpose cognitive systems and of our associative-learning capacities supports our sense that we are accountable for our actions. The absence of this capacity in other animals, and its limited presence in very young children, may support our sense that animals and children are not fully responsible moral agents.

In summary, the capacity to combine core concepts freely and productively may give humans a range of choice far beyond what either our learning history or our evolutionary history would seem to allow. Central to this capacity, I believe, is the faculty for natural language that Chomsky's work elucidates. If exercise of this faculty carries us beyond the limits both of innate and of learned concepts, then the study of language may speak to Chomsky's longstanding, distinct concerns for human knowledge and freedom.

REFERENCES

Biegler, R., and R. G. M. Morris. 1996. Landmark stability: Studies exploring whether the perceived stability of the environment influences spatial representation. *Journal of Experimental Biology* 199: 187–193.

Cheng, K. 1986. A purely geometric module in the rat's spatial representation. *Cognition* 23: 149–178.

Chomsky, N. 1975. *Reflections on language*. New York: Pantheon.

——. 1988. *Language and the problems of knowledge.* Cambridge, Mass.: The MIT Press.

Dehaene, S. 1997. *The number sense: How the mind creates mathematics.* Oxford: Oxford University Press.

Garcia, J., and R. A. Koelling. 1966. Relation of cue to consequence in avoidance learning. *Psychonomic Science* 123–124.

Gouteux, S., and E. S. Spelke. 2001. Children's use of geometry and landmarks to reorient in an open space. *Cognition* 81: 119–148.

Hermer, L., and E. S. Spelke. 1996. Modularity and development: The case of spatial reorientation. *Cognition* 61: 195–232.

Hermer-Vazquez, L., A. Moffett, and P. Munkholm. 2001. Language, space, and the development of cognitive flexibility in humans: The case of two spatial memory tasks. *Cognition* 79: 263–299.

Hermer-Vasquez, L., E. S. Spelke, and A. S. Katsnelson. 1999. Sources of flexibility in human cognition: Dual-task studies of space and language. *Cognitive Psychology* 39: 3–36.

Jeffery, K., ed. 2004. *The neurobiology of spatial behavior.* Oxford: Oxford University Press.

McGinn, C. 1993. *The problems of philosophy: The limits of inquiry.* Oxford: Blackwell.

Rozin, P., J. Haidt, and C. R. McCauley. 2000. Disgust. In *Handbook of emotions*, ed. M. Lewis and J. M. Haviland-Jones, 2nd ed., 637–365. New York: Guilford Press.

Scholl, B. J. 2001. Objects and attention: The state of the art. *Cognition* 80, nos. 1/2: 1–46.

Shusterman, A., and E. S. Spelke. Forthcoming. Language and the development of spatial reasoning. In *The innate mind: Structure and content*, ed. P. Carruthers, S. Laurence, and S. Stitch. Oxford: Oxford University Press.

Spelke, E. S. 2000. Core knowledge. *American Psychologist* 55: 1233–1243.

——. 2003a. Core knowledge. In *Attention and performance*, vol. 20: *Functional neuroimaging of visual cognition*, ed. N. Kanwisher and J. Duncan. Oxford: Oxford University Press.

——. 2003b. Developing knowledge of space: Core systems and new combinations. In *Languages of the brain*, ed. S. M. Kosslyn and A. Galaburda. Cambridge, Mass.: Harvard University Press.

Wang, R. F., L. Hermer-Vazquez, and E. S. Spelke. 1999. Mechanisms of reorientation and object localization by children: A comparison with rats. *Behavioral Neuroscience* 113: 475–485.

Wang, R. F., and E. S. Spelke. 2002. Human spatial representation: Insights from animals. *Trends in Cognitive Sciences* 6, no. 9: 376–382.

Woodward, A. L. 1998. Infants selectively encode the goal object of an actor's reach. *Cognition* 69: 1–34.

10

THE SCOPE AND LIMITS OF CHOMSKY'S NATURALISM

PIERRE JACOB

Fifty years ago, Noam Chomsky laid the foundations for a new scientific approach to the human language faculty (HLF), which he called "generative grammar." Furthermore, his argument that behaviorist explanations of human verbal behavior are inadequate was a major instigator of the "cognitive revolution" that took place in the 1960s and gave rise to the cognitive sciences.[1] Today, few contemporary analytic philosophers of mind or language would deny, I think, that Chomsky's work has deeply changed our scientific understanding of human language.

Many philosophers, however, have challenged one or another aspect of Chomsky's framework for investigating the HLF. Not only has Chomsky systematically responded to his critics, but he has also produced his own evaluations of their contributions to the understanding of the human mind and human language. As a result and as two recent publications demonstrate,[2] two gaps now separate Chomsky from the community of analytic philosophers. On the one hand, many philosophers cannot subscribe to what Chomsky calls "methodological naturalism." On the other hand, unlike some of the philosophers who do subscribe to methodological naturalism, Chomsky rejects what he calls "metaphysical naturalism."

My goal here is to explore the nature of these two divides. Over the years, Chomsky has come to make an important distinction between two versions of a naturalistic approach to human mind and language: methodological naturalism, which he accepts, and metaphysical naturalism, which he rejects. In the first section of this chapter, I will succinctly characterize the conceptual framework created by Chomsky for the scientific study of the HLF. In the second section, I will argue that many (if not all) of the criticisms directed by philosophers at Chomsky's scientific framework show that, whether intentionally or not, they embrace some version of methodological dualism, which is inconsistent with the methodological naturalist position advocated by Chomsky. But, in my view, the most unexpected and the most interesting divide is that which separates Chomsky from the program of some of the philosophers who subscribe to metaphysical naturalism and whose aim is to "naturalize intentionality." In the third section, I will examine the question of what prevents Chomsky from accepting metaphysical naturalism. In the fourth section, I will examine Chomsky's reservations about the program of naturalizing intentionality.

1. The Scope and Limits of Scientific Investigation of the Language Faculty

"Naturalistic" investigation, in the sense of Chomsky (2000), is nothing other than the scientific investigation of the world, whatever the aspect involved. Now, scientific investigation of the world, according to Chomsky (1980, 8; 2002), goes along with the acceptance of what, using Husserl's expression, the theoretical physicist Steven Weinberg called "the Galilean style" (in physics), i.e., "making abstract mathematical models of the universe to which at least the physicists give a higher degree of reality than they accord the ordinary world of sensation." Whatever aspect of the world is involved, what the scientific (or naturalistic) approach affords is an objective theoretical understanding of the world detached from ordinary human concerns and interests. Because it is based on strong idealizations, this theoretical understanding of the world is bound to be narrow and deep. It is bound to be narrow because the idea of a simultaneous objective theoretical understanding of all aspects of the world does not make sense.[3] It is bound to be deep because objective theoretical understanding of the world consists in discovering abstract principles that, like fundamental laws of physics, are inaccessible to the resources of human common sense alone, remote from observations and empirical evidence, but from which observations and empirical evidence may be inferred via long chains of explicit reasoning.

Theoretical understanding of the world is not the only kind of understanding accessible to humans. The world also offers us the possibility of artistic (or aesthetic) understanding: "the arts may offer appreciation of the heavens to which astrophysics does not aspire" (Chomsky 2000, 77). But if the goal we are seeking is theoretical understanding of the world, then the idealizations of scientific investigation are not dispensable.

The scientific (or "naturalistic") study of the HLF began in the 1950s, when Chomsky assigned generative grammar the task of providing an explicit and testable characterization of the computational properties of what is known by any person who is able to produce and understand the sentences of his or her native language.[4] The task is to describe the recursive procedures that allow the construction of a potentially infinite set of complex linguistic expressions out of a finite stock of simple lexical items. In Chomsky's more recent terminology, mastery of the recursive procedures that allow one to produce and understand a potentially infinite set of sentences from one's native language is an "internal" ("internalized") language, or "I-language," and the set of sentences generated by this "I-language" is an "E-language."[5] An E-language is composed of public E-expressions (including sentences); an I-language is composed of underlying mental I-constructions.

The fundamental task of theoretical linguistics is to understand how an I-language—a stable state of the HLF—allows infinite use of finite lexical resources. Chomsky (1980, 1986) calls this characteristic of the HLF "discrete infinity." In the framework of generative grammar, theoretical understanding resides in computational models of syntactic and semantic processes for constructing complex expressions from elementary constituents.

The goal of generative grammar is to discover the computational properties of the HLF—also called "universal grammar"—based on Chomsky's observation that one's ability to understand and produce sentences from one's native language raises three further questions:[6]

(Q1) What is the system of internal knowledge (I-language) that allows one to understand and produce the sentences of one's E-language?
(Q2) How does this system develop and stabilize over the course of ontogenetic development?
(Q3) How is this system exploited in verbal behavior (both in tasks of production and of understanding)?

Linguistic research shows that the I-language of an adult speaker consists of knowledge (partly explicit and mainly implicit or tacit) of a vast quantity of syntactic and semantic facts,[7] including the fact that, in the English sentences (1) and (3) but not (2), the proper name "Mary" can function as antecedent for the pronoun "she" or the possessive adjective "her" of the constituent "her daughter":

(1) Mary said that she would come.
(2) She said that Mary would come.
(3) Her daughter said that Mary would come.

In answer to (Q1), the task of generative grammar has involved a search for the basic computational principles from which we can infer the fact that in

(1) and (3) but not in (2) the noun can function as antecedent for, respectively, the pronoun and the possessive adjective. If an English speaker knows that the anaphor and its antecedent can be bound in (1) and (3) and not in (2), then question (Q2) arises: how does a human child learn this contrast?

The exploration of questions (Q1) and (Q2) was one of the major factors involved in the cognitive revolution that led to the shift from the study of human behavior to the study of the cognitive structures and processes that sometimes result in observable behavior. According to Chomsky, it would be a mistake to think that all interesting questions posed by the use of language can be handled by the scientific (or naturalistic) approach. Chomsky has repeatedly said over the years that the likelihood of reaching theoretical understanding or a scientific explanation of the "creative" aspects of language use is quite low. By contrast, according to him, questions (Q1) and (Q2) are well suited to scientific study. While the investigation of (Q1) is guided by a search for "descriptive adequacy," the investigation of (Q2) is guided by a search for "explanatory adequacy":[8] the latter should contribute to the explanation of how the child constructs her I-language (i.e., knowledge of the grammar of her native language) from the primary linguistic data provided by members of her linguistic community. In other words, (Q2) is the question: how do we characterize the initial state of the language faculty by means of which the child converts primary linguistic data into knowledge of the grammar of a particular natural language or E-language?

For forty years, Chomsky has maintained that inspection of primary linguistic data leads to what he calls the "poverty-of-the-stimulus argument," which in turn is a condition of adequacy on any purported answer to (Q2). This argument involves three complementary premises that could all be summarized by the proposition that the I-language of an adult speaker is vastly underdetermined by the totality of the linguistic data available to a child. First, grammatical knowledge is not the result of explicit learning or teaching: English-speaking parents do not teach their ten-month-old child that an English sentence is composed of a noun phrase followed by a verb phrase. Second, the utterances that a child encounters are a finite and fragmentary sample of the E-language spoken by members of her community (e.g., English). Third, children acquire knowledge of certain rules for which there are no clues in the set of utterances to which they are exposed: by hypothesis, the corpus of primary linguistic data only contains information of the category "*P* is a sentence of language *L*" and no information of the category "*P** is not a sentence of *L*."[9] On the basis of the poverty-of-the-stimulus argument, Chomsky concludes that a child could not acquire knowledge of the grammar of her language unless she was equipped with tacit knowledge of universal grammar, and this knowledge is a cognitive "module" specialized for the task of language acquisition.[10]

2. Methodological Dualism and Common-Sense Concepts

No contemporary philosopher of science would dream of subjecting theories of physics, chemistry, or biology to the authority of a priori conceptual reflection guided by mastery of such ordinary or common-sense concepts such as "matter," "movement," "air," "fire," "vegetable," or "life." Contemporary philosophers of the natural sciences take it for granted that only if theoretical scientific concepts can be freed from the constraints imposed by ordinary common-sense conceptions can scientific theorizing flourish.

Over the years, certain theoretical concepts of generative grammar have been disputed by philosophers of mind and language. In response, Chomsky has made the point that these criticisms tacitly rely on the presupposition that the naturalistic (or scientific) investigation of the HLF can be subjected to some kind of a priori conceptual analysis that accepts the authority of such ordinary common-sense concepts as "language," "languages," "knowledge," "mind," or "mental." If he is right—as I believe he is —then he is also right to conclude (Chomsky 2000, 112) that these philosophical criticisms rest on some kind of intellectual duplicity or double standard. In the natural sciences, the criteria of rationality are based entirely on explanatory success. But the criteria of rationality accepted in the natural sciences are supposed to be simply inapplicable to the study of human cognitive processes, for which the criteria of rationality are supposed to have an entirely independent source. This is the duplicity that Chomsky calls "methodological dualism," in opposition to methodological naturalism.

The concept of I-language has given rise to two sorts of philosophical perplexity. The first issue is whether an adult speaker of an E-language can truly be said to know the grammar of his or her language. A fortiori, can a human baby be said to know universal grammar? The second issue is whether the computational explanations of the HLF—or of any other human cognitive capacity—could or should be subjected to conceptual analysis based on the authority of ordinary common-sense conceptions of so-called mental phenomena.

When analytical epistemologists wonder whether an adult speaker really knows the grammar of his language, they decompose this question into two subquestions. First, given a particular E-language (let's say French), does there exist a unique, well-defined set of grammatical rules that generate all and only the sentences of the E-language in question? (This very question involves the contentious presupposition that E-language is given conceptual priority over I-language.) Second, is it appropriate to analyze the cognitive relation between an adult speaker of the E-language and this set of grammatical rules (if it exists) by means of the concept of knowledge?

I begin with the first question. In 1963, the philosopher Edmund Gettier published a short article in which he demonstrated that a person *S* could have a justified true belief in proposition *p* even though *S* could not be said to "know" that *p* in the ordinary sense of the word "know."[11] The majority of analytical epistemologists concluded from this that one ought to give up the traditional idea that having a justified true belief is a sufficient condition for knowing a proposition in the ordinary sense. Since 1963, analytical epistemologists have been wondering what other condition should be added in order to turn a true belief that *p* into genuine knowledge that *p*. They assume that not unless one truly believes that *p* can one know that *p*. They further assume that unless one were introspectively conscious of holding the belief that *p*—i.e., unless one could express its content verbally by uttering a sentence saying that *p* and recognize the belief as a belief—one could not believe (truthfully) that *p*. Since an ordinary speaker is not consciously aware of the grammatical rules of his E-language (which he cannot state), the relation between the speaker of an E-language and the grammatical rules cannot be the belief relation, let alone the knowledge relation.

To subject the scientific investigation of the HLF to the constraints of the ordinary concept of knowledge is to succumb to methodological dualism. Faced with the theoretical successes of molecular biology, philosophers of science would never subject the double-helix model of the DNA molecule to the authority of the ordinary concept of life. If the theoretical successes of generative grammar do not satisfy the requirements of the ordinary concept of knowing, what should we conclude? Only the intellectual duplicity inherent to methodological dualism can block the conclusion that the ordinary concept of knowing is inappropriate if we aim to satisfy the demands of the scientific investigation of the HLF.

The first question of analytical epistemologists was whether all and only the sentences of an E-language can rightly be said to be generated by a unique and well-defined set of grammatical rules. To justify a negative answer to this question, Quine (1972) developed an ingenious argument.[12] Obviously, if it is false that there exists a unique and well-defined set *R* of rules that generates all and only the sentences of an E-language, then the question of whether an adult speaker knows *R* is irrelevant. To discredit the idea that a speaker has tacit knowledge of a grammatical rule, Quine (1972) offers the distinction between the fact that (verbal) behavior conforms to a rule and the fact that it is actually guided by a rule. Quine (1972) conceives of the sentences of an E-language on the model of the "well-formed formulas" of the artificial languages of logic. According to him, it cannot be claimed that a single system of rules guides the verbal behavior of a speaker. He relies on two assumptions. First, he supposes that a speaker's verbal behavior cannot be guided by a rule unless the speaker can formulate and follow the rule consciously. Second, he supposes that, given a set of sentences belonging to an E-language, one can always imagine many rival systems of rules capable of generating the same set of sentences. Quine concludes from this that all a linguist can say is that the verbal behavior of a speaker conforms to many extensionally equivalent systems of rules.

In his response, Chomsky (2000, 78) observes, on the one hand, that the logical concept of a well-formed formula does not apply to natural-language sentences. On the other hand, comparison between rival systems of grammatical rules is not limited to extensional equivalence. Since Chomsky (1965), the methodology of generative grammar includes a distinction between weak extensional equivalence and strong intensional equivalence: two rule systems are weakly equivalent if they generate the same set of sentences, but two systems of rules are strongly equivalent if they assign the same structural descriptions to the generated sentences.

In a further move, Quine suggests that the relevant empirical evidence (or observable data) in linguistics is severely limited. He maintains that the evidence in favor of a syntactic or semantic hypothesis about the constituent structure of the sentences of an E-language is strictly limited to the observable verbal behavior of speakers of the E-language. As Quine puts it (1990, 37), "in psychology one may or not be a behaviorist, but in linguistics one has no choice. Each of us learns his language by observing other people's verbal behavior and having his own faltering verbal behavior observed and reinforced or corrected by others. We depend strictly on overt behavior in observable situations." Following Quine's behaviorist assumptions, observing the verbal behavior of monolingual speakers of Japanese surely would not help a French child learn French. Nor could, on Quine's behaviorist assumptions, monolingual speakers of Japanese enable a French child to learn French by observing, correcting, and reinforcing his verbal behavior.

Clearly, Quine's argument in favor of the restriction of relevant empirical evidence in linguistics presupposes that the child learning his native language and the linguist are confronted with exactly the same task. This is a dubious assumption. As Chomsky (2000, 54) emphasizes, acquisition of one's native language is a largely automatic process in which the child makes no conscious choice. The child applies his initial cognitive capacities (i.e., universal grammar) to the primary linguistic data made available by members of his community. On the other hand, the linguist makes conscious and laborious use of all possible empirical evidence relevant to discovering the structures, respectively, of the initial innate state of the HLF and the stabilized I-language of an adult speaker.[13] The scientific cues that the linguist can exploit are simply inaccessible to the child.

I will consider just three examples. First, certain discoveries about the neurological structure of the human brain (accessible to the linguist but not the child) are relevant to the comparison among competing linguistic hypotheses. Second, the linguist can systematically compare minimal pairs composed respectively of a sentence of a language and of an ungrammatical sequence composed of the same words. But the ungrammatical sequences constructed by the linguist cannot be part of the primary linguistic data available to the child.[14] Finally, as Chomsky notes (2000, 53–54), examination of what an adult speaker of Japanese knows may indicate that he has tacit knowledge of some abstract syntactic principle *P* for which no clue exists in the primary linguistic data available to a child learning Japanese. If so, then the generative linguist will have grounds for supposing that this abstract syntactic principle *P* belongs to universal grammar, the initial state of the HLF. Universal grammar is supposed to be common to children learning Japanese and to those learning French. Thus, even if there exist clues for this principle in the primary data accessible to a child learning French, knowledge of this principle by an adult French speaker might derive from universal grammar and not from the fact that he was exposed to utterances in French during childhood. It follows that description of the I-language of an adult speaker of Japanese can be relevant for determining what, in the I-language of a French speaker, is to be attributed to universal grammar and what depends on his personal linguistic experience.

Searle (1992) has offered another challenge (distinct from Quine's) to computational explanations of the HLF (or of any other human cognitive capacity). He maintains that any explanation of a genuine mental phenomenon must satisfy the constraint of the so-called connection principle, according to which a state or process cannot be genuinely mental if its content is not potentially accessible to the conscious subjective experience of the human agent to whom it is attributed. The very idea of discrediting computational explanations of human cognitive capacities by glorifying introspection will be judged harshly by those for whom the very task of the cognitive sciences is to unravel the functioning of mental processes whose very existence is inaccessible to mere introspection. Two answers are available in response to Searle's challenge.

On the one hand, the force of the connection principle is weakened by the fact that Searle fails to specify what counts as the potential accessibility of some content to the subjective conscious experience of a human agent. What makes the content of some unconscious mental process potentially conscious? Consider a human blindsight patient, who has lost the subjective visual experience of the form, contour, size, texture, and color of objects in the part of his visual field affected by a lesion in his primary visual cortex.[15] Should the visual attributes of an object count as potentially accessible to the conscious experience of a patient with blindsight on the grounds that they are fully accessible to the consciousness of a healthy subject (without blindsight)? Consider further the phenomenon of so-called subliminal perception, whereby a healthy human subject is shown a word (from his language) for such a short duration of time that, although he can visually process it, he cannot be aware of it. Nonetheless, it has been shown that, in such conditions, he can extract semantic information carried by the word, which, although visually processed unconsciously, facilitates recognition of a second, semantically related, word.[16] Should the content of the "subliminal perception" of a stimulus count as potentially conscious on the grounds that, had it been presented slower (for a longer period of time), the subject would have been aware of it? As long as Searle fails to specify what is potentially inaccessible to the subjective conscious experience of a human agent, the connection principle runs the risk of being devoid of empirical content, i.e., nonrefutable.[17]

On the other hand, as Chomsky notes (2000, 75, 106, 134), the connection principle is itself an answer to the question of what the criterion (or mark) of the mental is: on this criterion, accessibility to consciousness is what makes a mental phenomenon genuinely mental. By contrast, no philosopher of the physical sciences believes that he is expected to offer a criterion for what constitutes mechanical, optical, electrical, or chemical phenomena. But Searle could not apply the connection principle unless he took it for granted that computational theories of the HLF (or of some other human cognitive capacity) can be legitimately subjected to some a priori conceptual reflection based on the authority of the ordinary common-sense concept expressed by the word "mental." In accordance with methodological naturalism, Chomsky proposes to use the term "mental" on a par with the way physical scientists use such words as "mechanical," "optical," "electrical," or "chemical" to refer to different aspects of the world, without presupposing any problematic ontological or metaphysical divisions.

3. Chomsky and Metaphysical Naturalism

Unlike methodological naturalism, metaphysical naturalism is an ontological doctrine. To subscribe to metaphysical naturalism is to subscribe to physicalist monism, which stands in contrast to the ontological dualism between body and mind advocated by Descartes. To subscribe to ontological dualism is to suppose that mental entities are not physical entities, because the former are not reducible to the latter. To subscribe to physicalist monism is to assume that all chemical, biological, psychological, linguistic, or cultural processes are physical processes obeying the fundamental laws of physics. For a physicalist, the ontological problem of the relation between body and mind is to find out how to identify the latter with the former via an ontological reduction.

One might have expected Chomsky to appeal to the explanatory success of computational models of the HLF and argue for a physicalist ontology. If mental processes are computational processes, and if computational processes are in turn operations that can be carried out by a machine (built according to the laws of physics), then computational models of a human cognitive capacity show that a machine obeying the laws of physics can carry out operations characteristic of some fundamental human cognitive competence.

This argument has been offered by philosophers who, like Fodor (1975, 1987, 1994), subscribe to metaphysical naturalism. But no trace of it can be found in Chomsky's own work. On the one hand, unlike biological entities (including the HLF), machines, whether abstract or concrete, are artifacts: their functions depend on the intentions of their designers, whose contents in turn raise questions of interpretation that go beyond the limits of scientific investigation. Thus, according to Chomsky (2000, 44–45, 148), understanding the functioning of an artifact cannot contribute to the scientific understanding of any aspect of the natural world—including the HLF, whose structure and function (if any) is independent from the contents of the intentions of any designer.[18] On the other hand, Chomsky could not argue from the explanatory success of computational models of the HLF to physicalism because, on his view, for the past three centuries the ontological controversy between physicalism and substance dualism has been turned into a pseudoproblem between two theses equally devoid of content. This severe diagnosis itself calls for some explanatory comments.

According to Chomsky (2000, 83–84, 103, 108–109), the problem of the relation between body and mind was a genuine problem in Descartes' time—in the middle of the seventeenth century—when the physical universe was assumed to be governed by the principles of Cartesian mechanics. Since the mind's functioning did not seem to be governed by the laws of Cartesian mechanics, Descartes was rationally led to accept ontological dualism, according to which minds (or mental things) are distinct from bodies (or material things). However, the principles of the mechanical philosophy were swept aside by the explanatory success of Newton's introduction of universal gravitation (a force acting at a distance) as a unifying principle for both celestial and terrestrial mechanics. As Chomsky (2000, 84) puts it, Newtonian mechanics "exorcized the machine" but not the Cartesian conception of the mind.

Arguably, the principles of Cartesian mechanics govern our ordinary or naïve conception of a physical object.[19] But since it gave up the principles of Cartesian mechanics, theoretical physics no longer has a scientific concept of bodies or physical objects. Chomsky thus throws down a real challenge to physicalists: if the concept expressed by the words "physical object" has no assignable scientific content, then the ontological controversy between physicalism and substance dualism has lost its meaning.

Most contemporary advocates of physicalist monism subscribe to one or another version of the identity thesis between mental and physical entities. Some accept a reductionist version, others a nonreductionist version, of the identity thesis. A few advocate the elimination of mental entities. Because the concept of matter itself has been given up in theoretical physics, Chomsky rejects the very idea (which he takes to be devoid of any scientific content) of any purported physicalist reduction of mental things. Instead, Chomsky advocates the epistemological goal of unification (or integration) of scientific theories. But as the history of the physical sciences, often referred to by Chomsky, shows, unification between two scientific theories dealing with entities located at different ontological levels—such as the chemical theory of molecules and the physical theory of atoms—has often required a radical modification of the lower-level theory (e.g., the atomic theory) for it to be able to be integrated with a higher-level theory (e.g., the molecular theory).

Chomsky does not exclude the epistemic unification of computational explanations of (higher-level) cognitive competencies with (lower-level) neuroscientific explanations of the contribution of different brain areas to different human cognitive capacities. But the cognitive neurosciences must, he believes, undergo serious revision before their fine-grainedness equals that of computational theories.

Can an advocate of physicalist monism meet Chomsky's challenge by minimizing ontological commitments? For instance, consider the minimal physicalist distinction between two classes of things, both of which are physical but only one of which turns out to be also mental as a result of its internal structural complexity. Couldn't a physicalist meet Chomsky's challenge by appealing to such a minimal distinction?[20] The answer is no, and the reason is that this minimalist version of physicalism still faces Chomsky's (2003, 259) dilemma. Either contemporary physics supplies a complete description of the world or it does not. If it does, then mental entities are simply physical entities (and nothing more). If it does not, then we do not yet know what is (merely) physical.

Chomsky's challenge raises at least three fundamental questions. First, is the epistemological goal of theoretical unification itself really independent from any underlying ontology? Second, what authority should be granted, respectively, to physical theories and to neuroscientific theories in the adjudication of ontological controversies raised by the development of the cognitive sciences? Third, what authority should be granted, respectively, to scientific physical theories and to "naïve" (or common-sense) physics in the adjudication of such controversies?

Following Poland (2003), let's call "methodological physicalism" Chomsky's view that one should aim for epistemological unification between computational and neuroscientific approaches to the functioning of the human brain. Can methodological physicalism be justified on purely epistemological (or methodological) grounds? If theoretical unity (or simplicity) is a virtue at all, is it a purely epistemic (or even an aesthetic) virtue?

The reason one can (and ought to) be skeptical is that the epistemic goal being sought is a unification between theories belonging to different levels, and the very notion of a level is an ontological notion. We distinguish the chemical level of molecules from the physical level of atoms because molecules are things made up of atoms. The neurosciences investigate the structure and functioning of constituents of the human brain. Computational theories investigate so-called emergent properties of the brain, such as visual perception or the HLF. Now, Chomsky (2002, 55, 63, 65) treats the view expressed by the cognitive neuroscientist Vernon Mountcastle that "minds, indeed mental things, are emergent properties of brains" as a truism (and thus an obvious truth). But it is far from being a truism, since, for example, this is an ontological thesis that serves to justify the choice of methodological physicalism in favor of the epistemological unification between computational and neuroscientific theories of brain processes.[21]

Chomsky accepts the ontological assertion that human cognitive capacities are (emergent) properties of the human brain. But he rejects the claim that mental entities are physical entities. An opponent of ontological dualism will conclude that it is an error to subject ontological controversies in the cognitive sciences to the authority of the fundamental concepts of theoretical physics. For an advocate of physicalist monism, the computational properties of human cognitive capacities (including the HLF) depend on the neurological structure of the human brain. Even supposing that contemporary basic physical theories of elementary particles are fundamentally incomplete, there is still no reason to believe (as Fodor 2001 and Lycan 2003 have noted) that drastic revisions in fundamental theoretical physics would have a major conceptual impact on our scientific understanding of the molecular mechanisms involved in the firing of neurons, the transfer of information between neurons, and the functional roles of distinct brain areas.

Finally, the challenge thrown down by Chomsky to physicalists presupposes that only scientific concepts—not concepts of common sense—can arbitrate respectable ontological controversies.[22] Arguably, what physicalists should do is to reexamine their own conception of the role ascribed to concepts respectively from theoretical physics and from common sense in the adjudication of the ontological controversies underlying the development of the cognitive sciences.

Let us now consider one of the premises used by Davidson (1970) in support of his own nonreductionist version of physicalist monism, which he calls "anomal monism" and according to which any mental event is a physical event but no psychological concept or predicate is reducible to a physical concept or predicate. According to Davidson, there are, on the one hand, physical laws that subsume the relations between physical events. On the other hand, causal relations may also hold between pairs of mental events and even between pairs of events of which one is mental and the other physical. However, according to the premise that Davidson calls "the anomalism of the mental," there are neither genuine psychological laws able to subsume the causal relations between mental events nor genuine psychophysical laws able to subsume the causal relations between mental events and physical events. On Davidson's (1970) view, the anomalism of the mental (or of naïve psychology) is a reflection of the contrast between what he takes to be the "strict" (i.e., exceptionless) laws of physics and what he takes to be the "truisms" of either psychological or psychophysical correlations.

As Chomsky (2000, 88–89, 138) insightfully observes, the grounds for the anomalism of naïve psychology are also grounds for the anomalism of naïve physics.[23] The gulf between theoretical physics and common-sense or naïve psychology also separates theoretical physics from common-sense or naïve physics. There are no bridge laws linking the concepts of theoretical physics to either those of naïve physics or of naïve psychology. Davidson, who subscribes to a nonreductionist version of physicalism, seems to think that the relation between mental concepts and physical concepts raises a distinctive epistemological problem not raised by the relation between the concepts of naïve physics and the concepts of theoretical physics. But unless Davidson surreptitiously subscribes to some version of methodological dualism, acceptance of the "anomalism of the mental" should force him to accept the anomalism of naïve physics.

In short, Chomsky seems to be in a position that can be described by means of the following five propositions. First, he emphasizes the fact that there is no room in post-Newtonian theoretical physics for the concept expressed by the term "physical object." Second, on his view, the attempt to unify computational theories of human cognitive capacities with neuroscientific theories of the structure and functioning of the human brain is a legitimate research program. Third, he admits that human cognitive capacities are emergent aspects of the human brain. Fourth, acceptance of this ontological thesis grounds the search for theoretical unification. Finally, even if there is no room within the scientific theories of fundamental theoretical physics for the concept of physical object, this concept still plays a central role in naïve physics, of which the principles of the Cartesian mechanical philosophy are constitutive.[24]

The concepts of physical object and mental entity are not part of any scientific theory of the world. The ontological controversy about the nature of the relation between physical objects and mental entities is thus no part of any scientific (or naturalistic) investigation of the world. But the concepts of physical object and mental entity belong, respectively, to naïve physics and naïve psychology. Chomsky (2000, 90–91) himself calls "ethnoscience" the study of the common-sense conceptual resources by means of which humans, in all cultures, form their stable nonscientific representations of the world. The study of the relations between the concepts of naïve physics and those of naïve psychology thus pertains to naturalistic ethnoscience. If the conceptual representations of the world formed by common sense are features of the human brain, then they are themselves part of the world. If they are part of the world, then ethnoscience is a branch of the (naturalistic) scientific investigation of the world. If so, then the study of the relations between the concepts of mental entity and physical object is part of the naturalistic investigation of the ordinary conceptual system of representation of the world by human common sense.

4. Chomsky and the Naturalization of Intentionality

When in the late nineteenth century Brentano (1874) introduced the concept of intentionality into philosophy, he subscribed to the following three theses: first, he took it to be constitutive of intentionality, as it is manifest in such mental acts and states as love, hate, desire, hope, belief, judgment, perception, and many others, that these mental acts are directed toward objects distinct from themselves. Second, the so-called objects toward which the mind is directed by its intentionality have the property that Brentano calls "intentional in-existence." The notion expressed by Brentano's word "in-existence" has given rise to much exegetical controversy. Did Brentano mean nonexistence? Did he mean existence within the mind? Or did he mean both?[25] Third, intentionality is the mark of the mental: all and only mental acts and states have intentionality.

Let us assume (in conformity with Brentano's first thesis) that it is constitutive of intentionality that no one can be said to love, hate, desire, etc. unless there is something that is loved, hated, desired, etc. If this is true, then unless there were objects exemplifying the property of intentional in-existence, intentionality itself could not be exemplified (second thesis). But love, hatred, admiration, desire, and other mental acts are directed not only toward concrete objects but also toward abstract objects (like numbers), mythological constructions (Zeus), or fictional characters (Anna Karenina), which do not exist in space and time. Humans can even entertain thoughts about impossible numerical or geometrical objects such as the greatest prime number or squared circles. Thus, acceptance of Brentano's first two theses raises a number of fundamental ontological questions in philosophical logic: Are there intentional objects at all? Does recognition of the phenomenon of intentionality force one to postulate the ontological category of intentional objects? Could there be objects that fail to exist? These questions in turn have given rise to a major division within analytic philosophy. The prevailing (or orthodox) response has been a resounding "no." But an important minority of philosophers who subscribe to the "intentional objects theory" have argued for positive responses to these questions. Since intentional objects need not exist, on the intentional-objects theory, there are things that do not exist. According to critics of intentional-objects theory, there are no such things.[26]

Many contemporary philosophers of mind and language have responded to Brentano's introduction of the concept of intentionality by embracing a so-called externalist view of the content of psychological (or mental) states with intentionality. Externalism is the view that intentionality is not an intrinsic property of a cognitive system but rather a relation between a cognitive system and its environment. In other words, according to externalism, an individual's psychological states (e.g., his or her beliefs, desires, or perceptions) derive their contents from the relations holding between the individual (on physicalist assumptions, between the individual's brain) and properties exemplified in his environment. Externalist philosophers of mind and language, therefore, accept the burden of explaining what Soames (1989; cited by Chomsky 2000, 132) calls "the fundamental semantic fact of language . . . viz. that it is used to represent the world"—which presupposes that the function of human cognition is to represent the world. To simplify, we can distinguish two versions of externalism: a normative and a descriptive version. On the former but not the latter, intentionality is taken to arise from the norms obtaining in a linguistic community. Chomsky rejects both versions of externalism.

According to the normative version of externalism, the meaning and reference of linguistic expressions of a given E-language are constituted by the linguistic practices accepted by the community of people who speak the E-language in question. On this view, the intrinsic cognitive resources of a speaker are not sufficient to determine the meaning and reference of the words he uses and the beliefs he expresses by uttering them. If some individual did not belong to some linguistic community or another, then his mental states would fail to exhibit intentionality. On the normative version of linguistic externalism, the meaning and reference of the expressions of an external public language (or E-language) are given priority over the meaning and reference of the constructions of an I-language (in Chomsky's sense). Since they derive from the norms obtaining in a community, meaning and reference are normative properties of public linguistic expressions.

The normative version of externalism is incompatible with Chomsky's methodological naturalism for at least three reasons. First, on Chomsky's view, the expressions of some E-language are the derived products of the underlying psychological constructions that belong to a speaker's internalized I-language. By contrast, the normative version of externalism gives theoretical priority to the notion of a "shared public language" over a speaker's mental representations. Second, advocates of normative externalism assume that human verbal communication would simply be impossible unless there existed a single shared external public language (i.e., a set of public linguistic expressions each with a unique public linguistic meaning known to every speaker). Chomsky (2000, 30) rejects the normative externalist picture of verbal communication. He argues instead that verbal communication is a fallible inferential process and that resemblance—not identity—between the external products of different I-languages is sufficient to enable verbal communication.[27] Third, Chomsky (2000) constructs numerous examples aimed at discrediting the idea that there exists a fixed well-defined reference relation between the words of a language and nonlinguistic entities.[28] These examples at least cast doubt on the idea that the reference relation can serve the aims of naturalistic or scientific semantic investigation. I will consider three of these examples:

(4) The bank burned down and then it moved across the street.
(5) The book that he is planning will weigh at least five pounds if he ever writes it.
(6) London is so unhappy, ugly, and polluted that it should destroyed and rebuilt one hundred miles away.

Anyone who understands (4) knows both that the implicit subject of the verb "burn" refers to a concrete physical building and that the explicit subject of the verb "move" refers to an abstract institution—which can be physically instantiated as several concrete structures. In this sentence, the phrase "the bank" can, thus, refer at one point to a concrete building and at another to an abstract institution. In (5), the constituent "the book" is used to refer to an abstract content as grammatical object of the verb "write" and to a physical object as subject of the verb "weigh." In (6), the anaphoric pronoun "it" and its antecedent "London" are jointly used to refer to agents (its inhabitants), inanimate objects (the buildings), a place, and an abstract entity (which can be realized by different concrete entities, of which one can be physically destroyed and the other rebuilt elsewhere).

Chomsky does not use these examples to support the idealist (or nonrealist) metaphysical thesis that the world reduces to language and/or mental representations. What these examples are meant to suggest instead is that reference is an action carried out by human agents using words that are in themselves devoid of reference. Within one and the same utterance, a pronoun and its antecedent can subtly change reference as a function of the speaker's intentions, without the interlocutor experiencing the slightest difficulty in understanding this change of reference. Given the constant variability of referential intentions, the coordination between speaker and hearer transforms reference into a "mystery."

Unlike what a human being knows, what he does is, according to Chomsky, bound to remain a mystery. Generative grammar has shown the way to scientific understanding of an aspect of what a human knows: the HLF. But an epistemic divide separates the problems encountered in understanding what a human knows and the mysteries involved in explaining an intentional action. Although he does not endorse ontological dualism, Chomsky (1980, 79; 1988, 5–6) nonetheless accepts the Cartesian argument for the freedom of the human will, which says that, unlike the behavior of any machine (or mechanical device), human intentional action is "indeterminate" because a human agent is always *free* to choose between two distinct courses of action. A human agent can be "incited" to act, but he could always have acted differently. Now, reference (i.e., the use of a word to refer to something) is an intentional human action. Thus, because on Chomsky's view any act of reference (and what Chomsky also calls the "creative use of language") involves the freedom of the will, it is presently an epistemic mystery, not a scientific problem.

Most if not all of the philosophers who endorse the descriptive (non-normative) version of externalism also subscribe to metaphysical naturalism. Their basic goal is to domesticate intentionality within cognitive scientific psychology—to show that intentionality can be "naturalized" or that it obeys the laws of nature. The program of the naturalization of intentionality thus faces two complementary tasks: to show that intentionality has both respectable causes and respectable effects. A philosopher who subscribes to the program of naturalizing intentionality may be tempted to argue, as Fodor (1975, 1987, 1994, 1998) does, from the explanatory successes of computational models of the HLF to the computational-representational theory of the mind (CRTM). But, as I already noted, Chomsky does not. The following four features are distinctive of CRTM.

First, on this conception, all cognitive processes are computational processes. A computational process takes a representation (or symbol) as input and transforms it into a different representation following purely formal rules. CRTM thus presupposes the existence of a "language of thought" (or "mentalese") composed of symbols that are themselves endowed with syntactic and semantic properties. In virtue of their syntactic properties, the primitive symbols of the language of thought enter into combinations and form complex representations.[29] In Fodor's conception, the primitive symbols (or concepts) of the language of thought are supposed to possess "primitive intentionality," from which the "derived" intentionality of all other symbols flows (in particular, the derived intentionality of the linguistic symbols of external public languages).[30]

The thesis that the intentionality of symbols in the language of thought has priority over the intentionality of linguistic expressions is of course incompatible with the reversed priority granted by advocates of the normative version of externalism to the meaning of public linguistic expressions over the content of mental representations. Furthermore, the thesis that the intentionality of symbols in the language of thought is prior to that of public linguistic expressions is not immediately subject to Chomsky's objection against the possibility of a scientific understanding of reference. Suppose I hear a dog bark and the processing of this acoustic stimulus causes me to think of a dog. According to CRTM, the perceptual processing of the acoustic stimulus triggers in my language of thought an occurrence of the symbol "Φ," which just is my concept "dog."

According to CRTM, not all thinking amounts to some intentional (or voluntary) action. For example, the cognitive process whereby my auditory perception of the stimulus is turned into a conceptual representation of a dog is not an intentional (or a voluntary) action. When it results from my perception of a stimulus, my conceptual representation of a dog—the occurrence of my mental symbol "Φ"—is independent from any intention to refer to a dog.[31] If so, then the CRTM approach to the reference of mental symbols is not directly open to the neo-Cartesian argument based on the freedom of the will.

Second, according to the version of CRTM defended by Fodor (1994, 1998), the content (or semantic value) of a primitive symbol of the language of thought (i.e., of a primitive concept) results from the nomological correlations between this symbol and exemplifications of a property in the environment.[32] Thus, my primitive concept "Φ" draws its content or semantic value from the fact that it nomically covaries with exemplifications of the property of being a dog. In general, the intentionality of a primitive concept derives from psychophysical correlations between a cognitive system and parameters in the environment.[33]

Third, according to CRTM, psychological explanations are jointly intentional and nomological. The explanation of an action is intentional because what an agent does depends on the content of his intentions, beliefs, and desires. It is nomological because the psychological explanation of an action typically consists in subsuming the action under the psychological generalizations that cover the intentions, beliefs, and desires of human agents.

Finally, according to CRTM, what makes both psychological explanation causal explanation and intentional psychological generalizations causal laws is the computational thesis that psychological laws are implemented by underlying computational mechanisms. By virtue of the language-of-thought hypothesis, the contents of an agent's beliefs and desires reduce to the semantic values of symbols of the language of thought, but the underlying computational mechanisms transform mental symbols solely in virtue of their syntactic properties.

CRTM constitutes the most systematic contemporary effort to assign a role to intentionality in causal psychological explanations and to create a bridge between naïve psychology and computational models in cognitive science. More than anyone else, Chomsky has contributed to promoting computational explanations in cognitive science. But for at least two reasons he cannot endorse CRTM, which he takes to be a metaphysical project, not a scientific research program. On the one hand, Chomsky accepts the Cartesian argument against the possibility of offering causal explanations of human intentional actions based on the freedom of the human will. On the other hand, on Chomsky's view, purported explanations of human action that rely on the attribution of intentionality to an agent's psychological states cannot aspire to the same scientific explanatory value as computational models of human cognitive capacities.

However, even if causal explanations of intentional human actions are at present just protoscientific theories (not genuine scientific theories), an advocate of CRTM might well object to Chomsky that the Cartesian concept of freedom itself belongs to common-sense naïve psychology. So the following question arises for Chomsky: is not the neo-Cartesian argument from freedom of the will an instance of methodological dualism whereby protoscientific psychological theories of human behavior are subjected to some a priori philosophical reflection guided by the authority of the ordinary common-sense concept of freedom?

Because he thinks that the cognitive sciences should emulate the "Galilean style," which has proved so fruitful in the physical sciences, Chomsky takes it that, apart from computational models of a human cognitive capacity (such as the HLF or visual perception), there is presently no serious prospect for a genuine scientific (or naturalistic) investigation of the human mind. As his 2003 replies to Egan (2003) and Rey (2003) clearly reveal, he thinks that CRTM has failed to demonstrate either that intentionality has respectable causes or that it produces respectable effects.

As the history of twentieth-century philosophy testifies, the logical and ontological puzzles inherited from Brentano's intriguing definition of intentionality have given rise to much work in philosophical logic. Furthermore, the evaluation of Brentano's thesis that intentionality is the mark (or the criterion) of the mental, which has defined much of the landscape of contemporary philosophy of mind, has given rise to many important distinctions, including subtle distinctions between the concepts of intentionality and consciousness. Finally, some of the gap between the concept of intentionality and concepts as widely accepted in the natural sciences as lawful correlation and information has been filled by the work of philosophers working toward naturalizing intentionality.

It is one thing, however, to make the concept of intentionality metaphysically respectable by displaying some of its significant conceptual connections with concepts widely accepted in the natural sciences. (This much contemporary metaphysical naturalism can legitimately claim to have achieved.) It is another thing to show that the concept of intentionality can or should be a constitutive element of computational models of human cognitive capacities that can subsequently give rise to experimental inquiry. Clearly, Chomsky is skeptical about the prospects of a scientific research program into human cognitive capacities that would be based on the concept of intentionality. Furthermore, Chomsky (2003, 274) rejects Fodor's (1987, 1994, 1998) idea that the scientific investigation of human cognitive capacities is continuous with the generalizations of common-sense naïve psychology. The latter, which are intentional, derive from a priori conceptual reflection and are supported by no experimental confirmation. Nor could they be empirically disconfirmed.

In laying out his most recent version of CRTM, Fodor (1998) seems increasingly tempted to endorse a semantic view of computation according to which the preservation of the semantic relations among mental symbols is a constitutive feature of computational processes: "To a first approximation, computations are those causal relations among symbols which reliably respect semantic properties of the relata. . . . The essential problem . . . is to explain how thinking manages reliably to preserve truth. . . . Turing's account of thought-as-computation showed us how to specify causal relations among mental symbols that are reliably truth-preserving."[34] So it may seem as if it is constitutive of a mental representation (or symbol) to which computational processes are applicable that it has content: a semantic value or intentionality. If so, then it might seem as if a computational theory of a cognitive human capacity would be incomplete so long as it does not face the question: what do the symbols manipulated by the computational processes represent?[35]

As Chomsky's (2003) replies to Egan (2003) and to Rey (2003) testify, he would not endorse this semantic conception of computation and would rather keep the definition of a computational process apart from the problems raised by the semantic interpretation of mental symbols. In fact, the doctrine that in much recent work Chomsky (2000, 2003) endorses and calls "internalism" seems designed to support just the rejection of such a requirement for determining the content of the mental symbols to which computational processes apply.[36] One of the internalist principles he appeals to is a principle of symmetry (or parallelism), which (in accordance with his own minimalist program) governs the computational architecture of the grammars of natural languages. According to this principle of symmetry, the syntax of an I-language generates mental representations on which the rules of phonological and semantic interpretation operate in parallel. The mental representations generated by syntax thus constitute a twofold set of instructions for both the human sensorimotor system (which controls the articulation and perception of the sounds of language) and the human conceptual system (which controls inferences).[37]

Let *M* be the mental representation (or I-construction) associated with the proper name "London" by the syntax of an I-language. According to the principle of symmetry, *M* is a member of two distinct relations involving nonmental entities: on the one hand, *M* can be thought of as an instruction for the articulatory system enabling the pronunciation of "London." Thus *M* stands in relation *S*, the Sounds relation, with noises of category *N*. On the other hand, *M* is in the purported Refers relation *R* with some presumed nonmental entity *E* (e.g., a city).[38] Chomsky (2003, 271) observes that, unless the nature of entities *N* and *E* can be well defined, the relations *S* and *R* will remain totally indeterminate. In theoretical linguistics, it is widely taken for granted that a definition of relation *S* and entities *N* would be of no scientific interest. By parity, and contrary to most externalist philosophers of language, Chomsky (2003, 271) concludes that definition of the extrinsic relation *R* between *M* and *E* is scientifically futile.

On the basis of the computational principle of symmetry between the semantic and phonological interpretation of syntactic representations, Chomsky thus distinguishes two notions of representation: a pretheoretical relational notion and a theoretical nonrelational notion. The relational notion is intentionality in Brentano's sense: any representation is a representation of something. On Chomsky's view, the pretheoretical notion can play some auxiliary role in the informal or intuitive presentation of a computational theory, but not within the computational model itself. It is incumbent on the computational theory to explicitly introduce the operational (or formal) notion of representation, which is a purely syntactic notion obeying only the laws of the computational mechanisms. Thus, Chomsky's endorsement of an internalist (or logical syntactic) account of computation reveals the extent of the gap that separates him from most philosophers, who, following Brentano's legacy, subscribe to Soames's thesis (1989) that the most basic property of language is that it is being used to represent the world.[39]

NOTES

1. Interestingly, the psychologist George Miller (2003), who was also a major actor in the so-called cognitive revolution, argues that the revolution in question should rather be seen as a counterrevolutionary response to the behaviorist revolutionary denial of internal psychological intermediaries between sensory inputs and behavioral output.

2. Cf. Chomsky (2000), Antony and Hornstein (2003), and the interesting reviews of Chomsky (2000) by Stone and Davies (2002), Bilgrami (2002), and Moravcsik (2002).

3. As Chomsky (2000, 69) writes, for example, "the study of communication in the actual world of experience is just the study of the interpreter, but this is not a topic for empirical inquiry . . . : there is no such topic as the study of everything."

4. Cf. Chomsky's monumental "Logical Structure of Linguistic Theory," from which *Syntactic Structures* (1957) was drawn.

5. "I" stands for "individual," "internal," and "intensional." "E" stands for "external" and "extensional."

6. In fact, one could also consider the two further questions: (Q4) How did the human ability to acquire knowledge of some natural language or other arise in the course of phylogenetic evolution? (Q5) How is the grammatical knowledge (I-language) of an E-language stored in the human brain?

7. Cf. Chierchia (this volume) for a detailed presentation of these and other relevant syntactic and semantic facts.

8. This distinction was developed in Chomsky (1965). Cf. Boeckx and Hornstein (this volume) for clarification of the distinction.

9. This is the problem of "negative evidence." Cf. Boeckx and Hornstein (this volume) for detailed discussion.

10. Cf. Goodman (1968), Putnam (1968), and Quine (1969) for objections to Chomsky's conclusion.

11. Cf. Gettier (1963).

12. This Quinean argument for the indeterminacy of syntactic hypotheses is independent from Quine's semantic arguments in favor of his famous 1960 thesis about the indeterminacy of radical translation.

13. Incidentally, this critique of Quine's analogy between the child's task and the linguist's casts doubt on the version of the "theory-theory" of cognitive development defended by Gopnik (2003). Cf. Chomsky's 2003 response to Gopnik (2003).

14. Cf. the argument from the poverty of the stimulus.

15. Cf. Weiskrantz (1997).

16. Cf. Marcel (1983).

17. Cf. Block (1990), Chomsky (2000), and Jacob (1995).

18. This is an aspect of Chomsky's thesis according to which intentionality-based explanations (of, e.g., human action) outrun the limits of naturalistic (or scientific) investigation. I return to this point in section 4.

19. At least, they seem to govern human infants' "naïve physical" expectations about the behavior of physical objects. Cf. Spelke (1988).

20. Cf. Lycan (2003, 16).

21. An advocate of ontological dualism, e.g., Kripke (1972, 1982), could reject the view that mental properties are emergent properties of the brain.

22. According to Chomsky (2000, 139), any scientific investigation, including those above the level of fundamental physics, creates concepts that have no continuity with the ordinary concepts of common sense.

23. On behalf on Davidson, it might be pointed out that the generalizations of naïve psychology raise a problem not raised by the generalizations of naïve physics. Unlike the strict laws of basic theoretical physics, the generalizations of both naïve psychology and naïve physics are *ceteris paribus* and admit of exceptions. But in addition, Davidson (1970) has argued that, unlike the application of concepts of naïve (or common-sense) physics, the application of mental or psychological (either naïve or scientific) concepts to human agents presupposes some normative principle of rationality. Of course, the question is whether Davidson's arguments in this area presuppose some version of methodological dualism.

24. Cf. the work of Spelke (1988) on the cognitive development of naïve physics in human infants. The study of the cognitive capacities of newborns reveals the cognitive constraints on the diversity of human culture.

25. Cf. Jacob (2003).

26. Russell's theory of descriptions and Quine's theory of ontological commitment are intended to discredit the theory of intentional objects. But among others, the philosopher Terence Parsons has recently revived the Meinongian theory of nonexistent intentional objects. For some detailed discussion, cf. Jacob (2003).

27. This view is also that of Sperber and Wilson (1986).

28. According to Chomsky (2000, 130–31), the Fregean notion of reference (*Bedeutung*) is a technical semantic notion applicable only to the stipulated symbols of Frege's formal language to meet the demands of his logicist project of reducing arithmetic to logic.

29. A symbol of the language of thought (or concept) is said to be "primitive" if it does not result from the syntactic combination of other symbols (or concepts).

30. Cf. Jacob (1997).

31. Cf. Jacob (1997).

32. This follows from the informational semantic view of mental content to which Fodor (1994, 1998) subscribes.

33. Fodor (1994, 1998) himself subscribes to an atomistic version of conceptual content according to which the content of every primitive concept is independent from the content of any other primitive concept.

34. Fodor's (1998) endorsement of a semantic conception of computation is discussed and criticized by Damian Justo (2007) in his doctoral dissertation, in the context of an evaluation of Fodor's (2001b) arguments against massive modularity. As Justo (2007, chap. 2) rightly points out, it is one thing to recognize that it is a desirable property of a computational system that it preserves the semantic relations among symbols. It is another thing to build this feature into a constitutive feature of computation.

35. As Peacocke (1994), Egan (2003), and Rey (2003) observe, this question is raised both by computational theories of the HLF and by computational theories of vision.

36. Usberti (2002) develops Chomsky's internalism within the framework of an antirealist view of meaning partly rooted in Dummett's constructivism and in Meinong's theory of "intentional objects."

37. Example (6) was in fact intended to question the existence of the purported relation *R*.

38. Thanks to Richard Carter and Dan Sperber for their comments and also to Richard Carter for helping with the English version of this article.

REFERENCES

Bilgrami, A. 2002. Chomsky and philosophy. *Mind and Language* 17, no. 3: 290–302.

Block, N. 1990. Consciousness and accessibility. *Behavioral and Brain Sciences* 13, no. 4: 596–598.

Brentano, F. 1874. *Psychology from an empirical standpoint*. London: Routledge and Kegan Paul.

Chomsky, N. 1957. *Syntactic structures*. The Hague: Mouton.

——. 1965. *Aspects of the theory of syntax*. Cambridge, Mass.: The MIT Press.

——. 1968. *Language and mind*. New York: Harcourt Brace Jovanovich.

——. 1975. *Reflections on language*. New York: Pantheon.

——. 1980. *Rules and representations*. New York: Columbia University Press.

——. 1982. *The generative enterprise*. Dordrecht: Foris.

——. 1986. *Knowledge of language, its nature, origin, and use*. New York: Praeger.

——. 1988. *Language and problems of knowledge*. Cambridge, Mass.: The MIT Press.

——. 1994. Chomsky, Noam. In *A companion to the philosophy of mind*, ed. S. Guttenplan. Oxford: Blackwell.

——. 2000. *New horizons in the study of language and mind*. Cambridge: Cambridge University Press.

——. 2002. *On nature and language*. Cambridge: Cambridge University Press.

——. 2003. Replies. In *Chomsky and His Critics*, ed. L. M. Antony and N. Hornstein. Oxford: Blackwell.

Churchland, P. M. 1993. *A neurocomputational perspective*. Cambridge, Mass.: The MIT Press.

Davidson, D. 1970. Mental events. In *Essays on events and actions*, by D. Davidson (1980). Oxford: Oxford University Press.

——. 1984. *Essays on truth and interpretation*. Oxford: Oxford University Press.

Egan, F. 2003. Naturalistic inquiry: Where does mental representation fit in? In *Chomsky and His Critics*, ed. L. M. Antony and N. Hornstein. Oxford: Blackwell.

Fodor, J. A. 1975. *The language of thought*. New York: Crowell.

——. 1987. *Psychosemantics: The problem of meaning in the philosophy of mind*. Cambridge, Mass: The MIT Press.

——. 1990. *A theory of content and other essays*. Cambridge, Mass.: The MIT Press.

——. 1994. *The elm and the expert*. Cambridge, Mass.: The MIT Press.

——. 1998. *Concepts, where cognitive science went wrong*. Oxford: Oxford University Press.

——. 2001a. *Review of Chomsky's* New Horizons in the Study of Language and Mind. *Times Literary Supplement*.

——. 2001b. *The mind does not work that way*. Cambridge, Mass.: The MIT Press.

Gettier, E. 1963. Is justified true belief knowledge? *Analysis* 23: 121–123.

Goodman, N. 1968. The epistemological argument. In *The Philosophy of Language*, ed. J. Searle. Oxford: Oxford University Press.

Gopnik, A. 2003. The theory-theory as an alternative to the innateness hypothesis. In *Chomsky and His Critics*, ed. L. M. Antony and N. Hornstein. Oxford: Blackwell.

Jacob, P. 1995. Consciousness, intentionality, and function: What is the right order of explanation? *Philosophy and Phenomenological Research* 55, no. 1: 195–200.

——. 1997. *What minds can do*. Cambridge: Cambridge University Press.

——. 2003. Intentionality. In *The Stanford Encyclopedia of Philosophy*. Available online at http://plato.stanford.edu.

Justo, D. 2007. *La modularité de l'esprit revisitée*. Doctoral dissertation, EHESS.

Kim, J. 1993. *Supervenience and mind: Selected philosophical essays*. Cambridge: Cambridge University Press.

Kripke, S. 1982. *Naming and necessity*. Oxford: Blackwell.

Lycan, W. 2003. Chomsky on the mind-body problem. In *Chomsky and His Critics*, ed. L. M. Antony and N. Hornstein. Oxford: Blackwell.

Marcel, A. 1983. Conscious and unconscious perception: Experiments on visual masking and word recognition. *Cognitive Psychology* 15: 197–237.

Miller, G. A. 2003. The cognitive revolution: A historical perspective. *Trends in Cognitive Sciences* 7, no. 3: 141–144.

Moravcsik, J. M. 2002. Chomsky's new horizons. *Mind and Language* 17, no. 3: 303–311.

Peacocke, C. 1994. Content, computation, and externalism. *Mind and Language* 9: 303–335.

Poland, J. 2003. Chomsky's challenge to physicalism. In *Chomsky and His Critics*, ed. L. M. Antony and N. Hornstein. Oxford: Blackwell.

Putnam, H. 1968. The "innateness hypothesis" and explanatory models in linguistics. In *The philosophy of language*, ed. J. Searle. Oxford University Press.

——. 1975. The meaning of "meaning." In *Philosophical papers*, vol. 2, by H. Putnam. Cambridge, Cambridge University Press.

Quine, W. V. O. 1960. *Word and object*. Cambridge, Mass.: The MIT Press.

——. 1969. Linguistics and philosophy. In *Innate ideas*, ed. S. Stitch. Berkeley: University of California Press.

——. 1972. Methodological reflections on linguistic theory. In *Semantics of natural language*, ed. D. Davidson and G. Harman. Dordrecht: Reidel.

——. 1990. *Pursuit of truth*. Cambridge, Mass.: Harvard University Press.

Rey, G. 2002. Chomsky, intentionality, and a CRTT. In *Chomsky and His Critics*, ed. L. M. Antony and N. Hornstein. Oxford: Blackwell.

Searle, J. 1992. *The rediscovery of the mind*. Cambridge, Mass.: The MIT Press.

Soames, S. 1989. Semantics and semantic competence. *Philosophical Perspectives* 3: 575–596.

Spelke, E. S. 1988. The origins of physical knowledge. In *Thought without language*, ed. L. Weiskrantz. Oxford: Oxford University Press.

Sperber, D., and D. Wilson. 1986. *Relevance, communication, and cognition*. Cambridge: Mass. Harvard University Press.

Stone, T., and M. Davies. 2002. Chomsky amongst the philosophers. *Mind and Language* 17, no. 3: 276–289.

Usberti, G. 2002. Internalism and antirealism: A proposal. Mimeograph.

Weiskrantz, L. 1997. *Consciousness lost and found*. Oxford: Oxford University Press.

PART V

CHOMSKY AND THE INTELLIGENTSIA

11

CONSPIRACY

WHEN JOURNALISTS (AND THEIR FAVORITES) MISREPRESENT THE CRITICAL ANALYSIS OF THE MEDIA

SERGE HALIMI AND ARNAUD RINDEL

We do not use any kind of "conspiracy" hypothesis to explain mass-media performance.

EDWARD S. HERMAN AND NOAM CHOMSKY, *Manufacturing Consent*

Intentionalist and "conspiracy-oriented" thinking, from Chomsky to *PLPL*... tends to make the evil intentions of a few Powerful Ones in the shadows (closely connected to the media) the principal mode of oppression.

PHILIPPE CORCUFF, "Gauche de gauche: le cadavre bouge encore"

I confess that for some time I have occasionally been hearing and seeing things that no one has ever seen or heard.

NIKOLAI GOGOL, *Diary of a Madman* (1835)

To set out to recast any analysis of the structures of the economy and the reporting of the news as a "conspiracy theory" amounts to no ordinary falsification. It is part of a much larger design. During the last quarter of a century, the neoliberal counterrevolution, the crumbling of the "communist" regimes, and the weakening of the trade unions have all contributed, first to the renaissance, then to the hegemony, of individualistic thought. Collective institutions have been dismantled; those erected on their ruins privilege the atomized consumer, the "individual subject." The new dominant ideology that accompanies this grand transformation renders it more fluid and less detectable by labeling it "natural," produced by chthonic forces that no one can resist and that are supposed to bring numerous shared benefits in their wake.

Now perceived as "Marxist" and assigned the same low value as the regimes that actually called themselves that, structural analyses of history, politics, and the media are disdained. Any refusal to postulate that the "spontaneity" of actors and the impetuous upsurge of "human rights" are the essential principles driving globalization lays one open to the charge of being archaic, extremist, or paranoid. But of course people do have to be "informed," and informed in a way that runs counter to the institutional explanations that have been proscribed. So the major media present international news, and news about society, in the form of a moral fable based on binary clashes between Good (us) and Evil (them), portraits of great men (heroic or wicked), waves of emotion calculated to arouse unanimous compassion and tearful consensus.

The trap of "the end of ideology" condemns to the gallows the "grand collective narratives" built around a conflictual social history. Intellectual orientations such as these are depicted as having promoted "totalitarianism" and an ever-menacing "populism"—two terms whose generalization in the prevailing vulgate points clearly enough to the political profit expected from them. After all, they make it possible to launch a concerted attack on developments that had nothing in common except to challenge bourgeois democracy and its now sacrosanct system of representation.

In almost every domain of the social sciences, the search for and the exaltation of individualist and liberal traditions and the associated concepts of autonomy, complexity, liberty, and morality are becoming omnipresent.[1] In a parallel process, the historical movements that gave rise to a tradition that cannot yet be liquidated in its entirety (the French Revolution, socialism) are sifted by the new orthodoxy, which separates the wheat of the elements in them regarded as "salvageable" (in general, ones related to political liberalism, to the rejection of monarchical or feudal arbitrariness) from the chaff of those it would be better to get rid of (for example, when they offer more than merely verbal assent to an egalitarian project that is accused of having produced a "murderous utopia").

Cosseted by the media and by the intellectuals whom the media consecrate, a philosopher close to Nicolas Sarkozy's party, the UMP (Union pour un Mouvement Populaire), Monique Canto-Sperber, thus claims to have "saved" socialism by exhuming its liberal dimensions, and a historian, François Furet, undertook to split open the "bloc" of the French Revolution and cut out the tumor of Jacobin radicalization—so that the only thing left to celebrate was the promises of an enlightened aristocracy allied with a Girondin bourgeoisie. Finally, a thinker in the service of French business owners, François Ewald, is busy trying to ensure that everyone remembers Michel Foucault as a man supposedly drawn to the fluidity of a civil society set free of the constricting state and in harmony with the market.[2] Yet no one describes authors like these as being engaged in a "liberal plot."

Not even Marx or Proudhon have been spared this eagerness to hunt everywhere for fragments of individualist thought consonant with the current trend, this persistence in rescuing prestigious thinkers from their putative deviations.[3] Certain intellectuals and journalists with a distant past as participants in left-wing protest who now covet different, plusher political surroundings never seem to stop refashioning the references that structured them politically, so that they can then pretend to have kept faith with their own past commitments, once the latter are understood with a dose of maturity—and digested. Even Bernard-Henri Lévy, the friend and propagandist of the greatest capitalists in France, has no hesitation in appointing himself Sartre's biographer and the inheritor of his existentialist commitment.

Still, as long as the "classics" of protest are still alive, the deradicalized re-reading of their works demands their concurrence (Toni Negri, Jürgen Habermas). Otherwise, like the presumptuous commentator on McLuhan rebuked by the master himself in a scene from the film *Annie Hall*, you risk being disavowed by the author you have too freely interpreted. Whenever the latter's assistance has not been obtained—or when it is certain that his recuperation will be impossible—there has to be a switch from manipulative recycling to frontal opposition. Then it's open war: against the critics of "complexity," the end justifies the means.

This is the general context in which to properly frame the stigmatization of Chomsky or anyone else who undertakes to examine a set of systemic—hence collective—constraints, in order to try to deduce what the likely behavior of the agents in a given field (economic, cultural, or the media) will be.[4] The idea that the gravitational pull of social forces produces effects distinct from those that the will of individuals alone would generate is naturally highly disturbing to all who privilege more or less depoliticized private determinations: the choice of the individual, an aptitude for a redeeming "ethics" that protects the system against its "excesses." In this strictly individualist frame of reference, the critique of an institution can only appear to be a challenge to the morality of the individuals who compose it; a general dysfunctionality—as opposed to isolated cases—could not exist if it had not been intentionally and collectively planned. The numerous critiques directed at the major media have thus been treated, by these same media, as a putative "return of conspiracy theory." Their marked predilection for this theme has, as often happens, "spontaneously" coincided with a recrudescence of essays published in book form.[5] The thrust of this whole approach is seen in the ever more frequent comparison of any critique of the institutionalized media to imaginary conspiracies.[6]

I

The first thing to note is the paradoxical nature of most of the attacks on "Pierre Bourdieu" and "Noam Chomsky."[7] These two authors are blamed for failing to see the virtually spontaneous character of the economic transformation of the last quarter of the century, linked to phenomena regarded as "natural" and detached from any political or social intention ("globalization," "the technological revolution"). Nonetheless, they are simultaneously skewered for stating that collective determinations constrain the wills of individuals. When one juxtaposes these two charges, one gets the suggestion that people are autonomous vis-à-vis their social group but that human society is powerless against the market. To deduce the behavior of individuals from their socialization (including the information, more or less biased, that they receive from their environment) and to seek to explain economic policy on the basis of the interests of the class in power now obliges one to deny any sort of belief in an "order born of a conspiracy, like a new Protocols of the Elders of Zion."[8] This is to impute a very naïve, not to say infantile, social outlook to radical intellectuals. Rather than having sprung from the brain of "Chomsky" or "Bourdieu," it is more like something produced by the brain of an adolescent projecting onto Chomsky and Bourdieu fantasies based on too much time spent reading spy novels and watching television.[9]

In 1988, the economist Edward S. Herman and the linguist Noam Chomsky published a work that became one of the references for the radical critique of the media: *Manufacturing Consent: The Political Economy of the Mass Media.*[10] In it, the authors argue that media firms, far from shedding light on the reality of the social world, convey an image of it that suits the interests of the established powers (economic and political). The reaction of most high-profile journalists was similar to the comments of the American writer Tom Wolfe, the father of the "new journalism." He was trenchant and disdainful: "This is the—the old cabal theory that somewhere there's a room with a baize-covered desk and there are a bunch of capitalists sitting around and they're pulling strings. These rooms don't exist. I mean I hate to tell Noam Chomsky this."[11] Since then, the linguist has constantly reaffirmed that his analysis does not rest on any notion of conspiracy and that he was advancing "the exact opposite" of the caricature of him sketched by Tom Wolfe.

What might, at the limit, have passed for a misunderstanding in 1988 has become, generally at the hands of writers infinitely less talented and amusing than Tom Wolfe, the ordinary method of dismissal. It is all the more topical when those who utilize it, themselves situated at the progressive end of the political spectrum, perceive Chomsky as an obstacle to facile moralizing. In France, a country in which this campaign appears to know no respite, it is thus periodicals and intellectuals classified as belonging to the "left" who have most persistently falsified the work and the conclusions of the American linguist in order to keep alive his "bad reputation."[12]

From the time they were writing *Manufacturing Consent*, Chomsky and Herman had a presentiment of the coming storm. "Institutional critiques such as we present in this book are commonly dismissed by establishment commentators as 'conspiracy theories.'"[13] They tried to counter this in advance by pointing out in the preface: "We do not use any kind of 'conspiracy' hypothesis to explain mass-media performance." This double precaution ought not to have been necessary. In the eyes of a normally constituted reader, the analysis proposed in *Manufacturing Consent* has about as much chance of passing for a conspiracy theory as a field of sunflowers for a set of pots and pans. The point of the book is to emphasize that the forces that lead the media to produce information with a specific political and social orientation, far from being concocted by certain amoral individuals, depend above all on mechanisms inscribed in the very structure of the media institution—particularly its mode of organization and functioning.

According to Chomsky and Herman, these mechanisms operate in the form of "filters," "institutional factors that set the boundaries for reporting and interpretation in the ideological institutions."[14]

The first filter has to do with the nature of the ownership of the dominant media. Owned by private oligopolies, these channels of information and communication are driven by the logic of maximum profit.

The second filter is related to the sources of financing of the great media firms. Their revenue comes principally from their clients—their advertisers—to whom they sell audience shares.

The third filter concerns the sources of information privileged by journalists.

The fourth corresponds to the protests ("flak") addressed to those in charge of the press, which help mark out the limits of what can be published without risk.

The last filter is the dominant ideological presuppositions interiorized by journalists, which guide both their interpretation of current events and the account they give of them. In the 1980s, the dominant presupposition was the ideology of anticommunism. Its raison d'être has now virtually disappeared, but this "is easily offset by the greater ideological force of the belief in the 'miracle of the market' (Reagan)."[15]

This general frame of interpretation determines the way policies, governments, and "orthodox" commentators are treated. They are systematically valorized, and their errors, even their crimes, are always made to seem less atrocious than those supposedly committed by "dissidents" against the dominant belief system.

These five filters have a characteristic in common. They all favor the control and the orientation of news reporting in a direction that conforms to private interests. "In our view, the same underlying power sources that own the media and fund them as advertisers, that serve as primary definers of the news, and that produce flak and proper-thinking experts, also play a key role in fixing basic principles and the dominant ideologies."[16]

II

The film by Peter Wintonick and Mark Achbar, *Manufacturing Consent: Noam Chomsky and the Media*, which came out in Paris in 1993 under the title *Chomsky, les médias et les illusions nécessaires*, showed that, despite the precautions of the authors, the comparison of their work on the media to a "conspiracy theory" has been there from the start. Hardly a single public meeting filmed by

Wintonick and Achbar failed to include a participant who rose to his or her feet to demand that Chomsky account for the paranoid fantasies attributed to him. And in every debate and meeting, the linguist repeated: "Part of the structure of corporate capitalism is that the players in the game try to increase profits and market shares—if they don't do that, they will no longer be players in the game. Any economist knows this: it's not a conspiracy theory to point that out, it's just taken for granted as an institutional fact."[17]

But it was no use. Philippe Val is the owner of a satirical weekly long regarded as being on the left, *Charlie Hebdo*. In 2002, disturbed by the radical influence of Chomsky, he undertook to combat it in three successive editorials, all the more vindictive in that they swarmed with errors. The French translation of a lecture the linguist had given eleven years earlier[18] gave sufficient grounds for Val to deliver the following verdict: "For him [Chomsky], the news . . . is no more than propaganda."[19] And once again, Chomsky repeated: "I have never said that all the media were nothing but propaganda. Far from it. They offer a large quantity of precious information, and in fact they are better than they were in the past . . . but there is a lot of propaganda."[20]

The worst deafness is that of the person who refuses to listen. At the same time, Daniel Schneidermann, a journalist who specializes in covering the media, gave this "summary" of "the thesis . . . of the linguist Noam Chomsky": "Enslaved to the military-industrial lobby, unswervingly obedient to the political orders they receive, the media never stop retailing futilities by the kilometer, in order to prevent 'the imbecile masses' from concentrating on the essentials."[21]

Yet in 1996, Edward Herman had stated precisely how matters stood:

> We stressed that the filters work mainly by the independent action of many individuals and organizations; these frequently, but not always, share a common view of issues and similar interests. In short, the propaganda model describes a decentralized and nonconspiratorial market system of control and processing, although at times the government or one or more private actors may take initiatives and mobilize coordinated elite handling of an issue.[22]

And Chomsky noted: "if I was talking about Soviet planning and I said, 'Look, here's what the Politburo decided, and then the Kremlin did this,' nobody would call that a 'conspiracy theory'—everyone would just assume that I was talking about planning." But, added the linguist ironically, "nobody here ever plans anything: we just act out of a kind of general benevolence, stumbling from here to there, sometimes making mistakes . . . as soon as you describe elementary reality and attribute minimal rationality to people with power . . . it's a 'conspiracy theory.' "[23]

They were wasting their breath. In 2004, Géraldine Muhlmann published *Du journalisme en démocratie*. This book, which immediately earned her a flood of praise, sums up the four hundred pages of *Manufacturing Consent* in a dozen lines, learnedly attributing to its two authors "a schema of innocent public/wicked journalists, the former being held hostage by the latter."[24] Among the ovations she received for this particular effort was one from Philippe Corcuff.[25] This professor from Lyon has made "complexity" one of his trademarks (to the point where he is no doubt one of only a handful of people on the planet who have grasped the meaning, and the interest, of the concept—invented by him, and complex for sure—of "libertarian social-democracy"). So he too took his turn revealing a "rhetoric of conspiracy" that valorizes "the intentionality of a few 'powerful' actors" in Pierre Bourdieu and Noam Chomsky (as well as in Acrimed and *PLPL*).[26] It must have taken all of Philippe Corcuff's talents as a lecturer in political science, active member of the LCR (Ligue communiste révolutionnaire, a Trotskyist party), and member of the advisory council of ATTAC (L'Association pour la taxation des transactions financières et pour l'aide aux citoyens, an antiglobalization movement) for him to reduce the "structural" analysis of the media to the "rhetoric of a few actors" in this manner. His prowess was rewarded with acclaim from Edwy Plenel, the former editorial director of *Le Monde*, who himself was also becoming disturbed by "this vision of the world," in which "there is no place for anything but individual machinations."[27]

III

These repeated accusations have one thing in common: they all confine themselves to dismissing Chomsky's *conclusions* without ever discussing the evidence he adduces to back them up.

To examine the facts advanced by an author before presuming to assess the interpretation given them ought to constitute a prerequisite for any serious discussion. But one perceives immediately the problem that would arise from such scruples: the critics would have to come to grips with the mass of evidence—rigorous and detailed—presented by Chomsky and Herman to undergird their analysis. And if, having carried out such an examination, one were forced to admit that the deformations reported were real, it would no longer be enough to discredit the explanation advanced by the authors: one would be obliged to put forward a different one. The attempt to reduce this *ordinary* disinformation to *extraordinary* "moments of overexcitement" or "inevitable imperfections" would be seen as inadequate—especially after all and sundry have stipulated that the way the news is "reported" constitutes a fundamental demarcation between democracy and dictatorship.

So the denunciation of Chomsky's supposed conspiracy theory—and of those in France who adopt the same style of institutional analysis—is careful to limit itself to rehearsing the same baseless accusation. In the eyes of the mass of those who have not read the works under attack, the repetition of the same imputation amounts to confirmation that it must be true and an invitation not to waste time reading anything so crude.

These repeated assaults on the radical critique of the media reveal another intellectual contradiction. On one hand, the assailants reproach the two critics with defending the idea of a conspiracy hatched in the shadows. But in the same breath they point out that this "conspiracy theory" cites public documents and quotations from the press: proof that the critique is self-refuting and that media coverage of controversial subjects is in reality adequate.

But, given that the radical critique relies, as its adversaries themselves acknowledge, on public documents and interests that are transparent and publicly avowed, how exactly can it be said to be claiming that there is plotting going on in dark "rooms with baize-covered desks"? The answer is ready and waiting: since the critique of the media *is* a "conspiracy theory," the transparency of the logics that it illuminates . . . proves how inept it is! Nicole Bacharan, a lecturer at the Institut d'études politiques in Paris, explains slyly: "What in my opinion . . . totally defeats the conspiracy theory in this respect, is that it is very, very clear. The interests . . . are not hidden, there is no need to go looking in dark corners."[28]

It is worth noting here that our lecturer is likewise a commentator on American politics in the French media, as omnipresent as she is ignored by her university peers. One's degree of media visibility is for that matter generally proportional to one's propensity to disparage the oversimplifications of the radical critique of journalism. We trust we will not be suspected of engaging in "conspiracy theory" if we make the utterly ordinary psychological observation that a system that calls upon you (and remunerates and promotes you) is one you are never going to find entirely bad.

Odious when one imagines one has detected it in the writings of Chomsky (who never resorts to it), the everyday use of the lexicon of "conspiracy" arouses no protest when it turns up in the mouths, or in the texts, of the advocates of the established social system. Thus, when the strikes were occurring in spring 2003, the French minister of national education at the time, the media philosopher Luc Ferry, expressed indignation at the behavior of the teachers, repeatedly denouncing "a campaign of disinformation."[29] Two years later, it was his view that the angry secondary-school students were being "manipulated" by their teachers.[30] Curiously, no one reproached the essayist-minister for propagating a simplistic and homogeneous image of a professional body or a "conspiratorial" vision of a social movement.

Géraldine Muhlmann, likewise, is not often suspected of defending a conspiracy theory, of exaggerating the secret character and the unspoken motivating forces of social dynamics, when she expresses doubts about the utility of journalism and recommends that it should "go and look where no one generally goes, in the hidden spots."[31] Neither was this reproach directed at

Laurent Joffrin when he exclaimed that "obviously those in power conceal things,"[32] any more than it was at Daniel Schneidermann of *Le Monde* when he wrote that "to be a journalist is to believe nothing and nobody, to know that everybody lies, that you have always to check everything."[33] Disinformation, deceit, concealment of the truth, lies, complicity . . . what critique of the media, even a radical critique, could get away with using such a vocabulary?[34]

Since those who attack the "institutional analysis" have difficulty—understandably—in explaining exactly how it amounts to a "conspiracy-oriented" outlook, they adopt the tactic of lumping it in with unrelated phenomena. "If it wasn't you, it must have been your brother . . ." Thus, poring over the compost heap from which Thierry Meyssan[35] had emerged onto the media scene, Daniel Schneidermann put "the decades-old thesis of the linguist Noam Chomsky" into the same bag as *The X-Files*, Michael Moore, David Vincent (the hero of a 1960s television series about an extraterrestrial invasion), and Oliver Stone's film *JFK* (which blames the Kennedy assassination on a complex plot).[36]

If the criteria for setting the bounds of who is paranoid are broad, the hidden intentions attributed to the radical critique of the media are rather more narrow. For Géraldine Muhlmann, it "doesn't state clearly its implicit assumptions"[37] and has a tendency to "sometimes slip down an antidemocratic slope."[38] Nicolas Weill, a journalist with *Le Monde*, immediately congratulated the media-friendly philosopher for having spotted this "undeclared tendency to antidemocratism." He regretted, though, that Muhlmann's "analysis was not extended to the point, in the nineteenth and early twentieth centuries, when the critique of journalism converged with another phenomenon: that of anti-Semitism."[39]

As we see, intellectual and media elites blame the institutional analysis for everything and the opposite of everything, demanding that it respect principles that they themselves do not apply to others. They depict it reductively as a vague opinion, silently ignoring all the rational argumentation that grounds its legitimacy. They fail to inform their readers (and no doubt fail to inform even themselves) about this critique. Their "arguments" are limited to the infantile practice of lumping heterogeneous things together, defamatory insinuations, and imprecations to which there is no proper reply. These attacks, in fact, bear all the hallmarks of propaganda . . .

IV

Last and strangest: most of those who attack the radical critique of the media concede that it is, at least in part, well founded. *Le Figaro Magazine*, a widely read right-wing magazine, expresses concern, for example, at the "implacable economic logic" of the electronic communications sector.[40] Géraldine

Muhlmann concedes that when a journalist "flirts with those in power . . . he should be denounced."[41] Philippe Val agrees that Chomsky "is right" to criticize "the influence of the media."[42] And certain periodicals situated at the center of networks of intellectual and political connivance, such as the *Nouvel Observateur*, speak forthrightly about how urgent it is to "stick the pen in the wound"[43] in the face of the "danger for democracy" (no less) constituted by "the overlap between those who hold power and those who are supposed to check on them."[44]

But this concern soon melts away: behind superficial concessions to the radical critique there looms a fundamental denial. When they are pinned down, high-profile journalists and their stooges in the universities will concede that the criticism carries weight . . . only to add immediately that those who utter it are in the wrong. Their main goal seems to be to acknowledge that aberrations do occur—but they emphasize that this only happens within a confined space and does not in the least hamper the democratic mission of the media.

Laurent Joffrin, the former editorial director of the *Nouvel Observateur* who is now back at the helm of *Libération*, was also a regular contributor to France Inter, the presenter of a program on La Chaîne histoire, and a regular guest on France 5. In 2003, he edited a special *Nouvel Observateur* dossier on "The Hidden Face of Journalism."[45] In it he maintained that "investigation and common sense demonstrate that in the French media system, pluralistic for all its faults, the important news does get out, debates do take place (notably concerning the press itself), the powerful are criticized, and the ills of society are very widely exposed." So in exploring the "hidden face," we discover that it is both open to view and rather attractive.

"A journalism that is not part of 'the market,'" Joffrin continued with great subtlety on France-Culture, "is a journalism that is a journalism of the state. And when the state controls the press, that is unacceptable." The consequences of entrepreneurial capitalism as detailed by Chomsky are thus dispatched in a few words and reframed as simple "excesses": "Liberalism and journalism are intimately connected: but for the best! . . . There are excesses of liberalism. That is another thing. But the heart of the matter is that, if newspapers are free, they are enterprises."[46] Does such a conclusion not interdict one from then pretending to resolve a problem whose very existence one is denying? And, faced with the historical ignorance of a remark setting the market against the state in such a binary fashion, is it even worthwhile to suggest that Laurent Joffrin should try reading, if not Karl Marx (you can't expect miracles), then at least *Tintin au pays de l'or noir*?

But this refusal of any structural critique appears less disconcerting when one observes who its main proponents are: journalists occupying positions of power and prestige who draw substantial symbolic dividends (in the form of notoriety) from their media presence—dividends that can subsequently be converted into desirable financial advantages, through, among other things, free publicity for their books, albums, or concerts.[47] Their rejection of anything that might bring them face to face with their own responsibilities as deciders is all the more violent and sincere in that they have interiorized a neoliberal value system that accords a central place to the entrepreneurial drive. The questioning of the capitalist social order to which the structural analysis leads appears to them just as threatening as the evocation of an intentionality not immediately accompanied by precautions allowing one to feel that one is not oneself being called to account.

So to even register with the holders of media power, the critique has to give everyone the chance to exclude herself from it. Or to plead a collective short circuit (like the media's general "skid" at the time of the Gulf War), which unfailingly heralds the reestablishment of the pluralist radiance and the publication of conference proceedings in which the cadaver of the mistake is inspected and prepared for incineration.

So if the *Nouvel Observateur* rejects the idea of "automatic submission," that is also—in fact, primarily—in order to claim that, in the face of the "constraint [of advertising] that hampers the independence of journalists . . . some, including the *Obs*, resist, others cave in."[48] Likewise, if the "unilateralism" of the critiques irritates Philippe Corcuff so much,[49] that is mainly because it leads to the denunciation of the "false critiques . . . that supposedly throng the media," among them *Charlie Hebdo* (which employed Philippe Corcuff at the time), or because it might cause some to overlook "the differences" and, especially, to confuse "*Charlie Hebdo* with TF1." (Because it seldom calls upon outside journalists and essayists, TF1 has become the habitual target of all the institutional insolents who only appreciate insolence when it does not expose them to any discomfort.) While admitting "the justice of certain critiques," Thomas Ferenczi, a journalist with *Le Monde*, also introduces a distracting qualifier: the necessity to "recognize that *Le Monde* [is] precisely one of those that, to the extent their means allow, tried to resist this movement."[50]

All the "leading journalists" are thus ready to point to, and even condemn, "deviations," on condition that one simultaneously recognize (and celebrate) the status of exception to the rule, to which each lays claim. As far as they are concerned, conceding the existence of "structural" imperfections must not lead to any querying of the legitimacy of a system they have chosen to perpetuate. The *radical* analysts who wish to attack the evil *at its root*—simultaneously bearers of "bad news" and mirrors reflecting an ocean of everyday compromise—could hardly expect to be received with open arms by the diners whose party they were coming to spoil.

If, however, "the critiques . . . of journalism have often irritated journalists," that is not solely because journalists deny that they are serving particular interests or, as Thomas Ferenczi suggests, because "they seldom recognize themselves in the image of them given [by the radical critics]."[51] Journalists are not *that* unwitting. Some of them have clearly perceived the threat that the radical critique posed to the illusion of pluralism that grounds their social position and their privileges as a "countervailing power." They are also aware that the authors of articles and the suppliers of images are less and less the masters of what they produce, that editors' associations are disappearing or withering, that proprietors and advertisers are piling on more and more pressure all the time. Consequently, whoever wishes to preserve the illusion of journalistic liberty at a time when that liberty is on the wane has to keep on propping up the myth of the reporter without frontiers and without blinkers. For some years, daily newspapers have been undergoing a decline in readership and advertising revenue. At the same time, the public image of the media is degrading.[52] Aware of their growing discredit, against which the ongoing attempts to heroize the profession are unavailing, newspaper owners and editorial executives confess that there is "no more time to beat around the bush."[53] They nevertheless intend to hold on at any price to their monopoly on the critique of journalism, as a way of ensuring that it will never menace them. So they resort to various techniques.

The simplest one is to discipline their troops: article 3b of the national collective agreement of the journalistic profession restrains the freedom of public expression of journalists by stipulating that they must not "under any circumstances harm the interests of the press firm in which they work." Certain journalists have already felt the sharp edge of this provision.[54]

Ensuring group cohesion by sanctioning the bad elements and potential rebels is not a seemly approach for newspaper companies. A more singular method is to silence the most intractable critics in the name of a concern for democracy. This only works because of the double privilege of the press: the ability to celebrate the collective courage of the profession every time a particular journalist is kidnapped or killed in the exercise of her profession (which explains the formidably narcissistic echo aroused in all the media by the initiatives of Reporters Without Borders) and the ability to shut out or tamp down any criticism aimed at the press. Advantage of that kind is not something you come across every day. An affair reflecting little credit on truckers, masons, or druggists may resonate with public opinion, as long as it interests certain journalists, but the reverse, an affair reflecting little credit on journalists, has small chance of being communicated by truckers or masons to the entire population of druggists.

The press allows critical intellectuals few occasions to expound their views, *freely* (meaning without controls, intermediaries, or permanent interruptions), to the public at large, except via the diffusion of their works in bookstores (for those who succeed in getting published) and via samizdats in cyberspace. For the rest, the major media, not content with feeling no obligation to supply a forum to those whom they attack with all the more ease inasmuch as they distort their views, also apply very sparingly their legal obligation to publish the "entitlements to a reply" (*droits de réponse*) addressed to them.[55]

To justify this ostracism, media heavyweights claim that any radical critique coming from "outside" their own editorial departments would entail a new totalitarianism. This line of defense, the "journalistic preference," has been theorized by Jean-Marie Colombani. According to the editor in chief of *Le Monde* at the time, it is "impossible to regulate, control, or discipline the press from outside, without undermining that which grounds the very principle of freedom of expression."[56] The same changes were rung in the columns of *Charlie Hebdo*, where Philippe Val, annoyed by the creation of a French Media Observatory (Obsérvatoire Français des Médias, or OFM), attacked the "press cops" in an editorial. According to him, any outside scrutiny is equivalent to a "press militia," not to say "a morbid delation machine, feeding on the hatred of journalists in general."[57]

So are journalists the only ones authorized to criticize the media, comment on works dealing with the press, prescribe which ones are worth reading, and censure the ones that ruffle their feathers? If so, why not make soldiers the only ones entitled to investigate war crimes, and priests the only ones entitled to reflect on religion? Why not arrange things so that workplace accidents only hit the headlines when business owners start to get worried about them? In order to preserve their monopoly of legitimate criticism and guarantee its innocuousness, the media hierarchies avail themselves of a complementary strategy. This consists of filtering the external elements that are allowed to "enter," in other words, of ensuring that the commentators who are given a platform will be supportive.[58]

Two mechanisms give "media intellectuals"—both those who aspire to this status and those determined to keep it—an incentive to conform to the credo of the editors in chief and the proprietors of the press. These are lust for the carrot and fear of the stick. On one hand, praise of the dominant media and condemnation of the radical critique bring substantial dividends in terms of one's quota of visibility in the press and the invitations to conferences and facility in getting published that go along with it. Symmetrically, those who express agreement with the radical critique of the media find themselves reprimanded and ultimately blacklisted.

It is understandable that under these conditions the weaving of media laurels has become an expanding cottage industry. Its most zealous practitioners—who are often the ones with the most media visibility—are free to really let themselves go. In his journal *La Règle du Jeu* (May 2004), Bernard-Henri Lévy (414 appearances on television[59] and dozens of opinion pieces and articles published in *Le Monde*) commented on the "affaire *Le Monde*" (the paper's lawsuit against Philippe Cohen and Pierre Péan). He sang the praises of "the press, war reporters, and leading editorialists associated in the same struggle for truth, probity, and responsibility." But he also saluted the "grandeur of journalism . . . of journalists, of journals, and of intellectuals." In other words, his own grandeur along with the rest.[60]

To condemn the "radicals" automatically gains you a deferential hearing. Those with the best credentials to wield the sword, because of their own past as protesters, are celebrated by the media and invited into the studios to psychoanalyze the perversity of their former comrades, real or putative. After his volley of editorials against Chomsky, whose works it was plain he had barely skimmed, Philippe Val was welcomed with open arms on the set of *Culture et dépendances* by Franz-Olivier Giesbert, also the director of *Le Point* (a magazine of the right): "If we have invited you, Philippe Val, it is because we have been reading your editorials in *Charlie Hebdo* for a long time, which are exciting because you have a reputation for intelligence and moderation. Very good piece, by the way, on Noam Chomsky."[61]

As for Géraldine Muhlmann—may the reader pardon us for citing names almost forgotten already and that will soon be remembered for nothing except having once criticized Chomsky—as soon as her two books appeared (on March 9 and 17, 2004), she was showered with invitations. Her critique of the radical critique—which represents only a small part of her work—was systematically highlighted.[62] Muhlmann was also praised to the skies, and for the same reason, by a large portion of the media nomenklatura: Nicolas Weill (*Le Monde des livres*, April 2, 2004), Edwy Plenel (*Le Monde 2*, April 9, 2004), Philippe Corcuff (*Charlie Hebdo*, April 14, 2004), Roland Cayrol (*L'Esprit public*, on France Culture, April 18, 2004), etc.

V

Chomsky once explained the meaning of the attacks aimed at him: "The point is that any analytic commentary on the institutional structure of the country is so threatening to the commissar class they can't even hear the words. . . . So if I say there is no high cabal, what [Tom Wolfe] hears is there *is* a high cabal. . . . It's a tight-closed system of belief."[63]

It is still closed tight. In January 2005, Luc Ferry, the liberal philosopher and former minister of education, condemned the analysis of capitalism that he imputed to its challengers. He summed up his own contrasting vision of a social "system" that comes into being "automatically." In his analysis, he spread a pacifying and disarming description of spontaneous order over the whole global economy and used it as a platform to oppose the work of Chomsky and others on the media:

> The alterglobalists really go astray because they imagine that behind these globalized phenomena—the play of the financial markets, the plant relocations, the deindustrialization of some countries, the fact

> that cultural identities are swept away by an Americanization of the world that uniformizes modes of living and so destroys local cultures—there are people who are controlling the whole thing and pulling the strings. And that they were pretty much formed by the Chicago school, that they are neoliberals, that they are the bad guys. And we find the Marxist idea that behind the processes that govern the world, there are those with power. It is the myth of the two hundred families. We are back with cartoons by Épinal, financiers chomping on cigars and wearing top hats. Now the real problem, if you like, is exactly the inverse. If you look, for example, at plant relocations, what is striking is that nobody is in control, nobody is behind the scenes. *These processes are absolutely automatic*. There is no intelligence behind them.[64]

There is nothing in any case very new about this sort of portrayal of an "automatic" social order, on which the collective will is supposed to have no purchase. In 1932, Paul Nizan revealed the underbelly of this kind of analysis in *Les Chiens de garde*:

> When bourgeois ideas came to be regarded as productions of an eternal reason, when they lost the unstable character of historical productions, then they had the greatest chance of surviving and withstanding attack. The whole world lost sight of the material causes that had brought them into being and at the same time rendered them mortal. The philosophy of today is pursuing this attempt at justification.[65]

The attempt is still being pursued, and not just by philosophers. Its legitimizing function is so essential as to make it hard to imagine that one day argument and respect for texts will get the best of it. Thus Noam Chomsky and those inspired by his work of unveiling will probably never be done stating that their intellectual enterprise is "the opposite of conspiracy theory, it's just normal institutional analysis, the kind of analysis you do automatically when you're trying to understand how the world works."[66]

NOTES

1. In the crudest journalistic prose, the victory of neoliberalism is associated with the liberation "of the need for autonomy, for consumption, for creative and professional opportunities." Laurent Joffrin, *Marianne* (March 29, 2004). [Translator's note: Readers are reminded that in Europe the term "(neo)liberalism" means roughly the same thing as the term "free-market capitalism" does in the United States.]

2. The three names cited are merely examples selected at random, or what Pierre Bourdieu called "electrons" in a "field" that gives them their "minute force," "epiphenomena of a structure" that would be nothing without them. Pierre Bourdieu, *Sur la télévision* (Raisons d'agir: Paris, 1996), 63.

3. See, for example, the recent biography of Karl Marx by Jacques Attali, *Karl Marx ou l'esprit du monde* (Paris: Fayard, 2005), in which the essayist, a former adviser to François Mitterand and subsequently banker, contends that the author of *Capital* "writes continuously that he is for globalization, that he is for capitalism, that he is for economic liberalism, that he is for the bourgeoisie." *France Culture* (June 22, 2005). He hammered away at this "reading" of Marx throughout his interminable book promotion tour. [Translator's note: *France Culture* is a national radio network; see http://www.radiofrance.fr/chaines/france-culture/sommaire.]

4. For other examples, see "La critique des médias," *Pour Lire Pas Lu* (*PLPL*) 20 (June–August 2004); and also the column "Haro sur la critique des médias," Acrimed.org, rubrique 246.

5. The year 2005 saw a rich harvest of books, press coverage, and special dossiers devoted to this message. Among others: Antoine Vitkine, *Les nouveaux imposteurs* (Paris: La martinière, 2005); Véronique Campion-Vincent, *La société parano. Théories du complot, menaces et incertitudes* (Paris: Payot, 2005); Pierre-André Taguieff, *La foire aux illuminés. Esotérisme, théorie du complot, extrémisme* (Paris: Mille et une nuits, 2005); Frédéric Charpier, *L'Obsession du complot* (Paris: Bourin éditeur, 2005). The publication of these works (the first three, at any rate) was accompanied by laudatory articles in the main dailies and weeklies and multiple media appearances on the part of their authors. For its part, the magazine *TOC* (now defunct) dedicated a dossier to the "trendiness" of "belief in conspiracy" ("La mode des complots," March 2005). *Le Magazine Littéraire* followed suit with a dossier entitled "La paranoia" (July–August 2005), and *Le Monde 2* joined in with a feature on "the return of conspiracy theory" ("Le retour des théories du complot," November 5, 2005). [Translator's note: *Le Monde 2* is a weekly magazine supplement to the daily newspaper *Le Monde*.] The year 2005 also witnessed the appearance of a comic strip by Will Eisner on the history of the "Protocols of the Elders of Zion" (*Le Complot*, Grasset) and the cinematic release of a documentary by the American director Marc Levin on the same topic, *Protocols of Zion*, entitled *Les protocoles de la rumeur* in French.

6. The program *Arrêt sur images* recently supplied two examples. Faced with questions about the image of Islam that his stories were conveying, the journalist Mohamed Sifaoui immediately replied: "The general idea [advanced by his critics] seems to be: for the last thirty years all the French media have been in cahoots, because there is an enormous plot against the Muslim community . . . this intellectual tactic is totally senseless and idiotic, because it relies from the outset on the idea that there is a plot." *France 5* (October 16, 2005). Two months later, questioned about the threat by industrialists to punish journalists who dared to criticize their products, Alain Grangé-Cabane, president of the federation of perfume manufacturers, replied: "I wonder whether this isn't, a little bit, the conspiracy theory that is springing up all over the place. The horrible advertisers are supposed to be busy trying to smother the truth." *France 5* (December 4, 2005).

7. Often these names are used generically; this is a way of adopting a pose of preferring not to reply directly to other participants in the debate toward whom one wishes to feign contempt.

8. From the ultraliberal (in the European sense; it would be called ultraconservative in the United States) Web site http://www.chatborgne.com.

9. See the article by Laurent Joffrin, then editorial director of the newspaper *Libération* (of which he is now the overall director), likening the analyses of Pierre Bourdieu and his consorts to a "phantasmagoric, destructuring, and paranoid vision of the world," a "bizarre cross between *The X-Files* and Maurice Thorez." *Libération* (May 12, 1998). Thorez was general secretary of the French Communist Party from 1930 to 1964.

10. Fifteen years later, this work was translated (very loosely) into French under the incongruous title *La Fabrique de l'opinion publique. La politique économique des médias américains* (Paris: Le Serpent à plumes, 2003).

11. Mark Achbar, ed., *Manufacturing Consent: Noam Chomsky and the Media*, the companion book to the film by Peter Wintonick and Mark Achbar (Montreal: Black Rose Books, 1994), 61. Chomsky's response to Tom Wolfe's observation can be found at n. 63.

12. See Jean Bricmont, "Folies et raisons d'un processus de dénigrement. Lire Noam Chomsky en France," afterword to Noam Chomsky, *De la guerre comme politique étrangère des États-Unis* (2001; repr. Marseille: Agone, 2004), 227–257.

13. Herman and Chomsky, *Manufacturing Consent*, lx.

14. Peter R. Mitchell and John Schoeffel, eds., *Understanding Power: The Indispensable Chomsky* (New York: New Press, 2002), 26.

15. Herman and Chomsky, *Manufacturing Consent* (from the new 2002 introduction), xvii.

16. Herman and Chomsky, *Manufacturing Consent* (from the new 2002 introduction), xi.

17. Mitchell and Schoeffel, eds., *Understanding Power*, 26.

18. Noam Chomsky, *Propaganda* (Paris: Le Félin, 2002). The original lecture, given by Chomsky on March 17, 1991, was published as *Media Control: The Spectacular Achievements of Propaganda* (New York: Seven Stories Press, 1997).

19. Philippe Val, "Noam Chomsky dans son mandarom," *Charlie Hebdo* (June 19, 2002).

20. Noam Chomsky, *Télérama* (May 7, 2003).

21. Daniel Schneidermann, *Le cauchemar médiatique* (Paris: Denoël Impact, 2003), 121–122.

22. Edward S. Herman, "The Propaganda Model Revisited," *Monthly Review* (July 1996).

23. Mitchell and Schoeffel, eds., *Understanding Power*, 390.

24. Géraldine Muhlmann, *Du journalisme en démocratie* (Paris, 2004), 28.

25. Philippe Corcuff is a French university lecturer and was also a contributor to *Charlie Hebdo* when he attacked Pierre Bourdieu, Chomsky, and, in general, the radical critique of the media—of which *PLPL* (*Pour lire pas lu*), along with the Web site Acrimed (Action Critique Média), was one of the principal organs. Writing on behalf of Acrimed, the sociologist Patrick Champagne replied to the *Charlie Hebdo* columnist in "Philippe Corcuff, critique 'intelligent' de la critique des médias" (April 2004), http://www.acrimed.org/article1572.html. (The article from which Corcuff's epigraph to this chapter comes, "Gauche de gauche: le cadavre bouge encore" [October 2004], is available online at http://bellaciao.org/fr/article.php3?id_article=10032.)

26. Philippe Corcuff, "De quelques aspects marquants de la sociologie de Pierre Bourdieu" (October 2004), http://calle-luna.org/article.php3?id_article=136. Philippe Corcuff customarily plays fast and loose with quotation marks to make his point. In summarizing the ideas of certain authors, he regularly puts quotation marks ("scare quotes," as they are informally called) around terms such as "conspiracy" and "great liberal conspiracy," which these authors only ever use when stating that they reject them. For more detail, see Patrick Champagne, "Philippe Corcuff, critique 'intelligent' de la critique des médias."

27. Edwy Plenel, *Procès* (Paris: Stock, 2005), 78. A year earlier, Edwy Plenel was already casting vituperation on this "destructive and deathly critique" and saluting the publication of Géraldine Muhlmann's essays enthusiastically (*Le Monde 2*, 9, 2004).

28. Nicole Bacharan, "De quoi j'me mêle: Tous manipulés?" *Arte Television* (April 13, 2004).

29. Luc Ferry, "Pièce à conviction," *France 3* (June 5, 2003).

30. "Ferry/Julliard" on *LCI* [La Chaîne Info, i.e., "the news channel"] (February 12, 2005).

31. Géraldine Muhlmann, "Idées," *RFI* [Radio France Internationale] (May 16, 2004), 0h10 TU (Temps Universel).

32. *BFM* (October 15, 2004), 13h30.

33. Daniel Schneidermann, *Du journalisme après Bourdieu* (Paris: Editions Fayard, 1999), 41.

34. That's without counting all the periodicals that regularly carry headlines about conspiracies, hidden truths, and occult forces of every sort. On this, see the montage of front pages

and headlines assembled by *PLPL*: "Leur théorie du complot," *Pour Lire Pas Lu* (*PLPL*) 20 (June–August 2004): 12.

35. In his book *L'Effroyable imposture* (Paris: Carnot, 2002), Thierry Meyssan asserts, among other things, that no airplane crashed into the Pentagon on September 11, 2001.

36. Daniel Schneidermann, *Le Cauchemar Médiatique*, 121–122. The same tactic of "guilt by association" can be found in Antoine Vitkine's essay "Les nouveaux imposteurs." We note in passing that it is rather paradoxical to associate Chomsky with the conspiracy theory about the Kennedy assassination, since he has repeatedly and clearly rejected it, with detailed arguments; see Noam Chomsky, *Rethinking Camelot: JFK, the Vietnam War, and U.S. Political Culture* (Boston: South End Press, 1993). He has even described the Kennedy-assassination conspiracy theory as "extremely implausible," emphasizing that "nobody's even come up with a plausible *reason*." He adds that "the energy and the passion that goes [*sic*] into things like that is really extraordinary, and it's very self-destructive," and that "really there have to be serious efforts to get past this, I think." Chomsky, in Mitchell and Schoeffel, eds., *Understanding Power*, 326–331, 348–351. In a highly original manner (which says a lot about his command of the subjects he discusses), Antoine Vitkine sees the Kennedy assassination as "the only major conspiracy" that is "still not cleared up" ("Les Nouveaux imposteurs," 17). Yet it is Vitkine who charges Chomsky, among others, with helping keep alive conspiracy theories . . .

37. Géraldine Muhlmann, "La critique en reste au coup de gueule" (interview with Olivier Costemalle and Catherine Mallaval), *Libération* (June 19, 2004).

38. Géraldine Muhlmann, "Le travail du regard chez le journaliste" (interview with Raphaël Enthoven), *Lire* (May 2004).

39. "L'analyse n'a[it] pas été poussée jusqu'au point où la critique du journalisme au XIXe et au début du XXe siècle épouse un autre phénomène: celui de l'antisémitisme." Nicolas Weill, "Le journalisme au-delà du mépris," *Le Monde des livres* (April 2, 2004). For a more detailed analysis of this article, see Henri Maler, "Le Monde contre 'les critiques antimédias,' antidémocrates et antisémites" (April 26, 2004), http://www.acrimed.org/article1557.html.

40. Marc Durin-Valois, "L'implacable logique économique," *Le Figaro Magazine* (November 22, 2003).

41. Géraldine Muhlmann, "Idées," *RFI* (May 16, 2004), oh10 TU.

42. Philippe Val, *Charlie Hebdo* (June 19, 2002).

43. "La face cachée du journalisme," *Le Nouvel Observateur* (October 30, 2003).

44. Airy Routier, "Ces grands patrons qui tiennent les médias," *Le Nouvel Observateur* (July 1, 1999).

45. See the detailed analysis by Yves Rebours, "Le Nouvel Obs et les journalistes" (October 31, 2003), http://www.acrimed.org.

46. Laurent Joffrin, "Le Premier pouvoir," *France Culture* (November 6, 2004).

47. To be entirely fair to Philippe Corcuff and Géraldine Muhlmann, this charge does not, strictly speaking, apply to them, since their (more or less polished) works have won them no more than a rather confidential following, despite the stubborn persistence of these two authors in trying to get themselves noticed, either directly or with the help of their countless journalistic cronies.

48. *Nouvel Observateur* (October 30, 2003).

49. See Philippe Corcuff, "La critique gauchiste des médias," *Charlie Hebdo* (November 6, 2002); Philippe Corcuff, "De quelques problèmes des nouvelles radicalités en général (et de *PLPL* en particulier)," *Le Passant ordinaire* (September–October 2001).

50. Thomas Ferenczi, "Le journalisme critiqué et honoré," *Le Monde* (January 26, 2002).

51. Thomas Ferenczi, "Le journalisme critiqué et honoré."

52. Though one needs to bear in mind the unscientific nature of most opinion polls, the similarity of their results gives an idea of the suspicion that attaches to the media. A poll carried

out by Sofrès in January 2004 found only 6 percent of respondents who thought that "things really happen the way the papers say they do" (and 42 percent who thought that "things happen pretty much the way the papers say they do"). In *L'État de l'opinion 2003* (Paris: Le Seuil), only around 30 percent of those polled took the view that "the media are independent of political and financial pressure."

53. "La face cachée du journalisme," *Le Nouvel Observateur* (October 30, 2003).

54. The firing of Daniel Schneidermann in September 2003 by the management of *Le Monde*, for example, was justified by exactly this article 3b. But the case of a particular "personality" ought not to make us forget the "anonymous" firings or, above all, the chilling effect a threat of this kind has on the whole profession.

55. See Noam Chomsky, *Réponses inédites à mes détracteurs parisiens* (Paris: Spartacus, 1982); "Libération et Noam Chomsky" (April 2000), http://www.acrimed.org; Jean Bricmont, "Folies et raisons d'un processus de dénigrement"; "Pas lu dans Libé," *PLPL* (June 2000), http://www.homme-moderne.org/plpl/no/p3-1.html.

56. "Un tour du monde en 80 journaux," supplement to *Le Monde* (December 5, 2003).

57. Philippe Val, *Charlie Hebdo* (December 24, 2003). A number of *droits de réponse* sent to Philippe Val, including ones from OFM and the president of Attac, have never been published.

58. On this general topic, see Pierre Rimbert, "Les rapports entre journalistes et intellectuels: cul et chemise?" (November 8, 2004), http://www.acrimed.org.

59. See Philippe Cohen, *BHL, Une biographie* (Paris: Fayard, 2005), 13.

60. In *BHL, Une biographie*, Philippe Cohen explains in detail, with plenty of examples and totals, the public-relations technique of the *nouveau philosophe*. A chapter entitled "La République des lettres expliquée à ma fille" sets out some of his rules. Numbers 2, 3, and 4 are: "always return the favor if someone honors you," "never miss a chance to defend the media," and "no enemies among journalists." The results obtained by Bernard-Henri Lévy from following these three guidelines have exceeded his wildest expectations.

61. Franz-Olivier Giesbert, "Culture et Dépendances," *France 3* (November 26, 2003).

62. "La Suite dans les idées," *France Culture* (March 17, 2004), "Diagonales," *France Inter* (March 21, 2004); "Cultures et dépendances," *France 3* (March 24, 2004); "Répliques," *France Culture* (March 27, 2004); "Le Bateau Livre," *France 5* (May 2, 2004); "Idées," *RFI* (May 16, 2004). Additionally, there were two interviews in *Lire* (May 2004) and *Libération* (June 19, 2004), as well as a series of opinion pieces in *L'Humanité* (March 29–April 2, 2004).

63. Noam Chomsky, from a 1993 interview in Achbar, ed., *Manufacturing Consent: Noam Chomsky and the Media*, 61.

64. "Ferry/Julliard," *LCI* (January 29, 2005).

65. Paul Nizan, *Les Chiens de garde* (1932; repr. Marseille: Agone, 1998), 93.

66. Mitchell and Schoeffel, eds., *Understanding Power*, 26.

12

NOAM CHOMSKY AND THE UNIVERSITY

PIERRE GUERLAIN

> Intellectuals are both the main victims of the propaganda system and also its main architects.
>
> NOAM CHOMSKY, *Propaganda and the Public Mind*

> *La politique est l'art d'empêcher les gens de se mêler de ce qui les regarde* [Politics is the art of stopping people from interfering with what is of concern to them].
>
> PAUL VALÉRY, *Tel Quel*

Much ink and saliva have been spent to attack or justify Chomsky's political views, but few observers have dealt with Chomsky's place in academia or analyzed the various techniques used to marginalize or discredit him. Chomsky presents even his fiercest critics with a problem, for even those who call him "zany" know he is one of the most famous scholars in linguistics.[1] What I intend to do here is to comment upon a kind of catalog of the reactions to Chomsky's name within the academic world. I encountered these reactions both in articles written in English or French and during formal or informal discussions in academia. After my interview of Chomsky in 1997, I witnessed many such responses going beyond the anecdotal, which illustrate how the university operates.[2] I will not tackle the entirely political debates and reactions that have already been dealt with in other publications—debates about the Faurisson affair, a case in which Chomsky should not have let this Holocaust denier use his text in defense of freedom of speech as a preface to his book; and debates about Cambodia, in which opposition to Chomsky focused on statistics but misinterpreted Chomsky's denunciation of double standards. It is more difficult to comment upon silence, though silence is the most effective marginalization strategy both within the academic world and in the media. This silence about Chomsky's political work did not exist when he opposed the Vietnam War in the 1960s and 1970s. What follows should not be construed an uncritical endorsement of all of Chomsky's views; for instance, I am not sure I agree with his assessment of Walt and Mearsheimer's views of the Israeli lobby.[3] Rather, it is an analysis of how a major intellectual voice can be silenced or marginalized in what is claimed to be the very temple of free inquiry, the university.

Quite often, academics insist upon distinguishing Chomsky the linguist from Chomsky the political thinker for two sets of reasons. Linguists wish to discuss Chomsky within their field and do not necessarily refer to other aspects of his intellectual production. In the *New York Review of Books*, John Searle discusses, as a philosopher of language, Chomsky's linguistic theories, and Chomsky responds within the same field.[4] So there we see a scholarly debate between two language specialists. Yet, even within the confines of linguistics, it is quite common to hear debasing remarks such as "Chomsky is outmoded," "Chomsky is active mostly within the field of philosophy of language and therefore his work does not interest us as linguists," or, in France, "Chomsky's theories are of interest to Anglo-Americans but in France no one pays attention any longer." I am not competent to discuss Chomsky's work in linguistics, but this kind of remark can be analyzed sociologically to comment upon the way intellectual life operates. In France, when I point out to my colleagues in linguistics that Chomsky not so long ago published a book that was widely debated in the English-speaking world, they stress their lack of interest for his theories, which in their estimation do not deal enough with the linguistics of enunciation. The first remark mentioned here appears clearly as a wish to delegitimize Chomsky, for even when a scholar works on different objects or uses a different methodology, it does not mean that he or she is outmoded. Even a severe critic like Searle does not carry out this kind of intellectual assassination: he takes Chomsky seriously and reads him before criticizing him. Let us point out here that this neat division between "Anglo-Americans" and France has xenophobic overtones.

In the field of linguistics, this marginalizing attempt, which is more frequent in France, cannot go very far, as Chomsky's reputation is firmly established. Even his most virulent political critics acknowledge, for the most part, his competence in this field, which is why they insist on a neat separation between Chomsky the linguist and Chomsky the intellectual active in political debates. This separation is in any case justified: Chomsky's work in linguistics does not make his political stands more relevant. These stands are informed by his knowledge of political and historical phenomena. There is no need to resort to linguistics to understand the political phenomena Chomsky analyzes. Sometimes a parallel is made between Céline or Aragon and Chomsky; novelists or poets can make political mistakes but remain excellent writers. Thus Chomsky is presented as a significant scholar in linguistics who got lost in politics. This parallel ignores the fact that while Céline or Aragon did support a totalitarian ideology, Chomsky never has. Moreover, they proved their political incompetence. Céline and Aragon made political mistakes based upon their prejudices, lack of knowledge, or ideological preferences, whereas

Chomsky is extremely knowledgeable in history, philosophy, and political science. This is evidenced by his wealth of references and footnotes. Some of Chomsky's fiercest critics are far more tolerant of Heidegger or Carl Schmidt, who, contrary to the American linguist, did both express their support for a totalitarian ideology and collaborate with a murderous regime.

These comparisons with mistaken thinkers are problematic, for they presuppose a nonexisting similarity: Chomsky has always condemned Stalinism and Nazism in the most forceful terms. It is quite accurate, however, to state that competence in one field does not imply competence in others: Marguerite Duras the writer does not have the same status as Marguerite Duras the legal critic;[5] Ezra Pound the poet was not very gifted in political analysis (he supported Mussolini). It is therefore legitimate to distinguish between the linguistics professor and the political thinker or activist. His political writings must be tackled differently from his works in linguistics. However, linking him with dubious thinkers in the past is a way of denying Chomsky's intelligence and competence. The linguist may be considered bright, but the political thinker is characterized as "loony."

Another point made by Chomsky's critics is that his work in linguistics is published by university presses whereas his political work, which they call polemical, is often published by small marginal publishers or a dissident magazine, Z.[6] As a result, Chomsky is alleged to go through peer review in linguistics but is considered a militant in his other writings.[7] This criticism evokes the separation, frequent in the United States, between "academics" or "scholars" on the one hand and "intellectuals" on the other. This second category has almost disappeared from the public sphere. In France, this separation, used here in order to delegitimize, is strange, as "intellectuals," at least until recently, enjoyed a fairly good reputation in this country. Thus the distinction between the professor and the activist enables critics to restrict Chomsky to linguistics and to claim that he is not competent in political science.

When Chomsky's name was mentioned at a conference organized at the Institute for Political Studies (Institut d'Etudes Politiques de Paris)[8], in reference to the idea that the United States too is a "failed state," Tony Judt, a British historian who teaches at NYU, talked about Chomsky's "errors" in reference to the Cambodian genocide. These "errors" were supposed to be a definitive comment about Chomsky's competence in foreign policy. Judt is himself a courageous thinker who has been the victim of silencing,[9] but his statement at the Paris conference illustrates the strategy of silencing in the academic as well as in the media world. However, in academia there are a few unique features to this silencing. Chomsky is neither a political scientist nor a historian by training, and thus his writings are not considered legitimate by many in the field. So, in the name of peer review, Chomsky's thoughts are rejected by scholars in political science. This administrative division between academic disciplines acts as a barrier to the free pursuit of knowledge. Chomsky mentions the fact that he is sometimes invited to a conference of mathematicians, who do not require him to show a membership card before listening to what he has to say. Strangely enough, political scientists who work on the same topics as Chomsky have a different attitude. His name does not appear in most bibliographies, course descriptions, or mainstream political-science publications.[10]

Here is a specific example of the silencing technique: In his book entitled *Colossus: The Rise and Fall of the American Empire*,[11] Niall Ferguson, who taught at NYU and Oxford in Britain before moving on to Harvard refers (n. 9, p. 309) to Chomsky's and Said's "predictable comments" about U.S. foreign policy. These comments are so "predictable" that they are not quoted, not analyzed, and not refuted. Yet these comments are sometimes similar to those made by mainstream scholars or well-known professional historians. Chomsky is mentioned neither in the huge bibliography of this book nor in the index. Ferguson writes about the rocky relationship between the United States and Israel and the dysfunctional (by which he means Arab) culture of the Middle East, and he argues that the Shah of Iran "was not the worst of the despots installed and propped up by the United States during the cold war" (117) and, incredibly, that "the invasion of Iraq in 2003 was not without a legitimate basis in international law" (133). It is easy to see that the legitimate, mainstream historian could have learned a thing or two if he had read Chomsky, notably on the Middle East. The silencing or quasi-silencing technique used to delegitimize prevents the discussion of embarrassing ideas.[12]

Richard Posner's book is devoted to public intellectuals in the United States,[13] and he discusses Chomsky not as a widely recognized scholar in the field of political science but merely as a public intellectual. Posner marginalizes Chomsky's work in two ways. First, he states that Chomsky always changes the subject when he is embarrassed. Thus, according to Posner, when one talks of the positive aspect of military intervention in Kosovo, its prevention of genocide, Chomsky allegedly switches to Kurds and Turks, Timor, or the Palestinians (88). This statement is inaccurate, as readers of Chomsky's books on Kosovo know. Chomsky often compares the treatment of various situations in the media in order to show that the United States—or, more generally, the West—denounces the massacres committed by others but often forgets its own. The word genocide applied to Kosovo is highly problematic.[14]

Posner's second technique is to attribute to Chomsky what is in fact a statement made by an official that Chomsky comments upon. Thus when he quotes Chomsky on Cuba, he omits mentioning the fact that the idea that U.S. interventions did not primarily have to do with fighting communism comes from Arthur Schlesinger Jr., a historian who was a friend of John F. Kennedy. Chomsky merely comments upon this and feels it was indeed accurate but not made public (89). Posner is intent upon showing that public intellectuals in general make mistakes and do not check their sources, but he makes the very mistakes he denounces and caricatures Chomsky's work. When he levels the accusation of anti-Americanism at Chomsky, he seems to forget that the latter has often said that freedom of speech is best guaranteed in the United States.[15]

Chomsky and a Few International-Relations Scholars

One only needs to compare Chomsky's work with that of his neighbor at Harvard, Joseph S. Nye Jr.—who does not refer to him but who frequently quotes the columnist Thomas Friedman of the *New York Times* or articles by *The Economist*—to realize that the international-relations scholar is shallow and thinks like a "new Mandarin,"[16] without questioning his prejudices, the prejudices of his background, or the power struggles within his own country and discipline.

In *The Paradox of American Power: Why the World's Only Superpower Can't Go It Alone,*[17] Joseph Nye writes (4) that NATO's intervention in Kosovo stopped ethnic cleansing and that the promise of economic aid to Serbia after the war led the government to dump Milošević. It is now well known that NATO's intervention intensified the ethnic cleansing, and even General Wesley Clark has stated that the aim of this intervention was not to stop ethnic cleansing. Milošević's trial in The Hague, before he died, was not a model of respect for the law, either.[18] The famous Harvard professor does not do his academic work well: he accepts the official rhetoric of his own administration, to which he is loyal. In this context, one may wish to read Chomsky's piece entitled "Sovereignty and World Order" to assess who is more competent in history or political science.[19]

A bit further on, Nye writes: "the American presence is welcome in most of East Asia as a balance to rising Chinese power" (92). And then he warns: "If instead of the role of welcome balancer, the United States came to be seen as a threat, it would lose the influence that comes from providing military protection to balance others." Nye seems to be totally unaware that this military protection is imposed and therefore resented in many Asian countries. He talks about "efforts to promote democracy under the Reagan and Clinton administrations" (73), apparently without joking, probably in reference to El Salvador and Nicaragua of Iran-Contra fame (for Reagan) and to the Iraqi sanctions that led to thousands of deaths (for Clinton). His vocabulary betrays his prejudices: he talks about (77) "rioters confronting 'leaders'" in demonstrations against globalization, which he does not call "neoliberal," and he states that Americans will be able to take part in global Internet communities without ceasing to be "loyal Americans" (55). But loyal to whom or to what? This loyalty presupposes a national creed, and that is precisely what makes Nye's work narrow minded, hardly original, and incomplete (for him, opposition to McDonald's is anti-American). The famous, ideologically correct professor utters one platitude or falsehood after another, of the same type as those found in the media. His loyalty and his rather simplistic worldview are probably what facilitated his work for the Clinton administration as undersecretary of defense.

Writers such as Samuel Huntington ("The Lonely Superpower," *Foreign Affairs* [March–April 1999]) or Stanley Hoffmann ("Why Don't They Like Us?" *The American Prospect* [November 19, 2001]), who do not share Chomsky's views, could open Nye's eyes as far as the negative effects of his loyalty is concerned. His ideological bias stops Nye from understanding what he supposedly studies: the causes of negative images of the United States. Chalmers Johnson, who is not a Harvard professor, does this type of work much better,[20] as does the journalist William Pfaff, who writes: "The superiority of American political values and standards is taken to be self-evident."[21] Nye is a liberal whose objective is to ensure U.S. primacy thanks to *soft power*; he is absolutely sure that the United States is an open, democratic, progressive society that all other nations should emulate. This positive image, though not altogether wrong, is not counterbalanced by more negative aspects of American society. Nye wishes to appeal to the powerful, both within academia and in the world of politics, and he takes care not to antagonize them. He is a democrat and a liberal, but he praises Bush without ever making any legal or moral criticism of the president. Chomsky, on the other hand, following in the footsteps of Bertrand Russell, is committed to denouncing the misdeeds of people in his country or in the West. He is therefore not loyal to a particular group but to an ethics of truth.

As far as domestic issues are concerned, one could argue that Nye does not do his homework either. He states that in the U.S. voter participation has gone down to 50 percent, adding, "moreover, polls show that nonvoters are no more alienated or mistrustful of government than voters are" (122). His assertion comes from a newspaper article, and one can surmise that his belief that nonvoters are not alienated stops him from investigating the reasons why the citizens of a democracy do not wish to take part in the democratic life of their own land (in 2002, the voting rate for congressional elections was only 33 percent). Diplomas and the renown of the university where professors teach do not guarantee top-level research.[22]

Nye's ideological blindness and bias have their equivalents in France and other countries. Let us examine one specific example: Guillaume Parmentier responded to Robert Kagan's pamphlet "Power and Weakness"[23] in *Commentaire*, the French journal that published it in translation. He writes: "If one uses a tougher kind of vocabulary, which however expresses reality, can one say that the United States in 2002 has become more nationalistic? It would be somewhat paradoxical for the nation that has tried and succeeded beyond all hope in vaccinating Europe against the disease of nationalism to be contaminated by it."[24] How problematic this statement is! This specialist of international relations seems to think the United States stopped being nationalistic after World War II—where has he been for the last sixty years? He does not understand that fighting against the nationalism of others is itself a form of nationalism and has been a key feature of U.S. foreign policy since 1945, as Paul Wolfowitz et al. confirmed in the *Defense Policy Guidance* document in 1992.

Two other French researchers, Pierre Hassner and Justin Vaïsse, for their part, mention Chomsky in a fine scholarly book published in 2003.[25] This probably reflects the revival of interest in Chomsky's views after 9/11. They include an extract from his book *9/11* but err when they write: "This radical left-wing critique often may end up being a kind of isolationism, for every interaction between the US and the rest of the world is considered imperialistic or immoral: foreign adventures often leading to curtailment of liberties at home" (35). Well, for the curtailment of civil liberties the jury is in. The charge of isolationism does not really apply to Chomsky, a committed internationalist who does not distinguish between American and non-American immorality. On this count, Chomsky is the opposite of Raymond Aron,[26] who considered that morals and politics should not be confused, or of Machiavelli, who advised the Prince to resort to ruse and bracket moral considerations. As an "antipundit" scholar, Chomsky does not try to be a new Kissinger and influence the Prince; rather, he wishes to enlighten the general public.[27]

Ideas and Academic Structures

Appointing colleagues in higher education is a peer-reviewed process that is supposed to select the best. Historians choose their colleagues in accordance with some professional criteria, as do sociologists or linguists. This process, intended to wean out incompetents and amateurs, is often plagued by nepotism or ideological moves presented, of course, as the desire to appoint the very best. In a competitive market, such as the U.S. academic market, careerism and conformism go hand in hand; campus rhetoric and professional positions are often calculated in relation to tenure possibilities.

Each department or program has ideological preferences. Those of a women's studies department, for instance, will greatly differ from those of a business department. Cultural studies as a discipline has adopted different rhetorics and methodologies from history. Each academic discipline determines who will be appointed, shapes the legitimate discourses, and tries to prevent deviations from the dominant discourse within the field. In a sense, university departments create official disciplines by legitimizing some discourses and invalidating others. Each department and each discipline has a kind of border patrol in its area, which sometimes poses problems, as for instance in France, when a few sociologists legitimized the work of an astrologer who got a Ph.D.

without ever giving up the unscientific language of astrology.[28] Generally, administrative categories, personalities, and power struggles determine what is considered legitimate. A text of "genius" in a literature department might be deemed worthless in a sociology department. The methodologies of economics might appear quite naïve to psychoanalysts, and psychoanalysis itself is considered unscientific in many medical schools. Knowledge is made up of many little bits; legitimacy is often local. In the United States, historians and cultural-studies specialists often have major conflicts or disagreements. Sometimes the foundations of a discipline are challenged or utterly rejected by another discipline. *Homo economicus*, for instance, is not welcome in sociology departments.

Chomsky's work is outside the realm of such academic disputes and crosses disciplinary borders, something that cannot fail to displease the border patrols in various disciplines. Even some linguists try to eject Chomsky from their discipline, which is somewhat difficult, but such a move is much easier for historians or political scientists, who claim to be speaking on the basis of their professional competence. Chomsky's political thought is, of course, informed by his reading of many sources, most of them scholarly. He quotes his sources and offers interpretations that often go against the grain. The rhetoric of quality or professional training that is used not to take his work into account is often merely a pretext, for in his scholarly production he respects scholarly norms and standards, and his knowledge of philosophy and history is impressive. Everyone is entitled not to agree with Chomsky on this or that interpretation or on his general orientation, but in political-science departments in the United States, he is hardly discussed.

Chomsky's numerous articles on Vietnam, the Middle East, or U.S. foreign policy are, in Popper's terminology, falsifiable. Facts and sources are mentioned, so it should be easy to point out errors or misinterpretations. Yet, apart from the case of Cambodia, Chomsky is not discussed and refuted. Outside the United States, Chomsky figures in history and political-science syllabuses, and that includes Britain or Ireland. He is often invited by students or professors to give talks on various campuses. His book entitled *9/11*[29] has become a bestseller, but as far as academia is concerned, Chomsky remains marginal in the departments where he would be most relevant, such as political science. Edward Said, who before his death in 2003 taught literature at Columbia University in New York, was in a similar situation: he was an acknowledged thinker in his field but was not accepted as a political thinker. In the field of American studies, however, neither Said nor Chomsky are treated as pariahs.

Organized groups do not like free thinkers or mavericks, and recruitment committees in universities behave in exactly the same way. Appointments and official recognition, with few exceptions, go to men and women trained within the establishment. Chomsky dares to write about history or politics even though he was not officially trained in these disciplines. He is therefore an outsider, an "interloper" who has often read more extensively in the fields of history and politics than the average professor and is therefore embarrassing. For example, his knowledge of Adam Smith's writings is superior to that of the many liberals or conservatives who claim Smith as a mentor.[30] In other words, as a historian or a political scientist, he, to use Lacan's terminology, authorizes himself (*ne s'autorise que de lui-même*) and encourages other people to do the same. He goes even further: he thinks professors are often less clairvoyant and more under the spell of ideology than high-school students. Thus Chomsky is the quintessence of provocation for political scientists. Thinkers such as Lewis Mumford or Philippe Ariès also came from outside the boundaries of academic disciplines but were, in the end, accepted by their colleagues because they did not rock the boat of academic departments or disciplines. Chomsky both encroaches upon the territory of historians and political scientists and challenges their dependence upon the powers that be, their methodologies, and their blind spots. Besides his voluminous work in linguistics, he also produces much challenging work in political science. This no doubt irritates professional practitioners who may be protected by tenure but realize they are intellectually outclassed by an outsider.

As was the case for Pierre Bourdieu in France before his death in 2002, Chomsky has reached the very top in terms of academic recognition. In what Bourdieu called *la noblesse d'Etat* (state nobility), Chomsky is at the highest point: he is a university professor in linguistics in one of the most prestigious institutions in the United States and the world, the Massachusetts Institute of Technology, and a towering presence in his scientific field. This is the U.S. equivalent of Bourdieu, who went to the Ecole Normale Supérieure and taught at the Collège de France. Like Bourdieu, Chomsky refuses to play the game usually played by academic aristocrats: instead of praising the elite institutions that certify his competence and his individual genius, he criticizes the foundations of such institutions and takes part in public and street debates. Both Bourdieu and Chomsky are the very opposites of the *normaliens* who boast of their Ecole Normale Supérieure years all through their lives or of Harvard professors who use their professional affiliation as an unchallengeable label of intellectual worth. These two intellectuals deconstruct the conditions of production of specific discourses in elite institutions and show up the gaps in supposedly authoritative discourses.

Their social and political critique owes its strength, in part at least, to the fact that, as far as academic fame is concerned, they cannot be accused of being envious. Their dominant position in a given field gives them (or gave, in Bourdieu's case) an advantage and a freedom that "dominated" thinkers in the same fields do not enjoy. A journalist or artist who would formulate the same social and political critique would be accused of being envious: "he or she is jealous because he or she did not make the grade in academia." To top it all, Chomsky, like Bourdieu in the past, sometimes may find the rhetoric of some people in marginalized groups more relevant or better informed than the spin of pundits or experts. In Chomsky's case, the style in his political writings is easily accessible and therefore suspect in the eyes of academic aristocrats. This easy access does not stop Chomsky from explaining complex phenomena, and it often requires much historical or political knowledge about the topics at hand.

Chomsky is very tough with his academic colleagues, especially those that during the Vietnam War he called "the new Mandarins," because they chose to support power and think about their careers at a time when the United States was committing crimes in Asia. Political science is often in cahoots with power: professors go from Harvard or Stanford to the team of advisers of this or that president. Too sharp a critique of the powers that be would hinder their careers. Often in the United States these professors, past or future advisors to the Prince, share the ideological orientation of political leaders and can only be critical of their own country and its basic creed or administration within fairly strict limits. Chomsky points out those limits and denounces the lies of leaders and their supporters within academia (for example, Arthur Schlesinger Jr., in *American Power and the New Mandarins*).[31] He does so thanks to a tenured position at MIT and his personal fame. In effect, in academia as well as in the mainstream media, legitimate or loyal critique remains shallow, partial, and focused on individuals rather than on the system in general. Large sections of academia, under the influence of cultural studies or postmodernism, have adopted the language and analytical categories of marketing to express their critiques of the media or the political system. The singer Madonna thus becomes a symbol of political opposition or resistance. Thus "politics" leaves the "polis" and becomes a matter of personal preference.[32]

The posture of what one could call the "conformist rebel" or the "career dissident" is very fashionable in social-science and literature departments, where antiestablishment discourses are the norm among highly conventional academics caught in the usual rat race for positions, publications, and local privileges. The game here is to conform to an oppositional discourse deployed by those with local power. This dissidence is a pose; it is only discursive and hardly leaves the confines of the campus. Many of the "conformist rebels," who are supporters of multiculturalism and therefore supposedly in favor of cultural diversity and respect for the Other, are critics of Chomsky and what they call the "Chomsky left." Many supported the war in Afghanistan in 2001 and the so-called war on terror in its initial stages, commitments that hardly amount to "respect for the Other."[33] In politics, one talks of "limousine liberals." They have academic counterparts: professional denouncers whose rhetorical flamboyance is a career move and a ticket to publication in mainstream media. The Sokal affair revealed this split between a left of rhetorical flourishes and a left steeped in concrete issues.[34] Some pay the price for their dissidence; others play at being dissidents to advance their careers. Yet no one can blame a colleague who chooses conformity to avoid losing a job or a chance to be tenured. In brief, appointment procedures within universities have an impact varying from discipline to discipline, and greater conformity exists precisely in the disciplines that are the most politically relevant.[35]

Let us take a few specific examples. Chomsky thinks the United States achieved partial victory in Vietnam. In Chomsky's words:

> The destruction of Indochina ensured that it would not provide a model that others might follow; it would not be a "virus" that might "infect others," in the terms preferred by the planners. And the establishment of brutal and murderous military dictatorships in Indonesia, Thailand, the Philippines, and elsewhere ensured that "the rot would not spread." These too are considerable victories, enhanced by the U.S. stranglehold to prevent recovery since and U.S. support for Pol Pot, via Thailand, to ensure the more efficient "bleeding of Vietnam."[36]

Chomsky was one of the leading opponents of the war in Vietnam and does not minimize the influence of U.S. public opinion upon the administration, but his point that after 1968 big business wanted the United States to withdraw from a costly war is hardly ever mentioned in scholarly history books. Thus a French historian who is a well-known specialist about the United States does not mention the man the *New York Times* called "the most important intellectual alive" even once.[37] In historical research, it would seem appropriate to mention a view entertained by one of the most important actors in the protest movement, one whose scholarly erudition is staggering. This view could have been discussed, rejected, or parsed, but it quite simply has been silenced to this day. McNamara or Michael Lind's books have received more media and scholarly attention than Chomsky's articles published from the 1960s onward. Excluded from legitimate scholarly discussion, Chomsky reappears in public debates. His ideas are not totally absent in the academic fortress, but they have a kind of underground, *samizdat* status. Historical research is the poorer for it.

Chomsky's research on the Middle East, presented in *The Fateful Triangle*,[38] is also marginalized in U.S. political-science departments, probably because he criticizes the United States for its refusal to exert pressure upon Israel to find a fair settlement of the Israeli-Palestinian conflict. He rejects the idea that the pro-Israeli lobby is all powerful, something that has been hotly contested after Walt and Mearsheimer's essay published in 2006, and he offers ideas that could be used to fight anti-Semitism, but, once again, his interpretations do not cross over into university or mainstream-media discussions. His range of sources is remarkable, as is the masterful analysis of very convoluted and complex relationships between the various actors of the conflict. There are some similarities between his work and that of other researchers in international relations, such as Alain Joxe, for instance (*L'Empire du Chaos*),[39] but in scholarly discussions the same shallow questions as in the media are asked: Did Clinton do better than Reagan or Bush I? Why don't Palestinians grab the opportunities offered by Barak or Sharon or the United States? It is quite clear that discussing Chomsky's views, as well as those of other similar scholars, would foster historical research and the search for truth. Arguing that Chomsky is not a *bona fide* professional historian is a silencing move both in academia and the media.[40]

"Simplistic Views" and Other Smears

Chomsky is often blamed, as far as his political writings are concerned, for his allegedly simplistic views and for quoting too many journalistic sources; he is said to be more of a reporter than a scholar. His style is alleged to be too simple and easy to understand and thus not scientific or scholarly. One encounters here several typical academic categories. Chomsky's simplistic views are said to be related to his style, which is indeed easy to understand. A specialist needs specific concepts in his or her field; Chomsky does use such concepts or, as the case may be, deconstructs the ideological work they do. He has often made fun of "Parisian follies," that is, the use of abstruse language, and in Paris some intellectuals resent him for this. Yet there might also be a misunderstanding about his work. Chomsky does not wish to be acknowledged for his literary style; he is not a "literary philosopher" like Foucault or Derrida.[41] He is not a postmodern thinker. He believes in truth and in the possibility of revealing it. Of course, this causes some deconstructionists to snigger, but is closer to the methodology of hard science or the philosophy of Bertrand Russell and Bernard Williams,[42] among others including Orwell. Yet it is necessary to really understand his use of multiple points of view so as to not misinterpret Chomsky's style. Quite often Chomsky uses the vocabulary or arguments of those he criticizes to demonstrate their limitations or inanity. Some of his readers therefore think—or claim to think—that the vocabulary or style used ironically by Chomsky is his own.[43] Some political or philosophical texts in France are written in very elaborate, stylish, or poetic ways but also include factual errors and therefore can be refuted or falsified. Style cannot redeem a bad social-science article. The charge of oversimplification is often leveled at texts that have a pedagogical intent and that target various nonacademic audiences. Very often Chomsky repeats himself in different contexts, which can seem tedious to somebody reading many of his writings or the transcripts of talks. This is the fate of many academics frequently invited to talk to the general public about their work.

Blaming Chomsky for his use of journalistic sources is rather ironic when it comes from thinkers whose depth of reading is minimal compared to

Chomsky's. Chomsky, of course, also quotes books and reference works dealing with the topic at hand. He makes use of newspaper articles and the numerous cuttings friends or colleagues the world over send to him. Newspapers are a source of information when one knows how to read them. When he quotes the *Wall Street Journal*, he does not quote friends but, on the contrary, is referring to the state of thinking in the business world at a given time about a specific issue. Quite often he quotes the actors of the particular issue he is dealing with who publish articles in the press, and through meticulous cross referencing, he shows up the contradictions and what, in a Derridian fashion, one could call the aporias in these texts. Newspapers are an entry point into a discourse, but they are certainly not the only source of information for an avid reader like Chomsky.

Since mainstream media grant him very little space, even if the *New York Times* or *Le Monde* occasionally publish something by or about him, it is not surprising that he often chooses to write for the marginal or dissident press. (Although it is intellectually dishonest and difficult to disparage such a prominent linguist as Chomsky by calling him merely a journalist outside of the mainstream, this smear can be used against his supporters or against writers who develop similar ideas.) Of course, Chomsky's marginalization in the academic and intellectual world is far from total. Many academics in linguistics, philosophy, sociology, anthropology, media studies, and even political science take an interest in his work, use it, or get inspiration from it in their own work, even when they distance themselves from it. This small group of academics is itself criticized, and those who take inspiration from his work are said to worship without any critical distance a master or guru. This may well be true in some cases, but is Chomsky responsible for the flaws and mistakes of his disciples or worshipers? Anyone can provoke transference, and there is no doubt that Chomsky impresses many who meet him. As a man of conviction, great intelligence, and wide erudition in numerous fields, he of course causes transference. Some worshippers become zealots and betray his ideas, whether consciously or not. This has happened to major thinkers from Jesus to Marx. Chomsky is impressive but does not actively encourage transference, as some famous psychoanalysts in France have done—Lacan among them—or some flamboyant English professors in the United States, such as Stanley Fish. Freud, Foucault, or Derrida also caused massive transference. Chomsky always speaks calmly and does not keep silent in a Lacanian psychoanalytical fashion aiming at encouraging transference, and he does not hesitate to criticize or object to what he considers erroneous or dishonest in the discourse of others. His tough exchanges with Pierre Vidal-Naquet, whose views on anticolonialism and criticisms of Israeli policies were close to his own, show that he does not make concessions in political debates. A meeting with Chomsky can only be based on discourse and the search for truth. Chomsky is very demanding on this score; he does not yield in his determination to speak the truth (digging up the relevant facts), prompting some to say that he is rigid or naïve, or both. This determined search for truth explains in part why he is often the target of hostility. Freud before him also suffered personal denigration on account of his ideas, and he also chose his own ethics of truth rather than popular appeal.[44]

These criticisms amount to a refusal to tackle ideas and arguments on their merits. However naïve, inflexible, or unpleasant a writer may be, what matters is his or her ideas and not his or her psychology. The psychological approach to Chomsky's work is also an avoidance strategy designed to sidestep a discussion of ideas. Chomsky denounces a few examples of this in some of his responses to attacks.[45] Catherine Clément, for instance, accuses him of not understanding Jewish-American humor, which comes from Yiddish. Chomsky may not have a good sense of humor, but one has to know the private man to be sure of this. At any rate, is that really what should interest intellectuals reading political or historical texts? As Edward Said argues: "Least of all should an intellectual be there to make his or her audiences feel good: the whole point is to be embarrassing, contrary, even unpleasant."[46] Chomsky resorts less to humor than to irony, as an explicit reference to Jonathan Swift makes clear.[47] To discredit Chomsky, some critics resort to the sensationalism of the yellow press. They criticize the way Chomsky's ideas are presented in minor publications, without engaging in any discussion of these ideas. For outclassed academics or intellectuals, sometimes insult takes the place of debate. Such a pirouette is an evasion.

Chomsky is also often blamed for being a rationalist who thus has missed all the intellectual innovations of the twentieth century, notably psychoanalysis and postmodernism. It is quite accurate that Chomsky does not use Freudian concepts—Lacanian ones even less so. He is unimpressed by postmodernism, which does not figure in his scholarly work in political science or history. The Vietnam War can be tackled without analyzing Kissinger's or Johnson's unconscious (no doubt something that would interest a psychoanalyst). There is no need for Baudrillard's glitzy vocabulary to deal with Nicaragua or the Gulf War. For Chomsky, the Gulf War really *did* take place, people died in what he called "the Gulf Slaughter," and the video-game presentations of this war were an aspect of political propaganda.[48] There is indeed a divide among academics and intellectuals between those who think language enables access to truth, even if it is only partial truth, and those for whom everything is narrative and fiction. For Chomsky, the concepts of literary analysis cannot always and often should not be applied to political thinking. Understanding the words of a character in a novel, that is, a work of fiction that presents itself as such, differs significantly from understanding a political issue such as, for instance, the fraud organized by Republicans around the 2000 election, in which cronyism and violations of the law enabled their victory.

In the United States far more than in France, Chomsky is often presented as a Marxist or neo-Marxist. This is a particularly dishonest charge, as he has repeatedly stated that this is not the case. In fact, his critique of Leninism and Stalinism is severe.[49] He is not an anti-Marxist either. His judgment of political regimes claiming to be Marxist is complex: he criticizes these regimes and at the same time views the U.S. position in relation to them as dictated by the desire to eliminate any alternative to the dominant ideological system in the United States or in the West. Chomsky quotes Marx, but he prefers Rosa Luxemburg and Anton Pannekoek to Lenin, whom he ranks among totalitarians; he also quotes Bakunin against Marx. Ultimately, however, Chomsky is interested in ideas rather than idols or doctrines that have become rigid systems of thought. Chomsky's non-Marxism puts him in an awkward position among thinkers who claim to be followers of Marx; this applies even in the United States. Once again he is an outsider, and again this is somewhat similar to Bourdieu's position in France, for even though Bourdieu derived inspiration from Marx, he was not a follower or a disciple. He was vilified as the "last Marxist" so as to link him with the atrocities of Soviet regimes. All the opponents of the Vietnam War were accused of playing into the hands of totalitarians because they followed Marx. Some did indeed become totalitarian, but with the focus on them, the issue of the morality of the war was forgotten. In the United States, the charge of Marxism is leveled at those who talk of poverty or redistribution of wealth even when the targeted people have no connection with totalitarian regimes. This is reminiscent of McCarthyism and other witch hunts. Even the Catholic Church in the United States was accused of Marxism when it took a nonapproved stand over Latin America in the 1980s. In the academic world, Chomsky's non-Marxism did not make him more acceptable by the right (which, contrary to a widespread cliché, is quite strong on campuses, especially in law and business departments), nor did it endear him to the Marxist left.

Beyond Manichaeanism

In a Manichaean system, whoever is not with us is against us, as George W. Bush famously said about Islamic terrorism. Whoever does not side with the U.S. administration is a terrorist, a communist, or a Marxist. This label, "Marxist," is damning in the United States, and it enables those who use it to misrepresent the arguments made by the targeted thinkers. Chomsky has repeatedly criticized the so-called Marxist dictatorships, such as the Soviet Union, and has also attacked various political groups calling themselves "Marxist." He also criticized the PLO and Arafat, terrorist groups, and those responsible for crimes against humanity who make warmongering extremism in the West easier. He called the 9/11 attacks "crimes against humanity."[50] Manichean labeling saves the labelers the difficulty of dealing with complex positions. Academics may not be as brutal and blunt as Bush, but they do tend to put those they disagree with in a kind of infamous ragbag. After 9/11, this narrow-minded Manichaeanism led to unfounded accusations, for instance that Chomsky, like a lot of other left wingers, allegedly "blamed America first" and supported the terrorists, a blatant falsehood, as anyone who has read Chomsky's texts would know. By definition, Manichaeanism reduces everything to two opposite positions and as such should be banned in academic discourse. Academics should be aware of the various levels of interpretation and of the many facets of any political issue. Yet Manichaeanism is alive and kicking in the media as well as in academia. So although Chomsky denounced the "crimes against humanity" committed by the 9/11 attackers, he was presented as a supporter of terrorism because of his alleged hatred of America. Yet Chomsky never tires of saying the United States is the freest country in the world, meaning it has the best free-speech laws in the world—hardly a sign of hatred for his country.[51]

The other common charge against Chomsky is that he is "anti-American" or "un-American." He has answered on numerous occasions saying, for instance, "People who criticized Russian foreign policy weren't anti-Russian, they were pro-Russian. If you're criticizing U.S. crimes you're not anti-American, you're pro-American."[52] This discrediting accusation is sometimes leveled by people who are called leftists or left liberals like Gitlin[53] in the United States or Todd[54] in France and takes the place of a reasoned discussion of the arguments put forward. This is nothing new and is not specific to Chomsky; ideological character assassinations have always been common in intellectual life. In the 1950s among Stalinists in France or anticommunists in the United States it was even more common. After 9/11, a major split about the best way of fighting terrorism occurred within the left, between what some called the "Chomsky left" and the "liberal hawks." Chomsky is often considered as the leading figure of the so-called radical left or what the French call the "ultra left," which is supposedly blinded by its anti-Americanism and anti-imperialism and therefore makes common cause with the far right or even terrorists, called "islamofascists" by so-called liberal Paul Berman.[55]

Ad hominem attacks and ideological smears are nothing new in academia or in the public sphere, yet the choice of Chomsky as favorite whipping boy requires analysis. Chomsky is a prolific writer and frequent speaker on matters of foreign policy. He travels worldwide and is interviewed by non-American media outlets almost everywhere in the world. Debating him is a difficult task, as he has in-depth knowledge of events and publications and a fantastic memory. He can point out the contradictions between statements made by the same speaker decades apart; he can quote the writings and speeches of the Founding Fathers of the United States or of Israeli or British leaders at different historical moments. His "databanks," made up in part by a network of supporters and friends who send him articles, boggle the mind.

Few academics or reporters can compete with Chomsky as far as factual knowledge is concerned. In interviews he does not try to appeal to the interviewer or to the audience, he is not in cahoots with anyone, and he corrects factual mistakes of friends as well as unfriendly interviewers. What matters for him is not the identity of his interlocutor but his or her ideas. Very few top-level intellectual have exchanges with hundreds of ordinary people who are neither rich nor famous. Chomsky, the ultimate universalist, criticizes all the groups he belongs to: he is an American who is severe with the policies of the United States, a Jew who is critical of Israel and of its unconditional supporters in the United States, and a Westerner who denounces Western imperialism. But he is neither ashamed of being an American or a Jew or a Westerner. He dislikes Islamic fundamentalism as well as Christian or Jewish fundamentalism. He praises the diversity of public debate in Israel and denounces anti-Semitism. He praises freedom of speech in the United States and criticizes France for failing to understand what it amounts to. Like Edward Said, he denounced Arafat's corruption (though obviously not in the same way as did Sharon).

Chomsky is not easily put into a slot: his views on Palestine are not determined by ethnic considerations or a pro-Palestinian bias; his critique of U.S. imperialism is not made of a belief in communism or based upon admiration for a nation other than the United States. Thus, though he is no supporter of Fidel Castro he is a severe critic of U.S. policy toward Cuba. He is said to be a radical leftist, but some of his positive comments about the United States surely cannot appeal to xenophobic anti-Americans who claim to be leftists. This "left-winger" rejects the traditional divide between left and right; he sometimes quotes Pope John Paul II when he opposed war in Iraq or Latin America and sometimes praised the positions of the Catholic Church but, of course, he is not Catholic. All this is very simple: Chomsky is not Manichaean. He refers to the work of Amnesty International, the ACLU, and Human Rights Watch, all organizations in good standing among liberals (except when, as in the case of Human Rights Watch in 2006, they criticize Israel), but he may also criticize some aspects of their thinking and not share all their points of view. He often says positive things about the *Wall Street Journal*, a paper that he also sometimes calls "vulgar Marxist." He is neither faithful nor hostile to people but to ideas, so he may quote favorably someone he has also severely criticized (Arthur Schlesinger Jr., for instance). He is a scholar but also an intellectual, a public intellectual, though he rejects the label.[56]

Chomsky is the living proof that a determined intellectual can teach him or herself university disciplines by individual effort. Besides ideological opposition, this attitude toward intellectual life may explain some of the hostility targeting him. Others—for example, Tzvetan Todorov in a less directly political way—do the same thing and thus show that administrative boundaries between disciplines can be overcome or ignored. Of course, not every one shares Todorov's or Chomsky's genius or energy (he says he works about a hundred hours a week), and few people can read as much as he does, so multidisciplinary thinkers like him are rather rare. Yet such a thinker is a threat to all academic fortresses.

Chomsky hardly ever talks about television, although it is one of the major media used for indoctrination; however, his media analysis applies equally well to the mainstream press as to TV. He is sometimes said to be a moralist, which may mean that he believes in an ethics of truth and that he views his role as an intellectual in a traditional—that is, not postmodern—way: speaking truth to power, denouncing the lies of the powerful, and deconstructing media spin. Yet Chomsky rejects the idea of "speaking truth to power" because, in his words, "To speak truth to power is not a particularly honorable vocation. One should seek out an audience that matters—and furthermore (another important qualification), it should not be seen as an audience, but as a community of common concern in which one hopes to participate constructively. We should not be speaking *to*, but *with*."[57] An intellectual's job therefore is simple: to tell the truth as all decent people do. The intellectual must be active where his or her actions are likely to have an effect. That is why he or she must criticize mostly his or her country and not its enemies, as criticism in the latter case is both too easy and useless.

Chomsky's ethical approach leads him to refuse to reduce political or ethical issues to purely technical matters, something that sets him apart from many social or hard scientists. He is definitely not a "spin meister" whose job is to create positive impressions or images about a product or a policy. Chomsky, on the contrary, unwraps the packaging and points out the flaws of the product. As a scholar or an intellectual Chomsky has not followed the "cool," fashionable marketing approaches that privilege form over content, aesthetics over truth. That is why he is sometimes called "haughty." Once again, whether he is called "simplistic," "haughty," or "hysterical," insults have the same function: killing the messenger so as not to talk about the message. There are hybrid forms of this silencing technique: first a demeaning remark (A), then a distortion of his ideas (B), then a debate about the slanderous distorted idea (C). For instance: calling him hysterical (A) and someone who rejoices over the Twin Towers attacks (B), and thus someone who must be excluded from legitimate debate (C).[58]

Chomsky is not a poet, and his work is not literary. He is not read because of a "literary need"[59] by people who wish to dream or to embark upon an imaginary quest. His constant denunciations of crimes and misdemeanors, poverty and exploitation repel those who suffer from "compassion fatigue." Chomsky is neither a cheerful or pleasant writer nor an agreeable interviewee; he does not try to look good on TV and no one comments upon his shirt or his eyes or how big his smile is. These things have absolutely no intellectual relevance. He is very different from media stars or frequent talk-show guests who use their clothes and cool vocabulary to appeal to their audiences. Thus Chomsky is unfashionable; his characteristic mode of intervention is miles apart from the mush served by TV.

On campuses, Chomsky appeals to many students and some professors who welcome his dissident positions and do not put their careers at the center of their lives. For the academic establishment, he is a threat on account of his ideas, of course, but also because of his high cultural and scholarly capital and his extraordinary intellectual ability. For less gifted or more cautious scholars, his brilliance can only be fought with slander and marginalization. It is no surprise that political-science departments are the least welcoming of his ideas. Yet Chomsky enjoys a freedom of speech equaled only by major artists or novelists who, thanks to their fame, can free themselves of the administrative constraints or threats that weigh heavily upon those whose job or career depends upon the goodwill of higher-ups. Chomsky uses his fame to fight for his ideas, to oppose the powers that be with the power of truth. His ethics of truth is stronger than his ideological preferences and acts as a judge of careerists and opportunists. Away from the academic rat race, Chomsky the avid reader and free and original mind makes other academics insecure, especially those who have to yet to gain tenure and must accept conformity in order to be competitive. Still other scholars are stimulated by this intellectual maverick who does not mince his words. His fate within academia or the public sphere is not that different from Spinoza's in his time or any other dissident's in democratic societies. Indeed, compared to Norman Finkelstein, who has not been able to hold jobs or obtain tenure, Chomsky has been protected by his tenure and stature. Marginality goes with the territory, something that Chomsky is well aware of, but marginality also prepares the ground for future revisions and gives truth a chance.

The price to be paid when anyone crosses the line into serious dissent is steep. For example, Craig Murray, the former UK ambassador to Uzbekistan, was a typical and consistent liberal who took exception to Britain following the United States in its so-called War on Terror while supporting the Karimov regime, which kills and tortures dissidents. Murray strongly objected to Britain accepting information obtained under torture. Consequently, he was harassed, slandered, attacked physically and intellectually, and fired.[60] In other words, he is a decent person who objected to a vile and murderous system. He comes from a different intellectual tradition from Chomsky, but he made similar comments about the lies and deceptions of the leaders in power and had to be punished. This is the price of dangerous dissent: utter marginalization, silencing, and personal attacks. By remaining an intellectual force despite all this, by remaining outside the academic rat race and refusing to work for the powers that be, Chomsky is exceptional; he is a level-headed thinker who should be taken seriously and read in the same way Chomsky advocates the reading of Marx: not as a master who knows it all but as someone who makes one's personal thinking richer.

NOTES

1. Typical of the techniques used to discredit him is an article by Jeffrey Isaacs, "Thus Spake Noam," *The American Prospect* (October 16, 2001), http://www.prospect.org/print-friendly/web features/2001/10/isaac-j-10-16.html. The magazine gave Chomsky the right to respond on October 22, 2001 ("A Response"): http://www.prospect.org/webfeatures/2001/10/chomsky-n-10-22.html. Also of interest is the debate between Hitchens and Chomsky in *The Nation* (starting October 8, 2001). In an article devoted to literature and in particular to V. S. Naipaul's Nobel Prize entitled "Literature: The Humor and the Pity," *The American Prospect* (January 28, 2002), the writer Amitava Kumar links Chomsky to Eqbal Ahmad, who allegedly tried to kidnap Kissinger. This is the technique of guilt by association. The writer uses the word "comrade" to suggest a relationship between somebody who is presented as a criminal and Chomsky. Recently (September 14, 2007, on the radio station *France Inter*), Phillippe Val used the fact that bin Laden quoted Chomsky's name to link him to support for terrorism, advocacy of genocide in Cambodia, and so-called Jewish anti-Semitism, and he accused him of having no sympathy for 9/11 victims. These are all lies and slander, as anyone who reads or listens to Chomsky must know. Val's broadcast can be heard at http://media.la-bas.org/mp3/Val.mp3. For a relevant analysis of these techniques in the French context, see Jean Bricmont, "D'une mauvaise réputation . . . Lire Noam Chomsky en France," the preface to Noam Chomsky, *De la guerre comme politique étrangère des Etats-Unis* (Marseille: Agone, 2001); and "Folies et raisons d'un processus de dénigrement," the afterword to the second edition (2002). In Britain, a book analyzes the way left-wing dissidents are smeared: Scott Lucas, *The Betrayal of Dissent: Beyond Orwell, Hitchens, and the New American Century* (London, Pluto Press, 2004).

2. "From Aristotle to Adam Smith: What Is Left and What Is Right?" Interview at MIT, October 31, 1997. Published in *Sources* 4 (April 1998).

3. In "The Israel Lobby and U.S. Foreign Policy." The longer Kennedy School of Government version is online at http://ksgnotes1.harvard.edu/Research/wpaper.nsf/rwp/RWP06-011. Their recent book, *The Israel Lobby and U.S. Foreign Policy* (New York: Farrar, Straus and Giroux, 2007), is a scholarly study. Their views differ significantly from Chomsky's, but, after the publication of their article on the same subject in the *London Review of Books* and the Kennedy School of Government Web site in 2006, they were submitted to a similar smear campaign by the same sort of people as Chomsky was (for instance, Alan Dershowitz, whose scholarship is atrocious or nonexistent but who makes a specialty of attacking real scholars like Walt, Mearsheimer, or Finkelstein).

4. "End of the Revolution" (February 28, 2002), after the publication of Noam Chomsky, *New Horizons in the Study of Language and Mind* (Cambridge: Cambridge University Press, 2000). There was then an exchange of letters.

5. This is a reference to her public statements in the "Petit Grégory" affair, when she expressed her conviction that the mother had killed her child. Courts found the mother innocent.

6. Many articles, interviews, and excerpts from books can be found on the Web site of the magazine Z: http://www.zmag.org/chomsky/index.cfm. One can find, for instance, a February 16, 1970, talk on political philosophy, given at the Poetry Center of New York, which clearly shows Chomsky's competence in this field. See also Noam Chomsky, *World Orders Old and New* (New York: Columbia University Press, 1994). This book is a collection of talks delivered by Chomsky at Cairo University.

7. See Daniel Pipes, "Profs Who Hate America," *New York Post* (November 12, 2002).

8. "Eagle Rules? Foreign Policy and American Primacy in the Twenty-First Century" (May 29, 2001).

9. On October 3, 2006, Professor Tony Judt of New York University was scheduled to give a lecture titled "The Israel Lobby and U.S. Foreign Policy" at the Polish consulate in New York. He was prevented from doing so. See his article "Israel: The Alternative," *New York Review of Books* (October 23, 2003). He recently wrote, in an article entitled "Bush's Useful Idiots" in the *London Review of Books* (September 21, 2006): "In today's America, neoconservatives generate brutish policies for which liberals provide the ethical fig leaf. There is no other difference between them." On this he agrees with Chomsky.

10. As Edward Said writes: "There is little refutation offered his arguments; just the statement that he stands outside acceptable debate or consensus." *Representations of the Intellectual* (New York, Pantheon Books, 1994), 80. An example among many: James W. Ceaser, a professor of political science at the University of Virginia, published *Reconstructing America: The Symbol of America in Modern Thought* (New Haven, Conn.: Yale University Press, 1997), whose aim was the analysis of negative images of the United States so as to correct them. There was not a word about Chomsky, who is often accused of being the main American provider of such negative images. Silence in this case is akin to a professional mistake.

11. Niall Ferguson, *Colossus: The Rise and Fall of the American Empire* (London: Penguin Books, 2004). In a recent book, Amy Chua does the same thing: she ridicules Chomsky when she talks about the "antiglobalization movement" and calls him "one of the movement's high priests" who calls for more democracy and thinks that democracy is a panacea. But she does not quote him, she does not discuss his ideas about democracy—which are not as simplistic as she implies—and she does not include his name in her index. This censorship by mockery. Amy Chua, *World on Fire, How Exporting Free-Market Democracy Breeds Ethnic Hatred and Global Instability* (London: Heinemann, 2003), 12–13.

12. Ferguson could have found food for thought in some of Chomsky's works: "The U.S. Role in the Middle East," in *Pirates and Emperors, International Terrorism in the Real World* (New York: Black Rose Books, 1987), 151–174. He could also have found an analysis of the marginalizing technique he uses. A British publication, *Index on Censorship* (July–August 1986) had the gall to publish Chomsky. Elliott Abrams of the U.S. State Department had sent a letter to this journal to protest Chomsky's publication—Abrams wanted a journal devoted to the fight against censorship to censor one of its authors. In effect, Abrams, like Ferguson twenty years later, argued both that Chomsky should not be taken seriously and that he should be censored.

13. *Public Intellectuals: A Study of Decline* (Cambridge, Mass.: Harvard University Press, 2001).

14. For a comprehensive analysis see; Diane Johnstone, *Fools' Crusade, Yugoslavia, NATO, and Western Delusions* (London: Pluto, 2002). For reference, see Noam Chomsky, *A New Generation Draws the Line: Kosovo, East Timor, and the Standards of the West* (London: Verso, 2001). A typical example of Chomsky's style and his use of quotations: "It takes considerable effort not to recognize the accuracy of the report of the UN commission on Human Rights in March 2000 by former Czech dissident Jiri Dienstbier, now UN special investigator for the former Yugoslavia: 'The bombing hasn't solved any problems,' he reported: 'It only multiplied the existing problems and created new ones'" (41).

15. Thus in *Power and Terror, Post-9/11 Talks and Interviews* (New York, Seven Stories Press, 2003), 85: "One of the advantages of living here is that the United States has become, over the years, a very free country. Not as a gift from the gods, but, as the result of plenty of popular struggle, it's become an unusually free country, uniquely so in some respects."

16. See *American Power and the New Mandarins* (New York: Pantheon Books, 1969).

17. Joseph Nye, *The Paradox of American Power: Why the World's Only Superpower Can't Go It Alone* (New York, Oxford University Press, 2002).

18. See Diane Johnstone, "Great Power Meddling in Kosovo," *CounterPunch* (June 2–3, 2007), http://www.counterpunch.org/johnstone06022007.html.

19. See http://www.zmag.org/chomsky/talks/9909-sovereignty.htm.

20. See *Blowback: The Costs and Consequences of American Empire* (London: Little, Brown and Co, 2000), *The Sorrows of Empire: Militarism, Secrecy, and the End of the Republic* (London: Verso, 2004), and *Nemesis, The Last Days of the American Republic* (New York: Metropolitan Books, 2006).

21. "American Destiny: Safe for the Rest of the World?" *Commonweal* (May 27, 2002).

22. Professor Andrew J. Bacevich, at less prestigious Boston University, calls himself a conservative and shares some of Chomsky's ideas. In the preface to his book entitled *American Empire: The Realities and Consequences of U.S. Diplomacy* (Cambridge, Mass: Harvard University Press, 2002), he writes (vii): "Compared to the comprehensive horrors perpetrated by the regimes it opposed, the sins attributed to the United States during the Cold War—Washington's complicity with unsavory dictators, the conspiracies to overthrow unfriendly regimes, the mindless invasion at the Bay of Pigs, and the assassination of South Vietnamese president Ngo Dinh Diem—seemed venial." He does not quote Chomsky, although on some issues, like Kosovo, they agree or are very close. If one refers to the small book Chomsky published in 1986, *What Uncle Sam Really Wants* (Tucson, Ariz.: Odonian Press, 1986), one will see that the concepts he developed a long time before the emergence of so-called neoconservatives remain relevant. Chomsky, for his part, refers to Bacevich's work several times in *Hegemony or Survival: America's Quest for Global Dominance* (London: Penguin, 2003).

23. Originally published in *Policy Review* 113 (June 2002).

24. Guillaume Parmentier, "Force, faiblesse, puissance?" *Commentaire* 100 (Winter 2002–2003): 795.

25. Pierre Hassner and Justin Vaïsse, *Washington et le monde, Dilemmes d'une superpuissance* (Paris: CERI/Autrement, 2003). In Charles-Philippe David, Louis Balthazar, and Justin Vaïsse, *La politique étrangère des Etats-Unis, Fondements, acteurs, formulations* (Paris: Presses de Sciences-Po, 2003), Balthazar writes (47): "From Theodore Roosevelt to George W. Bush, and including Wilson, Franklin D. Roosevelt, Kennedy, Carter, Reagan and Clinton, liberalism, that is to say the promotion of basic liberties, has always been at the heart of the greatest policies of the US." One may wonder if this specialist confuses propaganda with scientific analysis. Further on in a book mostly aimed at university students, Vaïsse writes: "a few specialized mainstream newspapers, radio stations and TV networks give interested audiences, notably elites, rigorous and high quality news about what is happening in the world" (304). This was written after the war in Iraq, when mainstream media, far from offering rigorous information, instead bought the administration lies wholesale. This cozying up to the administration is analyzed by Michael Massing, "Now They Tell Us," *New York Review of Books* (February 26, 2004). The *New York Times* belatedly admitted it had blindly followed the erroneous leads provided by Ahmad Chalabi. See "The Times and Iraq," *New York Times* (May 26, 2004). Neither Fox News nor the major networks apologized for their unprofessional attitude.

26. *50 ans de réflexion politique* (Paris: Julliard, 1983). He also refers to his friendship with Kissinger, who symbolizes a Machiavellian approach to pragmatic politics. Kissinger deliberately prolonged the war in Vietnam and encouraged the invasion of Cambodia. See Howard Zinn, "Machiavellian Realism and U.S Foreign Policy: Means and Ends," in *Passionate Declarations: Essays on War and Justice* (New York: HarperCollins, 2003). On Aron and Machiavelli, see Robert D. Kaplan's very Machiavellian *Warrior Politics* (New York: Random House, 2002).

27. See Pierre Guerlain, "Robert Kagan and Noam Chomsky," *Comparative American Studies* 4, no. 4 (2006): 446–458.

28. See Bernard Lahire, "Comment devenir docteur en sociologie sans posséder le métier de sociologue?" *Revue européenne des sciences sociales* 40, no. 122 (2002): 42–65, http://recherche.univ-lyon2.fr/grs/index.php?page=9&fichier=textes/articles/ress122.htm&imp=2.

29. *9–11* (New York: Seven Stories Press, 2001).

30. In a text read during the second World Social Forum in Porto Alegre in 2002, "World Without War," Chomsky reminds his audience that contrary to what his current admirers think, Adam Smith warned about those who followed "the vile maxim of the masters of mankind," which is "all for ourselves and nothing for other people." Adam Smith was a liberal who probably would be closer to opponents of neoliberal globalization than to the so-called neoliberals themselves. (About neoliberalism, Chomsky writes: "what's called 'neoliberalism,' though it is not a very good term: the doctrine is centuries old, and would scandalize classical liberals.") See also the chapter entitled "Goals and Visions" in *Powers and Prospects: Reflections on Human Nature and the Social Order* (London, Pluto Press, 1996).

31. Yet Chomsky does not bear grudges. He may quote approvingly the same writer he criticized earlier. This is what he does in an August 2003 *Monde Diplomatique* article, in which he quotes Schlesinger's criticisms of the Bush administration.

32. See Thomas Frank's *The Conquest of Cool: Business Culture, Counterculture, and the Rise of Hip Consumerism* (Chicago: The University of Chicago Press, 1997) and *One Market Under God: Extreme Capitalism, Market Populism, and the End of Economic Democracy* (New York, Doubleday, 2000). The Sokal affair is also relevant here. See Alan Sokal and Jean Bricmont, *Impostures intellectuelles* (Paris: Odile Jacob, 1997); and Pierre Guerlain, "Affaire Sokal: Swift sociologue; les *cultural studies* entre jargon, mystification et recherche," *Annales du Monde Anglophone* 9 (1999): 141–160.

33. Michael Walzer and others at *Dissent* do not seem to have understood that bombing innocents in Afghanistan did not amount to punishing the criminals who committed the 9/11 atrocities.

34. In France, one can refer to Phillippe Sollers, who is not an academic but is typical of "conformist rebels," and in the United States to Michael Berubé, a multiculturalist who supported the bombing of Afghanistan and totally bought the lies of the Bush administration about Iraq (at least initially). See "Peace Puzzle: Why the Left Can't Get Iraq Right," *Boston Globe* (September 15, 2002).

35. In his major book about the historical profession in the United States, Peter Novick analyzes the case of David Abraham, who found himself without a job as a result of interference with the work of appointment committees. Peter Novick, *That Noble Dream: The "Objectivity Question" and the American Historical Profession* (New York: Cambridge University Press, 1988), 612–621. In 2007, Norman Finkelstein also became the victim of slander and collective punishment. He was refused tenure at De Paul University after his department approved him. See http://www.insidehighered.com/news/2007/06/11/finkelstein. Before that, he had written a Ph.D. at Princeton and criticized the work of Joan Peters in his work, which was later published. Finding a job proved extremely difficult for him. Chomsky deals with this in *Understanding Power* (New York, 2002), 244–248, and on pages found on the site Znet devoted to Chomsky: "The Fate of an Honest Intellectual," http://www.chomsky.info/books/power01.htm.

36. See "Memories," Z (July–August 1995), http://www.zmag.org/CHOMSKY/articles/z9507-memories.html. Chomsky here refers to his own book, *For Reasons of State* (New York: Pantheon 1973), 100–101.

37. Jacques Portes, *Les Américains et la guerre du Vietnam* (Brussels: Editions Complexe, 1993).

38. *The Fateful Triangle: The United States, Israel, and the Palestinians*, updated ed. (Boston: South End Press, 1999).

39. *Les Républiques face à la domination américaine dans l'après-guerre froide* (Paris: La Découverte, 2002). Chalmers Johnson's work, mentioned above, belongs to the same category.

40. The *New York Times* published an op-ed by Chomsky on February 23, 2004: "A Wall as a Weapon"—a rather rare occurrence.

41. Yet Edward Said, whose literary qualities cannot be doubted, was also criticized in a similar way.

42. See *Truth and Truthfulness: An Essay in Genealogy* (Princeton, N.J.: Princeton University Press, 2002). I cannot tackle the philosophical problems of truth in such a short piece. Williams does it in his comments about history and criticizes what he calls the "knowing evasiveness" of some academics (11).

43. This can be seen in the debate he has with Isaacs (see above). In a text prepared for the World Social Forum in Porto Alegre on 2002, Chomsky talks about the demonizing of opponents to neoliberal globalization, who are presented as "freaks." Later on, he specifies that this insulting word is not his own—something a good-faith reader had of course already realized.

44. In his book *A Godless Jew: Freud, Atheism, and the Making of Psychoanalysis* (New Haven, Conn.: Yale University Press, 1987), Peter Gay quotes a letter Freud sent to Charles Singer after the publication of *Moses and Monotheism,* on October 31, 1938, in which Freud explains that he was affected by the reaction of the Jews he had hurt in his book but that he had no other choice but to defend scientific truth as he had done all his life (152). Although Chomsky is neither Freud nor Freudian, there is a sociological similarity here. See also Edward Said, *Freud and the Non-European* (London: Verso, 2003).

45. See Chomsky's article "His Right to Say It," *The Nation* (February 28, 1981).

46. Said, *Representations of the Intellectual,* 12.

47. "A Modest Proposal," *New York Times* (December 3, 2002). Swift suggested in his "Modest Proposal" that parents should eat their children to put an end to famine in Ireland. Here, Chomsky suggests that Iran should invade Iraq and be encouraged to do so by the United States. Another example: in a recent article Chomsky writes, "The evidence that Iran is waging war against the US is now conclusive. After all, it comes from an administration that has never deceived the American people, even improving on the famous stellar honesty of its predecessors." An incompetent reader or one in utter bad faith could maybe imagine that this is Chomsky's take on Iran and miss the irony. *Znet* (August 27, 2007), http://www.zmag.org/content/showarticle.cfm?SectionID=11&ItemID=13629.

48. Baudrillard claimed in a *Libération* article in January 1991 that "the Gulf War would not take place," and then after it did take place, that it "had not taken place" (March 1991). Baudrillard argued it was a video game and therefore a virtual war. See *La Guerre du Golfe n'a pas eu lieu* (Paris: Galilée, 1991). Chomsky's expression "Gulf slaughter" can be found in his *Understanding Power: The Indispensable Chomsky* (London: Vintage, 2003), 328.

49. See Chomsky, *Understanding Power,* 224–230.

50. Here is a typical pronouncement: "I'm going to assume two conditions for this talk. The first one is just what I assume to be recognition of fact. That is that the events of September 11 were a horrendous atrocity, probably the most devastating instant human toll of any crime in history, outside of war. The second assumption has to do with the goals. I'm assuming that our goal is that we are interested in reducing the likelihood of such crimes whether they are against us or against someone else. If you don't accept those two assumptions, then what I say will not be addressed to you." From "The New War Against Terror," *CounterPunch* (October 24, 2001), based on a talk given at MIT.

51. See note 15, above. The *New York Times* published a piece by George Packer entitled "The Liberal Quandary Over Iraq" (December 8, 2002), in which Chomsky is accused of supporting Milošević out of anti-imperialism: "This thinking prompted Noam Chomsky to leap to the defense of Slobodan Milošević, and it dominates the narrow ideology of the new Iraq antiwar movement." This is to misread Chomsky's positions. Criticizing the United States for its manufacturing of a crisis situation does not mean that one agrees with the views of others. Opposing the war in Iraq did not mean supporting the Iraqi dictator but thinking about the number of victims among civilians. Subsequent events and even official congressional commissions

have totally discredited Packer's views. But where is Packer's new op-ed saying "I was wrong and Chomsky was right"?

52. "The Campaign of Hatred Against Us," Noam Chomsky interviewed by Ticky Fullerton (January 26, 2002), http://www.chomsky.info/interviews/20020126.htm.

53. Todd Gitlin, "Blaming America First," *Mother Jones* (January–February 2002).

54. Emmanuel Todd, *Après l'empire; Essai sur la décomposition du système américain* (Paris: Gallimard, 2002). Translated by Columbia University Press, in 2003, as *After the Empire: The Breakdown of the American Order*. This is a very interesting book in which the charge against Chomsky is marginal.

55. See his *Terror and Liberalism* (New York: Norton, 2003), where examples of historical confusions abound.

56. In Noam Chomsky and David Barsamian, *Propaganda and the Public Mind* (Cambridge, Mass.: South End Press, 2001), 166, Chomsky reminds his readers that "public intellectuals" approved of World War I, contrary to Eugene Debs or Bertrand Russell.

57. "Writers and Intellectual Responsibility" in *Powers and Prospects*. See http://www.chomsky.info/interviews/20010527.htm.

58. For a detailed study of such a case in France see "'Libération' et Noam Chomsky" (1), http://www.acrimed.org/article77.html; and "'Libération' et Noam Chomsky" (2), http://www.acrimed.org/article79.html.

59. Marc Chénetier uses this expression *besoin littéraire*, which was coined by Antoine Raybaud in his introduction to *Revue Française d'Etudes Américaines* 94 (December 2002).

60. See Craig Murray, *Murder in Samarkand: A British Ambassador's Controversial Defiance of Tyranny in the War on Terror* (Edinburgh: Mainline Publishing, 2007).

13
THE PRACTICE OF INTELLECTUAL SELF-DEFENSE IN THE UNIVERSITY

NORMAND BAILLARGEON

An anthology edited recently by C. P. Otero bears witness to the fact that, contrary to what a rapid survey of his work might lead one to think, Chomsky has written a great deal on education.[1]

Yet his ideas on this subject have not always received the sustained attention they deserve, and they are only very rarely discussed in the academic milieus concerned. This is all the more regrettable in that Chomsky's thought, in this as in so many other cases, diverges radically from the beaten path and is quite impossible to boil down into the kind of facile slogans and reductive dilemmas that too often set the bounds of debate about education. And perhaps this is precisely the factor that explains the relatively scant attention paid to his analyses.

As a matter of fact, Chomsky approaches education with the same principles and values that guide him in his approach to other social and political questions and problems—those of a libertarian socialist rationalist for whom, notwithstanding the extremely limited nature of our knowledge on this subject, human nature remains the essential point of reference in any vision of society and politics. The conjunction of these principles and values leads Chomsky to adopt a profoundly original critical stance[2] and allows him to keep alive what he calls a "hope." His hope might be formulated this way: nothing in what we know warrants us to reject the idea that, since human beings have a nature by virtue of which they aspire to liberty and a decent life, it is legitimate and reasonable for them to try to eliminate the illegitimate structures—

social, political, and economic—that in one way or another prevent or hamper the full unfolding of their potential; it is equally reasonable to hope that, once these structures are eliminated, a freer and more decent life will become possible.

With a modesty that does him credit, Chomsky has always refused to let the stance and principles he has set forth be called a "theory," and he has asserted that his contribution is limited essentially to the application of simple "Cartesian good sense" to certain human affairs. Despite that, the truth of the matter is that these principles have proved to be immensely fecund, including for our understanding of education—of what it is and what it could be.

In order to give the reader an idea of this, I shall examine a few concrete aspects of academic life in the university setting, a subject Chomsky has dealt with in several of his writings. The assault on the autonomy of the university being conducted on all sides at present makes them, alas, undeniably topical.

The importance of Wilhelm von Humboldt (1767–1835) in these writings, and in all Chomsky's thinking about education, should be noted at the outset. In Humboldt one finds, under the category of *Bildung*,[3] a strong articulation of rationalism and of the stance I noted above, which takes human nature as its point of reference:

> The true end of Man, or that which is prescribed by the eternal and immutable dictates of reason, and not suggested by vague and transient desires, is the highest and most harmonious development of his powers to a complete and consistent whole. Freedom is the first and indispensable condition which the possibility of such a development presupposes; but there is besides another essential—intimately connected with freedom, it is true—a variety of situations.[4]

It is not unimportant that Humboldt, a linguist and a theoretician of liberalism, was also the founder of the University of Berlin, and Chomsky has remained profoundly attached to the normative conception of the nature and function of the university, which these ideas dictated to Humboldt. This normative conception is "nothing other than the spiritual life of those human beings who are moved by external leisure or internal pressures toward learning and research."[5] The capacity to permit these human aspirations to be satisfied, Chomsky notes, is an indicator of the level of development attained by any given civilization.

However, and without neglecting the importance of these ideas, hopes, and ideals, the libertarian would be unwilling to rest content with them. Higher education, Chomsky notes, is a powerful instrument for the perpetuation of social privilege. Social pressure requires the universities to impose norms and standards grounded in criteria that are biased against those who do not possess the character traits, the attitudes, and the background typical of the upper middle classes. These criteria ensure that the values of these classes are reinforced. Inequality is maintained subsequently, since only those who possess university training can pursue further training as adults (keeping their skills up to date or receiving ongoing training), and that despite the fact that it is persons of more modest status who would derive the greatest benefit from such opportunities—who, indeed, would be most deserving of them.[6] In this sense, "it is pointless to discuss 'the function of the university' in abstraction from concrete historical circumstances."[7] The ones that shaped his own vision of the university were those of the activism and the movements—especially the student movement—of the 1960s, in which he played an important part[8] and which enflamed American campuses until the turning point, which came when, as Chomsky put it in 1973, "I think these [ideological] controls are being reasonably effectively reestablished."[9]

Chomsky often pays fitting tribute, in numerous writings and interviews, to the militancy of those years, a rare phenomenon in the history of universities, which produced a good number of "islands of . . . independent critical thought."[10] The student movement, he said in 1969, had given a salutary shakeup to "the complacency that had settled over much of American intellectual life" and brought to the fore the necessary idea of "university reform . . . democratizing it, redistributing 'power' in it, reducing constraints on student freedom, as well as the dependence of the university on outside institutions."[11]

But, he immediately added, "I suspect that little can be achieved of real substance along these lines." To understand this reservation, it should be noted that Chomsky, the reader of Humboldt and a standard bearer of the ideals described above, with his devotion to the idea of a free place dedicated to the formulation, criticism, and free discussion of ideas, to the values of objectivity, truth, and intellectual honesty, remained highly critical even at the time of some of the positions vis-à-vis the university then typically being adopted both on the conservative side and in some areas of the "new left."

Some tended to think that what was needed, above all, was to change the administrative structure and the top decision-making personnel of the universities, since it was they who were forcing the institution and its professors to carry out certain shameful and antiacademic functions for the benefit of the dominant powers in society. Others, faced with the militarization and instrumentalization of the university, recommended working to abolish it altogether.

Chomsky argued that the university remains, despite everything, one of the most open and free institutions in our society—because freedom and creativity are the indispensable conditions for the accomplishment of its particular mission and because some of the results, or the spinoffs, of what is accomplished there are potentially useful, even vital, for the dominant powers. The university is, in this sense, "ultimately a parasitic institution from an economic point of view,"[12] and its loss would be a dramatic detriment for all who aspire to a radical social transformation—the more so in that the relative openness of the university ensures that these activities take place there, up to a point, in the light of day, whereas they would still take place outside the universities if the universities were abolished, but in much greater concealment and with no opportunity for the public to exercise real control over them.

Besides, and without in the least wishing to deny that the universities effectively are institutions in which such functions are being carried out, Chomsky notes the crucial fact that nothing is dictated there from on high: in other words, that those in the universities, freely and of their own accord, do exactly what the dominant powers expect of them, without the university administrations having to compel them in any way whatsoever. It is of their own accord that these "secular priests," these "experts in legitimation," profoundly indoctrinated themselves, become agents of indoctrination in turn.

Chomsky for his part proposes to work at bringing about change that will be more than cosmetic and that will target, first and foremost, "the cast of mind and the general intellectual and moral commitments of very substantial segments of the faculty themselves,"[13] in the hope that the university can, without denaturing itself, contribute in its own fashion to social change.

Before examining some of the methods advanced by Chomsky toward this end, here are some examples of these mechanisms, which act as so many filters and make it possible for the university to exert this ideological control.

The first might be called "nitpicking methodologism." The following anecdote, told by Chomsky, will illustrate what this means. He is what is called an "Institute Professor" at MIT, and that means he is authorized to give courses in all departments—which he has done during all the years he has been at MIT, just as he has sat on supervising and examining committees for Ph.D. candidates in all sorts of disciplines. With one exception: he is never welcome in the department of political science and has almost never sat on Ph.D. committees in this discipline. Each time he has, he explains, the candidates who requested his presence were women from the third world, and the probable reason the department made an exception in these cases is that it feared being accused of sexism or racism.

When a young woman who wanted to prepare a doctoral thesis in political science on the media in southern Africa asked for Chomsky to be on her committee, the fact that he had done so much work on the media meant that he could not be excluded for not being knowledgeable about the subject. Nonetheless, as Chomsky recounts, a way was found to sideline him that reveals mechanisms of control that I think many people who work in certain sectors of the social sciences and humanities will recognize.

First, a memo was circulated to make sure that every member of the department turned up for the meeting at which the young woman was to present her proposal. Then, as soon as she began, she was bombarded with methodological questions from all sides, asked by the exceptionally high number of professors who had shown up for the occasion. "What's your hypothesis?" "What's your methodology going to be? What tests are you going to use?" And so on. Chomsky relates the upshot:

> And gradually an apparatus was set up and a level of proof demanded that you just can't meet in the social sciences. . . . But she fought it through, she just continued fighting. They ultimately required so much junk in her thesis, so much irrelevant, phony social-scientific junk, numbers and charts and meaningless business, that you could barely pick out the content from the morass of methodology. But she did finally make it through—just because she was willing to fight it out.[14]

Chomsky has often mentioned an important disparity, to which he refers briefly in the foregoing example, between the intellectual work done in the sciences (natural and mathematical) and that done in disciplines with a strong ideological coloration. Success comes to a researcher in the former from mastering a body of established knowledge (*un savoir*) and then showing proof of independence of thought and creativity, qualities that are strongly encouraged because they are indispensable. In the latter, however, where the body of established knowledge does not amount to much, conformism is typically encouraged, and creativity, audacity, and so on are repressed. That potentially constitutes another ideological filter.

Chomsky relates that for many years he has been invited to speak before gatherings of researchers in the numerous fields in which he has worked—from mathematics and the theory of automation to philosophy and the history of ideas. He has noticed that physicists, mathematicians, and so on never ask him what degrees he holds, preferring simply to assess the intellectual value of what he has to say. But in political science, on the other hand, the reaction is quite often (Chomsky points out that there are exceptions) to contest his very right to speak because he holds no degree in that field. The explanation, he suggests, is simple: "in the sciences, there is no need to worry about credentials, since the fields have intellectual substance and integrity. In the humanities, where a substantial number of practitioners are people with tiny minds and limited understanding, it is necessary to keep outsiders from prying in (they might have ideas, which would be terrifying)."[15]

The trajectory of this self-described "child of the Enlightenment"[16] could not fail to intersect with that vast family of ideas that has proliferated since the last decades of the twentieth century within the North American university and elsewhere, under widely varying denominations: postmodernism, deconstruction, strong program in the sociology of science, radical constructivism, poststructuralism, and so on. To a large extent, these ideas also appear to Chomsky to act as a filter, diverting attention away from important questions and conferring on those who practice these "disciplines" a status and an intellectual capital that Chomsky—I do not think I misrepresent him in saying this—essentially regards as counterfeit coin.

The most substantial piece by Chomsky in this respect is, I think, his contribution to the debate on postmodernism, science, and rationality organized by Michael Albert under the auspices of the magazine Z, to which readers are referred, since a summary cannot really do it justice.[17] In it they will find a remarkable defense of classical rationalism and the ideals of the Enlightenment. But in reading what he has published on the "postmodern" movement in the universities, I sometimes seem to detect Chomsky adopting a darker, more ominous tone than usual. There is nothing particularly surprising in that: if one compares Chomsky's perception of the typical approach of these authors to the academic and university ideal he has always defended, the contrast is striking. And one grasps the malaise of the man who thinks that the university presents, for the left, a chance to acquire indispensable intellectual capabilities, the intellectual and moral virtues of deliberation, honesty, and reflection, in sum, that it "permits the political life to be an adjunct to the academic one, and action to be informed by reason"[18]—his malaise, I say, faced with approaches that seem to him, in certain cases at least, to signal its abandonment. In his contribution to the debate mentioned above, Chomsky writes:

> It strikes me as remarkable that their left counterparts today [that is, today's left intellectuals as opposed to those of yesterday] should seek to deprive oppressed people not only of the joys of understanding and insight, but also of tools of emancipation, informing us that the "project of the Enlightenment" is dead, that we must abandon the "illusions" of science and rationality—a message that will gladden the hearts of the powerful, delighted to monopolize these instruments for their own use.

And so I now come to several paths of resistance evoked by Chomsky.

First, he has suggested, apparently with perfect seriousness, that the things that go on in the university should be given their real names. Are we dealing with military research or, more generally, with the development of dangerous or death-dealing technologies? Then let them be identified on campus, under an banner so explicit that everyone will be well aware what is going on in the departments concerned. Let us have a Department of Death and perhaps a Department of Justification of Pollution as well. In Chomsky's own words:

> universities ought to establish Departments of Death that should be right in the center of the campus, in which all the work in the university which is committed to destruction and murder and oppression should be centralized. They should have an honest name for it. It

shouldn't be called Political Science or Electronics or something like that. It should be called Death Technology or Theory of Oppression or something of that sort, in the interests of truth-in-packaging.[19]

In addition, and to return to the anecdote recounted above about the doctoral thesis disfigured by sterile and unattainable methodological prerequisites, Chomsky has suggested, with good reason it seems to me, that there ought not to be such rigid insistence that doctoral theses be solely the product of work done by an individual on her own, or that they be completed within a fixed timespan, or that they should have narrowly set aims and deal with conventional topics of research that limit the scope for speculation. Such constraints, in his view, produce trivial research and can cause scholars to spend their careers making minor adjustments to their own previous body of work.[20]

Arguing that impermeable disciplinary boundaries and the very division of the university into departments may be contributing to the screening out of certain questions and certain problems, he has proposed various forms of what might be called transdisciplinarity. For example, graduate students should be encouraged to defend the relevance of their field of research in light of a critical perspective that does not automatically accept the premises and limits that individual fields set for themselves. Within the university, philosophers seem to him particularly well prepared to accomplish this task, and they should be encouraged to do so, particularly with students in the human and social sciences.[21]

Chomsky has likewise insisted on the importance of the university's mission to educate, emphasizing that "to restore the integrity of intellectual life and cultural values is the most crucial task that faces the universities."[22] He adds that a university "should be a center for . . . radical social inquiry in the pure sciences, . . . permit a richer variety of work and study and experimentation," and "should provide a home for the free intellectual, for the social critic, for the irreverent and radical thinking that is desperately needed if we are to escape from the dismal reality that threatens to overwhelm us."[23]

For his part, as a means to fulfill this ambitious program, he has for many years offered a not-for-credit course at MIT, open to all, which has proved to be extremely popular. And he has of course also expressed a wish to see members of the university take part in the public debates and contests of their time—sometimes even expressing nostalgia for a not-so-distant past: "many scientists, not too long ago, took an active part in the lively working-class culture of the day, seeking to compensate for the class character of the cultural institutions through programs of workers' education, or by writing books on mathematics, science, and other topics for the general public."[24]

Indeed, there is a pedagogical dimension to all Chomsky's work that I wish to emphasize. In his writings and lectures—on matters of general interest and addressed to the general public—Chomsky really speaks to people as intelligent human beings. He speaks to them about subjects that are important and often complex, but he does so clearly and soberly, in such a way as to be understood by the men and women he is addressing. He adduces the facts and the data indispensable for understanding how matters stand, but without overtheorization, especially artificial overtheorization—and without abusive oversimplification either. It is arguable that much of Chomsky's success as an author and lecturer derives from these qualities. It is also arguable, alas, that they are too rare in the academic world.

Chomsky has suggested many other means of intellectual self-defense to university scholars. They are all aimed at making it possible to preserve the intellectual independence of the academic community and defend the independence of the university against the factors that might lure scholars into a betrayal of academic freedom—including access to money and power; ideological monolithism; the fact of concentrating on trivial problems that interest the professions alone; and the tendency, particularly in certain behavioral sciences, to engage in experiments on anything and everything, without worrying about the consequences for human beings.[25]

This brief overview will make it possible, I hope, for readers to get a sense of the originality of Chomsky's analyses and the practical steps he recommends. Their value, as his commentators have noted, is to allow "children of the Enlightenment to be optimists of the will, without condemning themselves to being irrationalists of the intellect. It is Chomsky's insistence on this point, his commitment to both reason and moral hope, that we take to be his signal contribution to social thought."[26]

I subscribe unreservedly to that judgment.

NOTES

1. Noam Chomsky, *Chomsky on Democracy and Education*, ed. C. P. Otero (New York and London: RoutledgeFalmer, 2003).

2. On the principles of Chomsky's social and political thought, see the excellent synthesis in Alison Edgley, *The Social and Political Thought of Noam Chomsky* (London: Routledge, 2000). On their implications for a critical theory of education, see, in addition to Chomsky's own texts, C. P. Otero's introduction to *Chomsky on Democracy and Education*.

3. It is well known that the German language possesses two sets of terms referring to education: *bilden* (verb) and *Bildung* (noun) on the one hand, and *erziehen* (verb) and *Erziehung* (noun) on the other. The former places the emphasis on the result of education, the second on the process. Humboldt employs the word *Bildung* when speaking of culture and education.

4. Wilhelm von Humboldt, *The Limits of State Action*, ed. J. W. Barrow (Indianapolis, Ind.: Liberty Fund, 1993), chap. 2, para. 1, p. 10.

5. Wilhelm von Humboldt, cited by Chomsky in his "The Function of the University in a Time of Crisis," in *Chomsky on Democracy and Education*, 178. This text, which dates from 1969, was first anthologized in Noam Chomsky, *For Reasons of State* (New York: Pantheon Books, 1973; repr. New York: New Press, 2003, with a new introduction by Arundhati Roy).

6. Chomsky, "The Function of the University in a Time of Crisis," in *Chomsky on Democracy and Education*, p. 181.

7. Chomsky, "The Function of the University in a Time of Crisis," 179.

8. For more on this, see Robert F. Barsky, *Noam Chomsky: A Life of Dissent* (Cambridge, Mass.: The MIT Press, 1997). Chapter 4, "The Intellectual, the University, and the State," 119–163, is particularly illuminating in this regard.

9. Noam Chomsky, "The Universities and the Corporations" (May 1973), in *Language and Politics*, expanded 2nd ed., ed. C. P. Otero (Oakland, Calif.: AK Press, 2004), 137. In the first edition, *Language and Politics*, ed. C. P. Otero (Montreal: Black Rose Books, 1988), this quotation is at 161.

10. Noam Chomsky, "Language Theory and the Theory of Justice" (October 13, 1977), in *Language and Politics* (2004), 218. In the first edition of *Language and Politics* (1988), this quotation is at 250.

11. Chomsky, "The Function of the University in a Time of Crisis," 178–179.

12. Chomsky, "The Function of the University in a Time of Crisis," 189.

13. Milan Rai, *Chomsky's Politics* (London: Verso, 1995), 128.

14. Noam Chomsky, in Peter R. Mitchell and John Schoeffel, eds., *Understanding Power: The Indispensable Chomsky* (New York: New Press, 2002), 244.

15. Noam Chomsky, from "Exchanges on Reconstructive Knowledge Between Noam Chomsky and Marcus G. Raskin, October 1983 to February 1985," in Marcus G. Raskin and Herbert J. Bernstein, eds., *New Ways of Knowing: The Sciences, Society, and Reconstructive Knowledge* (Totowa, N.J.: Rowman and Littlefield, 1987), 149.

16. Noam Chomsky, "Helping People Persuade Themselves" (February 15, 1988), in *Language and Politics* (2004), 657. In the first edition of *Language and Politics* (1988), this quotation is at 773.

17. Chomsky, "Rationality/Science and the Post-This-or-That," in *Chomsky on Democracy and Education*, 87–97. This text and all the others from the debate organized by Z are available at http://www.zmag.org/ScienceWars/sciencetoc.htm. Chomsky's text is there given the slightly different title "Rationality/Science—from Z Papers special issue," http://www.zmag.org/ScienceWars/sciencechomreply.htm. It has been published in French under the title "Le Vrai visage de la critique post-moderne," in *Agone* (Marseille) 18–19 (1998), http://atheles.org/agone/revueagone/agone18et19.

18. From an important text of the student movement of the 1960s entitled "The Port Huron Statement," cited by Chomsky at "The Function of the University in a Time of Crisis," 192. See http://www.tomhayden.com/porthuron.htm.

19. Rai, *Chomsky's Politics*, 129.

20. Chomsky, "The Function of the University in a Time of Crisis," 179–180.

21. Chomsky, "Remarks Before the MIT Commission on MIT Education (November 1969)," in *Chomsky on Democracy and Education*, 294. Chomsky, "Philosophers and Public Philosophy," *Ethics* 79, no. 1:: 1–9.

22. Chomsky, "Philosophers and Public Philosophy," 9.

23. Chomsky, "The Function of the University in a Time of Crisis," 192.

24. Chomsky, "Rationality/Science and the Post-This-or-That," 96.

25. See especially Chomsky, "The Function of the University in a Time of Crisis," and "Scholarship and Commitment, Then and Now," in *Chomsky on Democracy and Education*, 195–201.

26. Joshua Cohen and Joel Rogers, "Knowledge, Morality, and Hope: The Social Thought of Noam Chomsky," *New Left Review* 187 (May–June 1991): 27.

14
CHOMSKY, FAURISSON, AND VIDAL-NAQUET

JEAN BRICMONT

More than thirty years ago, Chomsky got dragged into what is known in France as the "Faurisson affair." At the end of the 1970s, Robert Faurisson, a lecturer in literature at the University of Lyon, was subjected to all sorts of intimidation: his lectures were disrupted, he was physically attacked, and finally he was suspended from teaching by the university. In addition, he was taken to court by "antifascist" organizations such as LICRA (Ligue internationale contre le racisme et l'antisémitisme), MRAP (Mouvement contre le racisme et pour l'amitié entre les peuples), and associations of persons who had been deported or been active in the resistance during World War II. These lawsuits were successful.[1] The reason for these actions was that in some of his writings Faurisson had denied the existence of the gas chambers during the war.

At the time, Chomsky defended Faurisson's right to free expression. This affair did much to discredit Chomsky's writings in France, and it still does.[2] Pierre Vidal-Naquet (1930–2006) was one of Chomsky's most fervent critics in this debate. Yet he was also one of those who signed the appeal launched by nineteen historians in December 2005 for the abrogation of various laws, including the Gayssot law criminalizing holocaust denial, that they regarded as injurious to liberty.[3] In fact, Vidal-Naquet was always opposed to laws criminalizing holocaust deniers and to the suits that were brought against them. How does his appeal against the Gayssot law sort with his harsh criticism of Chomsky's stance in favor of Faurisson's freedom of expression? The remarks that follow attempt to shed light on this question, on the one hand by elucidating Chomsky's position and more generally the logic of those who defend freedom of expression, and on the other by analyzing the position of Vidal-Naquet and other opponents of Chomsky at the time of the Faurisson affair.

What Really Happened?

Chomsky's involvement in the Faurisson affair began in 1979, when, along with five hundred others, he signed a petition demanding that "university and government officials do everything possible to ensure his safety and the free exercise of his legal rights." Chomsky has always made it clear that he did not share Faurisson's views and that his "support" of Faurisson was strictly limited to the defense of his freedom of expression. Since Vidal-Naquet's position on this is, in principle, identical to Chomsky's, we may legitimately wonder exactly what Vidal-Naquet was reproaching him with.

In an article published in *Esprit* in September 1980,[4] Vidal-Naquet declared that "what is scandalous about the petition is that it never raises the question of whether what Faurisson is saying is true or false, that it even presents his conclusions or 'findings'[5] as the result of a historical investigation, one, that is, in quest of the truth."[6] The "scandal" provoked by the petition compelled Chomsky to engage in a protracted exchange of letters not just with Vidal-Naquet but also with a certain number of other French intellectuals and periodicals.[7] Chomsky summarized some of the arguments from these letters in a short text that he gave to a friend of his at the time, Serge Thion, telling him to do with it as he wished. A specialist on Cambodia and a long-time militant in the anticolonial movement, Thion had drawn close at this time to Faurisson and saw fit to attach Chomsky's text as a preface to the *Mémoire en défense* published by Faurisson in 1980 in reply to the lawsuits against him.[8] Although Chomsky has always affirmed that he never had any intention of publishing his text as a preface to a book by Faurisson, and although he attempted, too late, to prevent it from appearing, it offered Vidal-Naquet and many others an opportunity to renew their attacks on Chomsky.[9] Rumors flew, Chomsky's name came to be associated with Faurisson's for almost a quarter of a century, and his writings, particularly his remarkable works of the 1980s and 1990s on U.S. policy in Central America and the Middle East, became *de facto* untranslatable into French. Only at the end of the 1990s did this situation begin to change.[10]

Freedom of Expression: What Principles?

Chomsky often remarks that in France the very concept of freedom of expression is not understood. We might refine that a little by saying that it is the logic of freedom of expression that is often not understood, or that those who claim to defend freedom of expression often do so in an incoherent fashion. Many people take it "that there are some things you can't say," but just what is to be prohibited and on the basis of what principles is rarely discussed with clarity, although it is the fundamental question.

In the first place, it should be recalled that law is not morality, and even if it is based on moral principles, law cannot regulate everything: there are actions that one may judge to be immoral but that no law can prevent. In consequence, one can perfectly well take the view that propositions are odious or scandalous without wishing to prohibit them on that account.

What is more, in every society and at every time, freedom of expression has existed—for some people at least. No one ever prevented the pope or the king of France from expressing himself freely. Censorship is always by definition exercised by those who hold power, and in particular by those who enjoy freedom of expression, against those who do not. Hence the sole question to be asked about censorship is: in the name of what *principles* do those who are able to express themselves have the right to prevent others from expressing themselves too?

The proponents of censorship might retort that censorship exercised through the courts and under the control of an elected parliament in a democratic state is not the same thing as "totalitarian" censorship. This is true, but such a defense of "democratic" censorship runs up against difficulties that are insurmountable, in my view. These I will explain, starting with a few general considerations, then dealing with the fundamental juridical problem, and concluding with a few observations of a pragmatic nature.[11]

The first observation one might make is that censorship of the minority by the majority entails obvious risks. It might muzzle Faurisson, but it will also reduce Galileo, Darwin, and Einstein to silence, along with all the creative artists who at one time or another have adopted a minority stance. Is the game worth the candle? Proponents of censorship in democratic societies are often conscious of problems of this sort, and will therefore state that they want to limit their censorship to ideas that are "really" odious or "really" dangerous, leaving the Einsteins and the Picassos to work in peace. But the fundamental juridical problem this kind of thinking runs up against lies in the fact that modern democratic law relies on rules stated in general terms that are meant to be applied impartially. And in fact, if you do not accept this premise, you lapse back into arbitrary power and *lettres de cachet*. So how do you combine this stipulation with censorship?

The criteria most often invoked to justify censorship are the manifestly false and mendacious character of statements, their odious and wounding character, or their harmfulness. Let us examine each of these criteria in turn, to show why none can be maintained or applied in an impartial fashion.

The first criterion is clearly inapplicable: what are we to do with thousands of weird or pseudoscientific doctrines? Are they not, in some cases at least, manifestly false? But who will be rash enough to try to legislate about that? Would a censorship commission that undertook to do so not run the risk straightaway of prohibiting new scientific ideas? As for censoring mendacity, that presumes that one can assess not just the falsity of statements but also the intentions of those who make them, which renders the problem even more complicated.

As for odious statements, to try to censor them is to reinstate the crime of blasphemy. But almost all believers feel themselves wounded by the utterances of atheists or followers of other religions. Is it reasonable to entrust legislators with the task of distinguishing between "really" odious statements and statements that only appear odious from the vantage point of a particular religion? The late Oriana Fallaci, whose books sell like hotcakes, wrote that Muslims breed "like rats." The French comedian Dieudonné compares religion to "farting in one's bath" (meaning that it is a private matter) without arousing any protest. If these statements are tolerated, on what basis would others not be? And if one does wish to prohibit them, how is that to be done in an impartial fashion without also prohibiting quantities of others?

Next we come to the question of the "effects" of speech, the idea being that one ought to censor it when it is harmful or dangerous. The problem is that, by definition, censorship always reinforces the power of those who already possess it and never allows the most harmful aspects of the dominant discourse to be censored. It only allows the silencing of those on the margins, whose speech, precisely because it emanates from marginalized persons, cannot have important direct consequences. It is highly probable that the discourses whose effects are, in practice, the most harmful are the "sacred" texts of the "major" religions, which all contain appeals to make war on infidels and which are sometimes taken very seriously by their believers. But who wants to censor the Bible or the Koran? Besides, if we consider political speech in the strict sense, that which justified or justifies the Vietnam War or the Iraq embargo is far from being prohibited, though the victims are counted in the millions: indeed, it is well rewarded for the most part. In this respect, Chomsky cites an interesting example of "revisionism": Americans were asked what they thought the number of Vietnamese killed during the Vietnam War was. The median reply was around 100,000, in other words, about 5 percent of the American government's official estimate. As Chomsky says, what would one think of German political culture if similar replies were given there about the massacre of the Jews?[12] This revisionism, moreover, unlike that of Faurisson, is upheld by the core of the intellectual and media elite. Do we really want to reinforce it by giving this elite the additional power to censor what it doesn't like?

It is sometimes suggested that we should prohibit marginal discourse precisely when it is marginal, before it becomes dominant and dangerous—Hitler's discourse before he came to power, for example. But seizures of power do not result purely from the expression of ideas (there still exist Nazis in our societies, and some express themselves—so why do they not take power?); they are linked to all sorts of socioeconomic and historical circumstances. Whether or not the latter can be acted on is a complex problem, but supposing that remedial measures had been taken (for example, a solution to the 1930s economic crisis, or an alliance among Hitler's opponents): it is hard to see how censorship of Hitler would then have been necessary. Vice versa, it is hard to imagine that in the absence of such measures, censorship would have been enough by itself to stop Hitler.

Actually, one can respond in general terms to arguments of this type by invoking an idea of John Stuart Mill: let truth and error confront each other on equal terms. Which do you think will win? The proponents of censorship will no doubt answer: error (invoking, for example, the idea that the masses are easy to manipulate). But, if that is your reasoning, what hope do you have that censorship will not ultimately be controlled by those who are in error, and the chance for truth to prevail not be eliminated once and for all? To avoid that, censorship would have to be entrusted to "safe" hands and stay there for good. But why should those who hold the truth wield the arm of censorship any better than the others? This type of reasoning in favor of censorship is widespread, at least implicitly, but what it testifies to is both a radical pessimism about democracy and an unrealistic optimism about autocracy.

Freedom of Expression: What Limits?

It is true that certain words are not protected by freedom of expression: if one person is holding a revolver to the head of another person and a third person says "fire," that is not regarded as an expression of opinion in any country. The difference here is that this is a direct incitement to an immediate (illegal) action. From the viewpoint of the defense of freedom of expression, the fundamental distinction is between word and deed—words are free but deeds evidently are not, and in certain circumstances words such as "fire!" can be likened to deeds. To artificially provoke a panic (by shouting "fire!" in a crowded theater when there is no fire) is not covered by freedom of expression either, for the reason that in this case too, the word is a form of action with immediate effects.

I note in passing that, as regards Faurisson, it is hard to accuse him of inciting the commission of a fresh crime, since he denies that a past crime was possible (his main thesis being that the gas chambers were technically not feasible).

Clearly it is impossible, as it often is in legal questions, to supply mechanical and universally applicable criteria that would make it possible to distinguish between the expression of opinion and words comparable to deeds—

but the idea that only words that give incitement to *immediate* action are blameworthy is a good barrier against opinions being censored on account of the long-term consequences they are thought to have. "Incitement to racial hatred," for example, is not, except in very particular circumstances, incitement to immediate action, and its criminalization must be seen as a hindrance to freedom of expression. If we were to exercise censorship of incitement to racial hatred impartially, we would quickly find ourselves banning a good many sacred books, as well as a large part of Western thought, which abounds in apologies for war, slavery, and racism. I come back to the point that when incitement to racial hatred is criminalized today, it is only marginal discourse that gets targeted. As Chomsky notes, there are really no "laws against hatred," only laws against persons hated by those with the power to make the law, which is a very different thing.

To illustrate this idea, one might contrast the righteous indignation displayed by the European press in February 2006 in the wake of reaction in the Muslim world to the Danish cartoons of the prophet Mohammed and the almost complete silence of these same "defenders of liberty" when a British historian, David Irving, who was traveling in Austria, was arrested and sentenced to three years in prison for statements he had made in 1989 denying the Holocaust, for which he has apologized and about which he declares that he has changed his mind.[13] On this last point, it is worth noting that the prosecutor maintained that this retraction was false and was intended only to avoid a lengthier sentence. According to that logic, the Holy Office ought not to have shown the clemency it did to Galileo, whose repentance was assuredly feigned.

Clearly, there are also legitimate limits that may be imposed on "words" in connection with defamation, insult, or the right to privacy. But, aside from the fact that problems of this kind fall under the purview of civil rather than criminal law, it is important not to allow powerful institutions or individuals to silence those who criticize them by suing them, in abusive fashion, for slander. In the Faurisson affair, nobody could regard themselves as being *personally* defamed, and to put forward notions such as the defamation of memory or the responsibility of the historian is once again to abuse the notion of slander in order to reduce certain opinions to silence.[14]

Aside from these questions of legal principle, there exist pragmatic problems that are ignored by people who think "that there are some things you can't say." The first is the efficacy of censorship. Anticommunist ideas of all kinds spread through the socialist countries despite censorship. But the same thing happened with republican ideas under the monarchies or the idea of independence in the colonies. And all that was before the Internet. Since the advent of this new medium, all you can say to the censors is "good luck." Besides, to believe that laws criminalizing the expression of racism have any positive effect at all is to seriously delude oneself about the reality of racism in our societies. As for Holocaust denial, the tenor of statements made by the president of Iran in 2005 suggest that such ideas are relatively widespread in the Muslim world.

Another pragmatic problem encountered by censorship is that human thought is something so subtle, and capable of so many nuances, that when you try to put shackles on it (what can and cannot be uttered), it almost always finds a way to evade them or to say the very thing the censors want to prohibit "in a different fashion," "with other words," and so on. Only an absolutely totalitarian system can prevent this from happening, and then only for a while. Laws such as the Gayssot law, which try to prohibit certain thoughts in a society that is democratic, are especially absurd. To prevent the drift into totalitarianism, such laws are formulated so as to aim at very precise targets (the rejection of the Nuremberg decision on . . .), but in doing so, they contradict one of the very bases of law in democratic societies: the fact that laws are general in nature.

A final pragmatic problem is that of the slippery slope: where will censorship end? If we censor Mr. X but authorize Mr. Y to defend Mr. X's freedom of expression, it will be very difficult to keep the opinions of Mr. X from becoming known. But to keep that from happening is probably one of the main purposes of censorship. So it becomes necessary to silence Mr. Y as well. Then along comes Mr. Z, who agrees with the condemnation of Mr. X but believes that Mr. Y should be able to express himself. These are not purely theoretical concerns. The Gollnisch affair illustrates the problem. Gollnisch is a professor of international law and Japanese civilization at the University of Lyon. He is also a prominent member of the National Front, and he was suspended from teaching for having said that historians ought to be able to work freely and that a debate on the crimes committed during World War II ought to take place.[15] These statements are not, in themselves, illegal (yet). It is only in light of Gollnisch's position in the National Front that his statements were interpreted as being implicitly denialist. But, even assuming that this interpretation is correct, the juridical problem posed by this suspension is not just that it reintroduces "thought crime" but that it also puts a person's intentions on trial, since it is his *intention*, "what is going on inside his head," and not the fact itself that constitutes the infraction here.

The question is not, by the way, whether or not one "likes" Gollnisch (or Faurisson, or Irving, etc.) but how far one is willing to sacrifice the most elementary principles of justice and law to silence people whom one does not like.

It is paradoxical, come to that, how frequently in France it is the left or the extreme left that encourages censorship, against its enemies at least (no freedom for the enemies of freedom, for "fascists," etc.),[16] without realizing that the logic of censorship entails that sooner or later it will be utilized by the powers that these political movements wish to criticize and thus will turn around and bite them. The conviction of "radical" critics of Zionism on charges of anti-Semitism illustrates the phenomenon perfectly.[17] You often hear it said on the left that the defense of freedom of expression, when it comes to Faurisson or Gollnisch, "serves the purposes" of the extreme right. But, to the extent that that is true (which is debatable), it is only because the bulk of the left refuses to defend freedom of expression in principle, meaning even for its enemies.

Finally, there is often confusion between access to the media and the opportunity to express oneself in a private capacity. There is no doubt that, in the United States as in France, access to the media is a great deal easier for proponents of the "free market" and militarism. Chomsky is the last person who would gainsay that, having done a great deal to analyze the phenomenon. And such bias is a serious problem in a society that calls itself democratic. But whether or not someone has the right to express her ideas in a private capacity, by expounding them orally, writing letters, or sending articles to newspapers, is an entirely different matter. No one has ever insisted that Faurisson should be able to present his ideas on the evening television news: the adversaries of censorship simply ask that he should not be punished for expressing himself in purely individual fashion. The answer to the problem of bias in the media comes, once again, from the ideas of Mill: those who criticize the social order ought to be insisting that the dominant media give more space to real debate, where their ideas can clash with those of others "on a level playing field," not demanding that the owners of these media should indirectly be provided, through the reinforcement of censorship, with the opportunity to suppress dissent of any kind.

Vidal-Naquet's Critiques of Chomsky

Vidal-Naquet might perhaps have agreed with everything that has just been said, but the rebukes he directed at Chomsky would seem to indicate that he did not grasp that while one obviously ought to ascertain, in defending freedom of expression, that what is at stake really is the expression of opinion and not, for example, incitement to commit illegal acts, one should not pass judgment on the veracity of the writings that are attacked or the intentions of those who composed them. The fact is, if we accepted the obligation to assess the truth of statements made subject to criminal charges, then any defense of freedom of expression would become impossible in practical terms, if only because of time constraints: at a minimum you would have to study the file, scrutinize the exact statements incriminated, and have them translated if necessary. (Chomsky does in fact read French, but it is in no way an obligation, when one is defending someone's freedom of expression, to read the language in which he writes.)[18] In addition, you would have to make a balanced assessment, distinguishing the true from the false in the incriminated writings (and as we shall see below, even Vidal-Naquet acknowledges certain merits in those of Faurisson). In fact, Vidal-Naquet gives a good argument in Chomsky's favor inadvertently when he writes "Ought one to refute Butz?[19] It would be possible, of course, and even easy, assuming one knew the file, but it would be long and tedious."[20] That is exactly why the defense of freedom of expression must not be concerned with finding out whether the victim of censorship "is telling the truth or lying."

Whether the statements of those accused of thought crime were true or false is a question that was obviously never put to Chomsky when he was defending Soviet dissidents who were, as he himself said, "advocates of ongoing U.S. savagery in Indochina, or of policies that would lead to nuclear war, or of a religious chauvinism that is reminiscent of the dark ages."[21]

Finally, there is the question of intentionality: determining the intentions of an author is even more difficult than determining whether what he says is true or false. It is evident that Faurisson's research was "historical" in nature, rather than, let's say, sociological or philosophical. Qualifying it with that term implies no judgment on the value of his research. As a physicist, I often receive nutty letters and papers written by "independent" researchers and dealing with topics in physics. I have no way, as I read these texts, of divining the intentions of their authors, but I see no reason why I should not refer to them as works on "physics," even if they are without value.

Let us now examine in more concrete detail the problems encountered by whoever wishes to go beyond the simple defense of freedom of expression apart from content, and in order to do so, let us simply highlight everything that Vidal-Naquet concedes to Faurisson.[22] He acknowledges that Faurisson supplied the "demonstration that Anne Frank's *Diary*, as it was first published, is, if not a 'literary hoax,'[23] at least a document that has been tampered with."[24] Returning to this point, he adds in a note: "For the sake of completeness, I will say that in his new book there is material on gas chambers that were either imaginary or did not function in the western camps, Buchenwald and Dachau."[25] Vidal-Naquet thus concedes a great deal more to Faurisson than Chomsky ever has.

Vidal-Naquet also wrote a preface to a book by his friend Arno Mayer, published in French in 1990.[26] What do we find in this work? Mayer writes that:

> Sources for the study of the gas chambers are at once rare and unreliable. . . . No written orders for gassing have turned up thus far. . . . Most of what is known is based on the depositions of Nazi officials and executioners at postwar trials and on the memory of survivors and bystanders. This testimony must be screened carefully, since it can be in-

> fluenced by subjective factors of great complexity. . . . In the meantime, there is no denying the many contradictions, ambiguities, and errors in the existing sources.

It is true that Arno Mayer adds (contrary to Faurisson) that "such defects are altogether insufficient to put in question the use of gas chambers in the mass murder of Jews at Auschwitz."[27] Still, he expresses himself in a manner that might lead to confusion. If a physicist or a biologist says that the sources that support a given idea are "rare and unreliable," one will naturally be inclined to think that this scientist has doubts about the idea in question. It is true that Vidal-Naquet criticizes this phrase of Arno Mayer in his preface. But he begins by declaring: "that today we are able to state that the numbers given in such and such a piece of testimony have to be reduced by 75 percent is a gain for scholarship that we would quite wrong to snub."[28] But he does not state precisely which numbers are meant, which once again risks causing the reader to assume that the historical data are not, in general, as certain as they are thought to be. He then states that, "after the book of Jean-Claude Pressac,"[29] it will no longer be possible to refer, in relation to the gas chambers, to "rare and unreliable" sources, as Mayer does.[30] Fair enough. But Pressac's work was published in 1989. If we follow Vidal-Naquet's logic to the letter (after Pressac's book we can no longer refer to rare and unreliable sources), then before his book we could—and the Faurisson affair took place in 1980. Obviously, I do not think that is what Vidal-Naquet intended to say, but since the allegations he originally made against Chomsky bore essentially on the use of slightly ambiguous expressions that he took to be scandalous ("findings," even if taken to mean "*découvertes* [discoveries]" and "historical"), I merely wish to point out here that one can turn the same technique back on him.

In sum, we have two contrasting attitudes: Chomsky defends Faurisson's freedom of expression, says nothing about his writings, and does not debate with him; Vidal-Naquet reproaches Chomsky with being too favorable to Faurisson and undertakes a critique of Faurisson's writings. But as I see it, Vidal-Naquet thus does much more to propagate Faurisson's ideas—inadvertently of course—than Chomsky does. The lawsuits brought against Faurisson and others (for which Vidal-Naquet was not responsible) have, as Chomsky remarks, guaranteed maximum publicity for the Holocaust deniers, especially in the Muslim world.

But what is really "scandalous" (to use his own word) in Vidal-Naquet's stance is that he denies that the attacks on Faurisson constitute an infringement of freedom of expression. He takes the view that "the conditions under which Faurisson was brought to request leave of Lyon . . . were certainly regrettable, and I have said as much, but his freedom of expression, subject to extant laws, has not been threatened at all." He likewise considers that the suits launched by LICRA and other groups against Faurisson "do not prevent him from writing or being published." If one follows Vidal-Naquet's line of reasoning here, the Soviet state did not prevent a certain number of dissidents from being published (in the West), it only punished them when that occurred. Even if the sanctions here are less drastic than they were there, the fact of suing an individual, harming his career, or intimidating him in any fashion on account of his opinions *is* a form of repression of freedom of expression. In the United States, groups such as the American Civil Liberties Union (which is just as "antifascist" as the corresponding French organizations, but with a totally different conception of what "the struggle against fascism" is) did not hesitate, at around this time, to launch appeals against court orders that prohibited the Ku Klux Klan and the Nazis from demonstrating. Moreover, they demanded that the demonstrators be protected by the police, which the University of Lyon did not do in the case of Faurisson.[31]

The height of incomprehension is attained when Vidal-Naquet asks rhetorically whether Chomsky is requesting that Faurisson's works "be advertised and sold at the entrances to synagogues" while affirming that "the principle he invokes [freedom of expression] is not what is at stake."[32] Alain Finkielkraut displayed the same incomprehension at the time of the Faurisson affair, writing in *Libération* (January 16, 1981) that the lawsuits brought against Faurisson were part of "the normal exercise of freedom of expression" in "all democratic countries, including the United States." Finkielkraut claims to regard freedom of expression as "the highest and least debatable of principles" but seems not to realize that such lawsuits would in fact have no chance of succeeding in the United States, or in many other countries. In the same article, Finkielkraut wrote: "For our Chomskians, the Nazi genocide no longer forms part of reality, it has ceased to exist concretely [*il a cessé d'avoir lieu*]. It has now entered another domain: that of tastes and colors. One can believe in it or not, as climate, caprice, digestion and temperament dictate, the way one believes in Father Christmas or the victory of the left." This is exactly the opposite of what Chomsky thinks: he has described the massacre of the Jews as "the most fantastic outburst of collective insanity in human history,"[33] and he has never concealed his total opposition to relativism, moral or cognitive. Readers will note that rhetoric of this kind is the best way to sow confusion in people's minds, by mixing up the question of the defense of freedom of expression with the question of assent to the content of a given discourse.

An Affair That Is Not Over

In fact, one can go much farther in defense of Chomsky. Let us suppose that, contrary to what is the case, he had quite deliberately written a preface to Faurisson's book in order to symbolically defend his freedom of expression. In what respect would this gesture have been scandalous? After all, the text of the preface is very clear about the fact that he limits himself to the question of freedom of expression and does not in the least address Faurisson's ideas. In addition, the "book" by Faurisson to which Chomsky's preface is attached is a memorandum in his own defense, responding to the accusations directed at him. If "antifascist" organizations had not brought liberticidal lawsuits against him, this book might never have existed. Why do historians, among them Vidal-Naquet, who denounce lawsuits against thought crime in principle, never feel obliged to make the slightest gesture in defense of the victims of these lawsuits and indeed condemn those who do make such a gesture? What is antifascist, or even just humanly decent, about attacking Faurisson when he is the target of aggression, including the physical sort, from all sides and is prevented from giving courses (on subjects that have nothing to do with the Holocaust), all the while demanding that he be prevented from replying?[34] (In his *Mémoires*, Vidal-Naquet recounts that when he published "A Paper Eichmann" in *Esprit*, he presented its editor, Paul Thibaud, with just one condition: "In no case would *Esprit* publish a response from Robert Faurisson.")[35] Is Chomsky's stance, which is that in an affair of this kind the foremost priority is to defend freedom of expression, even at the risk of incurring discredit (which he himself has done, thanks in part to the attacks of Vidal-Naquet), not the most noble and at the same time most lucid one?

In an article in *Marianne* (December 24, 2005) favorable to the petition of the nineteen historians, Philippe Cohen writes: "Condorcet, Voltaire, Zola, awaken, we are all about to go mad! In the last several weeks the public debate in France has taken on a warped, sometimes asphyxiating, tone [*une tournure viciée, parfois irrespirable*]. Freedom of speech, and even of thought, are in danger." But the problem did not date from just "several weeks" ago. What happened recently is that some groups marginalized in the past (Africans, West Indians, Arabs) intervened in the debate, particularly about colonialism and slavery, and attempted (wrongly, in my view) to utilize for their own ends the logic of state truth and state censorship. But this should not hide the fact that this logic took root at the time of the Faurisson affair and that it is precisely this logic that leads to the drift that one deplores today—or are we to marginalize the groups in question *ad vitam aeternam*?

Vidal-Naquet entitled the book in which he assembled his writings against Holocaust denial *Assassins of Memory*. His systematic and clumsy attacks on Chomsky have, on account of their considerable impact (flowing from the moral authority he derives from his double status as a victim of the Holocaust and as one who denounced torture during the Algerian war),[36] turned Vidal-Naquet himself, even if this was not his intention, into one of the main assassins of the radical social critique incarnated by Chomsky, in particular the critique of American foreign policy. At a time when American imperialism, supported *mezzo voce* at least by a large part of the French intelligentsia, is lashing out not just in the Middle East but also, if only in words, against France and the republican ideals it represents, one may wonder if Vidal-Naquet and all those who have followed him have any reason to be proud of having so effectively caused to vanish from France—and this at a time when they were spreading everywhere else—the ideas of one whom even the *New York Times* calls "the greatest living intellectual."[37] One may also wonder if the ideals of Voltaire, as well as those of laicity and the République Française, were not better embodied during the Faurisson affair by an anarchist American Jew than they were by the entire French intelligentsia, which chose to despise and ignore him.

NOTES

1. Chomsky responded vigorously to his French critics. See Noam Chomsky, *Réponses inédites à mes détracteurs parisiens* (Paris: Spartacus, 1984). This remarkable book contains unpublished letters, or ones published only in truncated form, addressed to publications such as *Le Monde, Le Matin de Paris*, and *Les Nouvelles Littéraires*. It includes as well the full text of an interview originally prepared for publication in *Libération* in 1981 but which he refused to allow to appear there in mutilated form. This interview also appears in the original French edition of this book, *Cahier Chomsky*, under the title "Réponses à mes détracteurs parisiens." In it Chomsky states: "The French courts have now condemned Faurisson for having, among other pieces of knavery, failed to show the 'responsibility' and the 'prudence' of a historian, for having failed to use probative documents, and for having 'let others take charge of his discourse (!) with the intent of furnishing an apologia for war crimes or incitation to racial hatred.' In a show of moral cowardice, the court then claims that it is not restraining the right of the historian to express himself freely, only punishing Faurisson for having made use of it. With this shameful judgment, it gives the State the right to determine what is official truth (the protestations of the judges notwithstanding), and to punish those guilty of 'irresponsibility.' If this does not trigger widespread protests, it will be a black day for France" (222–223). Naturally, there were no protests. All this took place before the Gayssot law, the purpose of which was to supply a legal foundation for proceedings of this type, was passed. The text of the Gayssot law, and other relevant documents, were printed as appendices to *Cahier Chomsky*.

2. See my afterword, "Folies et raisons d'un processus de dénigrement. Lire Noam Chomsky en France," to Chomsky's book *De la guerre comme politique étrangère des États-Unis* (Marseille: Agon, 2002) for a fuller discussion of the denigration of Chomsky in France.

3. "Liberté pour l'histoire," *Libération* (December 13, 2005), published as one of the appendices to *Cahier Chomsky*. [Translator's note: The generic meaning of the French term *révisionnisme* is simply "revisionism." But in the present context, it bears the specific meaning of "Holocaust denial," and that, or a variant, is how I translate it throughout.]

4. Pierre Vidal-Naquet, "Un Eichmann de papier," *Esprit* (September 1980). Reprinted with revisions in Vidal-Naquet, *Les Assassins de la Mémoire* (Paris: La Découverte, 1987). This book appeared in English as *Assassins of Memory: Essays on the Denial of the Holocaust*, trans. Jeffrey Mehlman (New York: Columbia University Press, 1992). The 1980 article is entitled "A Paper Eichmann." [Translator's note: All citations here of this and other writings assembled in *Assassins of Memory* are from Mehlman's English translation, which I have adapted in one case.]

5. The term "conclusion" is the correct French translation of the English word "findings," which Vidal-Naquet had initially and improperly translated as *découvertes* [discoveries] and which he therefore took to be an illustration of Chomsky's support for the theses of Faurisson. In *Assassins of Memory*, he acknowledges this error, but asserts that this problem of translation is

"insignificant" ("On Faurisson and Chomsky [1981]," 68), even though it lies at the origin of his quarrel with Chomsky. In fact, if the petition does adopt a tone slightly favorable to Faurisson's ideas, that is no doubt because it was drafted by persons close to him, which is unfortunately the inevitable result of the totalitarian climate that prevails among the French intelligentsia. If there existed in France persons or associations that really defended the freedom of expression, which is to say the freedom even of people who have opinions different from theirs, petitions on behalf of Faurisson that were more rigorously neutral in their wording would be possible, or might not even be necessary, since in a less totalitarian climate Faurisson's rights would be respected. As for Chomsky, he was faced with a simple dilemma, one faced by everyone presented with a petition: to sign or not. Rewriting the petition is generally not an option. Given his constant commitment in favor of freedom of expression for people who defend positions contrary to his own (apologists for the Vietnam War, for example, whom some in the student movement at the time wanted to silence), one can perfectly well understand why he signed this petition without hesitation. The French text of the petition appears in the appendices to *Cahier Chomsky*.

6. "A Paper Eichmann," in *Assassins of Memory*, 58.

7. See Noam Chomsky, *Réponses inédites à mes détracteurs parisiens* (Paris: Spartacus, 1984). That an intellectual with the international renown of Chomsky should be unable to respond to attacks against him in the French press and should only be able to find a marginal publisher to publish his responses seems not to disturb Vidal-Naquet, who contents himself with a mention of the "small book published, alas, by Editions Spartacus" (*Assassins of Memory*, n. 18 to "On Faurisson and Chomsky," 162). It would no doubt have been better, as far as Vidal-Naquet was concerned, if Chomsky's responses had not been published at all.

8. Robert Faurisson, *Mémoire en défense. Contre ceux qui m'accusent de falsifier l'histoire. La question des chambres à gaz* (Paris: La Vieille Taupe, 1980). Chomsky's preface is entitled "Quelques commentaires élémentaires sur le droit à la liberté d'expression." See *Cahier Chomsky* for the French text. For the English text, see Noam Chomsky, "Some Elementary Comments on the Rights of Freedom of Expression" (October 11, 1980), http://www.chomsky.info/articles/19801011.htm

9. See in particular Vidal-Naquet, "On Faurisson and Chomsky" (1981; with a 1987 postscript), in *Assassins of Memory*, 65–73.

10. With, among other things, the publication of Chomsky's *De la guerre comme politique étrangère des États-Unis* (2002).

11. To be precise, the following discussion will limit itself to the situation in rich, powerful, and internally stable countries such as France and the United States. The question of freedom of expression in countries with a very different socioeconomic situation (Cuba, Iran, China, Rwanda, etc.) would require a separate discussion, which will not be undertaken here.

12. See Noam Chomsky, *Media Control: The Spectacular Achievements of Propaganda* (New York: Seven Stories Press, 2002); and R. W. McChesney, *Corporate Media and the Threat to Democracy* (New York: Seven Stories Press, 1997); published together in French translation as Noam Chomsky and R. W. McChesney, *Propagande, médias et démocratie*, new augmented ed. (Montréal: Écosociété, 2004).

13. One notes in passing that when the Western press congratulates itself because it dares, without running any real risk, to publish these cartoons, it forgets to congratulate certain publishers in the Arab countries when they, in their fashion, "combat censorship" by publishing the writings of Faurisson, Thion, or Garaudy.

14. During his first trial, Faurisson faced civil suits launched by "antifascist" groups (and lost). This was before the Gayssot law, which penalized Faurisson's "crime." The fact that certain associations have the right to consider themselves as coming under attack from the "racist" opinions, for example, of an individual is obviously a serious hindrance to freedom of expression.

15. See, for example, http://www.phdn.org/negation/gollnisch2004.html, for quotations from Gollnisch, from an article that is hostile to him.

16. All the while proclaiming itself "antifascist" and "anti-Stalinist" with a perfectly good conscience but apparently without realizing that censorship can only be defended coherently within a totalitarian logic.

17. Examples include the "affairs" of Morin-Sallenave-Naïr, Dieudonné, Mermet, Boniface, Ménargues, and a few others. These cases certainly differ among themselves, but they all have in common the "repression of anti-Semitism" (real or imaginary), which the "antifascist" left has done so much to encourage. Mention may also be made of the pressure exerted on the French historian Annie Lacroix-Riz, who was accused of "denial" with respect to the Ukrainian famine of the 1930s. See http://www.historiographie.info/menu.html.

18. It is a fair presumption that all those who defend the freedom of expression of "dissidents" writing in Russian, Chinese, Farsi, and so on do not have detailed first-hand knowledge of their writings.

19. Arthur Butz is an American author, a Holocaust denier whose works preceded those of Faurisson.

20. Vidal-Naquet, "A Paper Eichmann," in *Assassins of Memory*, 51. [Translator's note: I have adapted Jeffrey Mehlman's translation here, substituting the word "file" where he writes "archives." The word used by Vidal-Naquet is "*dossier.*"]

21. From Chomsky, "Some Elementary Comments on the Rights of Freedom of Expression."

22. Not being a historian, I will not pass judgment on the pertinence of the concessions Vidal-Naquet makes to Faurisson, only on the internal logic of Vidal-Naquet's approach and on the risk he runs, in some of what he says, of causing unsophisticated readers to take on board ideas opposite to the ones Vidal-Naquet himself wishes to encourage.

23. Faurisson's expression was "*supercherie littéraire.*" See document II in Serge Thion, *Verité historique ou verité politique?* (Paris: La Vieille Taupe, 1980).

24. Vidal-Naquet, "A Paper Eichmann," in *Assassins of Memory*, 17. Vidal-Naquet regards this concession as unimportant. But that is debatable: the diary has become very well known, after all, and to have shown that it was "tampered with" might seem to an unsophisticated reader to be not such a slim result for an amateur historian—Faurisson—who is, moreover, accused of being a charlatan.

25. Vidal-Naquet, "On Faurisson and Chomsky," in *Assassins of Memory*, 69–70, and n. 12, p. 162. Vidal-Naquet does, however, go on to state in the same note that Faurisson's new material is "poorly analyzed" and "hard to utilize."

26. Arno J. Mayer, *Why Did the Heavens Not Darken? The "Final Solution" in History* (New York: Pantheon Books, 1989). The French edition of Mayer's book with a preface by Vidal-Naquet is entitled *La "Solution finale" dans l'histoire* (Paris: La Découverte, 1990). This preface appears in an English translation entitled "On an Interpretation of the Great Massacre: Arno Mayer and the 'Final Solution'" as chapter 11 in Pierre Vidal-Naquet, *The Jews: History, Memory, and the Present*, trans. David Ames Curtis, with a foreword by Paul Berman and a new preface by the author (New York: Columbia University Press, 1996).

27. Mayer, *Why Did the Heavens Not Darken?* 362–363.

28. "Que l'on puisse dire aujourd'hui que tel témoignage doive être affecté, quant aux nombres, d'un coefficient de division par quatre, est une conquête scientifique que nous aurions grand tort de bouder." From Vidal-Naquet's preface to Mayer, *La "Solution finale" dans l'histoire*, viii–ix. [Translator's note: Vidal-Naquet is referring to the fact that in some cases the numbers of Jewish victims given in certain testimony had been revised downward in light of historical research. The translation is my own. David Ames Curtis offers the following transla-

tion: "It is a scholarly success that today we are able to say that such important evidence is to be assigned a divisor coefficient of four, and we would be greatly wrong to go around sulking about it." Vidal-Naquet, *The Jews: History, Memory, and the Present,* 160.]

29. Jean-Claude Pressac is the author of the book *Auschwitz: Technique and Operation of the Gas Chambers* (New York: The Beate Klarsfeld Foundation, 1989). Vidal-Naquet characterizes him as "once a strong Holocaust denier."

30. From Vidal-Naquet's preface to Mayer, *La "Solution finale" dans l'histoire,* viii–ix. An English version appears in Vidal-Naquet, *The Jews: History, Memory, and the Present,* 160.

31. Freedom of expression may not be better respected overall in the United States, the land of McCarthyism, than it is in France. In my introduction to Chomsky's "Réponses à mes détracteurs parisiens," I describe not only the editorial manipulations on the part of *Libération* that caused Chomsky to withdraw the interview from publication in that newspaper but also a case of "private" censorship in the United States involving Chomsky and Edward S. Herman; see *Cahier Chomsky,* 219. But, contrary to the situation that prevails in France, the notion of freedom of expression is generally understood in the intellectual world (where Chomsky's stance on the matter shocks almost no one), and there exist organizations such as the American Civil Liberties Union, which defend this freedom "360 degrees," meaning for their enemies too. This is what escapes Philippe Val (in *Charlie Hebdo,* June 19, 2002) when he ridicules Chomsky for paying tribute to the Americans when they authorize "Nazi marches and Holocaust-denial publications." In fact, almost all Americans, including those who on other questions would share Val's own political stance, would agree with Chomsky on the question of freedom of expression. There exists, on this topic, a difference in political culture between France and the United States, even if mentalities in France are showing signs of evolving.

32. Vidal-Naquet, "On Faurisson and Chomsky," in *Assassins of Memory,* 71–72. The rhetoric about the sale of Faurisson's works at the entrance to synagogues exemplifies the confusion, discussed above, between the private expression of opinion and access to channels of diffusion.

33. Noam Chomsky, "His Right to Say It," *The Nation* (February 28, 1981), http://www.chomsky.info/articles/19810228.htm. Chomsky is quoting himself from his book *Peace in the Middle East.* For the French text, see "Il a le droit de le dire," in *Cahier Chomsky.*

34. The fact that it is the "antifascists" who consistently refuse to engage in the rough and tumble of debate with the "fascists," at any rate on questions of this kind, shows that things have changed a lot in France since the end of the war. Vidal-Naquet, by the way, although he says he does not wish to debate the *révisionnistes* (Holocaust deniers), has in fact debated Serge Thion quite a bit, as shown by the correspondence between them published in Serge Thion, *Une allumette sur la banquise. Écrits de combat* (1980–1992), a text available only on the Internet. I note that Serge Thion was expelled from the CNRS (Centre nationale de la recherche scientifique), where he was a specialist on Southeast Asia, for the crime of Holocaust denial.

35. Pierre Vidal-Naquet, *Mémoires, 2. Le trouble et la lumière* (Paris: Le Seuil/La Découverte, 1998), 271. In the same work, he complains (272) that Chomsky has "covered him with insults in the press all over the world." Since he gives no reference, one is left to wonder where it was that Chomsky could have published these insults, access to "the press all over the world" not being easily afforded him in general. In fact, Chomsky was not even able to get the American magazine in which the English translation of the article in question by Vidal-Naquet had appeared to publish the response he sent them. According to Vidal-Naquet, "it was a sad business" (*ce fut triste*), and there were others besides himself to express this tristesse, "Alain Finkielkraut for example." As we have seen above, Finkielkraut shares with Vidal-Naquet a very particular conception of freedom of expression.

36. See, among other titles, Pierre Vidal-Naquet, *La Torture dans la République* (Paris: Éditions de Minuit, 1972); *La Raison d'État. Textes publiés par le Comité Audin* (Paris: La Découverte,

2002; a new edition of a book first published in 1962 by Éditions de Minuit). [Translator's note: Some library catalogues list Pierre Vidal-Naquet, *Torture: Cancer of Democracy, France and Algeria, 1954–62*, trans. Barry Richard (Baltimore, Md.: Penguin Books, 1963).]

37. Of course, the *New York Times* adds immediately that it does not understand how this great intellectual can write the things he does about the foreign policy of the United States. But at least it awards him this status. Chomsky was also chosen as the most important contemporary intellectual in 2005 by the readers of the British magazine *Prospect*, far ahead of Umberto Eco, Vaclav Havel, or the pope. See http://www.prospect-magazine.co.uk/intellectuals/results.

15

CHOMSKY AND BOURDIEU

A MISSED ENCOUNTER

FRÉDÉRIC DELORCA

During the 1990s, the figures of Noam Chomsky and Pierre Bourdieu emerged as major points of reference for movements "on the left of the left," which were then enjoying a renaissance and heading toward planetwide unification—what has come to be called the "alterglobalization movement." Yet no encounter, either between these two men or between their ideas, ever really took place, a failure that is perhaps symptomatic of the difficulties in dialogue not so much between Europe and America as between two traditions of thought present on both continents.

Two Different Relationships to Factual Objectivity

The main obstacle to dialogue between the worlds of these two intellectuals lay in an essential, one might even say transcendental, point: the epistemological relationship to reality. Whoever peruses either Chomsky's political writings or his work in linguistics is struck by his attachment to the objectivity of facts, which he perceives as univocal and empirically verifiable by scientific means—a stance that clearly places his research at the Massachusetts Institute of Technology in the area of the "hard" sciences (especially the cognitive sciences).

Bourdieu, on the other hand, coming from the constructivist tradition of the social sciences (Berger and Luckmann 1996) and partly inspired by the heritage of Émile Durkheim (Barkow, Cosmides, and Tooby 1992) and Ernst Cassirer (Bourdieu 2001a, 202; 1997, 28), always held to the definition of the object of scientific knowledge as a construct, such that its objectivity depends principally, for him, on the intersubjective agreement of the social actors qualified to produce it (scholars and scientists). He clearly expresses this conviction in his final course at the Collège de France (Bourdieu 2001b, 151) and sees it as a way of overcoming the opposition between "naïve realism" and the "relativist constructivism" of Bruno Latour (Bourdieu 2001b, 151)—an idea we also find, albeit in a form that tones down the agonistic dimension of knowledge, in Jürgen Habermas.

Bourdieu's constructivism, like that of most sociologists of his generation for that matter, goes along with a polemical stance against the notion of human nature, which is seen primarily as an element that legitimates political conservatism (Bourdieu 1997, 114). Chomsky, in contrast, has always emphasized the importance of human nature as the site of the innate emergence and reproduction of a universal generative grammar that makes language possible, a human nature conceived in a more open manner than it is in "sociobiology," as a bearer of contradictory tendencies—see, for example, his critique of the Lorenzian vision of nature grounded exclusively in competition (Chomsky 1996, 137). Thus the Chomskian theory of natural language inscribed in the genes rejects the structuralist heritage (which Bourdieu shares), for structuralism assigned a radical autonomy, in relation to the biological substrate, to the sphere of representations (see, for example, Lévi-Strauss's completely antinaturalistic interpretation of the incest taboo).

There is no doubt that Chomsky's attachment to objectivity and a positivist tradition entails a form of intellectual humility that is absent in Pierre Bourdieu, especially in his manner of approaching disciplines other than his own. Hence, while Chomsky has always been prudent enough to make it clear, particularly in his correspondence, that he was not really familiar with the sociology of Pierre Bourdieu (which might also be taken as a sign of disdain), Bourdieu hazarded an unanticipated attack on Chomskyan linguistics. "Chomskyan 'competence' is simply another name for Saussure's *langue*. . . . The shift in vocabulary conceals the *fictio iuris* through which Chomsky, converting the immanent laws of legitimate discourse into universal norms of correct linguistic practice, sidesteps the question of the economic and social conditions of the acquisition of the legitimate competence," he wrote in *Language and Symbolic Power* (1991, 44). In his *Méditations pascaliennes*, Bourdieu charges Chomsky with making the principle of discourse "grammar, a typical product of the scholastic point of view" (1997, 67). The implication here is that the hypothesis of a universal grammar, far from answering the concrete questions raised by the study of language, is the projection of a university professor who is intellectualizing the practice he observes. These charges—purely gratuitous and apparently supported by nothing except the forcefulness with which they are stated, unburdened by any technical justification—are so misplaced that they rebound against sociology. Is this hubris on the part of an overachiever from the École Normale Supérieure who fancies himself authorized to hold forth on any topic without having studied it seriously,[1] or is it a manifestation of a certain tendency in the social sciences to confuse circular reasoning with rigorous argument on empirical grounds? It is likely in any case that this lapse did nothing to make dialogue between the two men any easier.

Vice versa, though, one could make the case, from a Bourdieusian point of view, that Chomsky's refusal to study critical sociology has deprived him, especially in his political analyses, of conceptual resources for apprehending the relations between Western institutions and the dynamic that drives them and prevented him from grasping the mechanisms that condition language and representations, including even bodily practices, in richer and more problematic terms than the Chomskyan vocabulary affords, trapped as it often is in a certain functionalism.

Bourdieu, Chomsky, and the Dominant Political Ideas in the French Intellectual Field

Yet, divergences, misunderstandings, and reciprocal ignorance apart, the shared features of their intellectual itineraries are striking. Bourdieu, born in 1930 in a modest rural milieu in Béarn in south-west France, influenced by a father who aided the peasants in his village and frequented Spanish republicans in exile (Bourdieu 2004, 111), and Chomsky, a child of the "Jewish ghetto in Philadelphia—not a physical ghetto . . . but [a] cultural ghetto,"[2] two years his elder, who wrote his first "political essay" on the war in Spain at the age of ten, are marked by the same libertarian relationship to institutions and especially by a shared mistrust of the chic intelligentsia, especially when it adopts a phony left-wing rhetorical pose. The two men both reject the Marxist vulgate that was the *koinè* of the Western (and especially French) intellectual elites in the 1950s and 1960s. Allergic to fashion, both of them also refused, with the same determination, the political reaction that began in the 1970s with the appearance of the *nouveaux philosophes* ("new philosophers"), the avant-garde of triumphant neoliberalism and humanitarian neocolonialism. Chomsky, who knew the French intellectual milieu almost as well as Bourdieu, must have taken the rise to power of these media intellectuals almost as badly as did Bourdieu. For their part, they covered him with opprobrium on account of his preface to the Faurisson book—a polemic in which Bourdieu took no part.

Chomsky's oppositional sensibility, like Bourdieu's, is rooted in a profound openness to alterity, especially that of the third-world peoples who have suffered from Western imperialism. When Bourdieu was a young man doing military service in 1957, Kabylie and the fate of the Algerians amid a full-scale war of decolonization made a deep impression on him, just as the Vietnam War, of which Noam Chomsky was one of the leading opponents in the United States, was a major milestone in his intellectual trajectory.

This has caused both men to be ostracized equally by the institutions (notably the mainstream media) in their respective countries. Chomsky, throughout his career, has been relegated to the "Siberia of American discourse," as *Business Week* put it (April 17, 2000), and on the French media and political scene, Bourdieu, when he supported the general strike in 1995, was derided as an archaic mandarin and accused of putting his scholarship in the service of corporatist interests (like the unions representing minor bureaucrats) or outmoded revolutionary utopias.

On the terrain of politics, Chomsky—who has always been strongly inclined toward the study of international relations—doubtless has a more global vision of the overall stakes for humanity. While Bourdieu, along with Michel Foucault, Yves Montand, and Bernard Kouchner, signed a very *bien pensant*[3] article published in *Libération* on December 15, 1981, which reproached the French socialist government for its passivity in the face of the coup d'état in Poland (Bourdieu 2002, 164), Chomsky for his part drew attention to more uncomfortable truths: if a priest was murdered in Poland at the start of the 1980s, dozens were dying at the same time in Central America, along with hundreds of peasants, at the hands of paramilitaries armed by the CIA. Faithful to the philosophy of Bertrand Russell, for whom true courage lies in criticizing one's own side, Chomsky compared intellectuals who attacked the Soviet bloc to Soviet journalists who spent their time denouncing the West rather than opposing Brezhnev's policies in Afghanistan (Chomsky and Peck 1987, 223–227). By focusing the thrust of his and Edward Herman's book *Manufacturing Consent* on the propaganda of the mainstream media concerning Nicaragua and El Salvador, rather than on the already waning Soviet empire, the linguist took greater risks than did the sociologist and at the same time saw further than the latter did, perceiving that the repression of the small peoples of Central America by the dominant military power bore within itself a dynamic of the destruction of progressive values far more dangerous for humanity than the coup d'état in Warsaw.

The Aborted Encounter

During the second half of the 1990s, when Bourdieu decided to assume, within France, a role comparable to that once played by Jean-Paul Sartre in order to resist the rise of neoliberalism, the names of Bourdieu and Chomsky often appeared in the same vicinity. For example, they both signed two petitions against the policies of NATO in the Federal Republic of Yugoslavia in 1999–

2000 (the "Appeal for a Just and Durable Peace in the Balkans" of 1999[4] and the Bruxelles Appeal of 2000),[5] which caused them to be listed among the "objective accomplices of Milošević" by a major French weekly (the equivalent at the time of being put on the proscription list under the dictatorship of Sulla). The two thinkers have also been active in support of the Palestinian cause—shortly before his death, Bourdieu, along with Derrida and Vidal-Naquet, signed a petition of solidarity with the imprisoned Arab-Israeli parliamentarian Azmi Bishara, whom Noam Chomsky also supported[6]—and in the setting up of alternative media (in France, associations like ACRIMED acknowledge the influence of both Chomsky and Bourdieu).

These convergences led some to formulate the project of organizing a debate between the two thinkers, like the ones that took place between Michel Foucault and Noam Chomsky, or between Chomsky and Jean Piaget in the 1970s. The death of Pierre Bourdieu in January 2002 prevented the fulfillment of this project. The task remaining today, despite the academic barriers that render it difficult, is to carry out a thoroughgoing comparison between their two bodies of thought, on the political plane as well as on the epistemological and scientific ones.

NOTES

1. This is especially the case in everything touching the philosophy of language, where Bourdieu often makes ill-judged use of approximate allusions that bend theories out of their initial context in order to adapt them, in a highly artificial manner, to the intuitions of his sociology, and decorate his references with dropped names—especially when he deals with the works of Ryle (Bourdieu 1994, 228; 2001a, 320; 1997, 42, 74, 177), Strawson (Bourdieu 1997, 42, 161), and Kripke (Bourdieu 1994, 84), whose analyses he nevertheless proposes to "put to the test" (see the blurb on the cover of *Raisons practiques*). In result, Bourdieu's concepts with respect to the English-language debates on language can probably only be rigorously discussed on the basis of the secondary literature. See, for example, Shusterman (1999) or the way in which Searle makes use of Bourdieu in the service of his own notion of "the Background" against Chomsky and Fodor (Searle 1998, 172).

2. http://www.chomsky.info/interviews/20020322.htm.

3. Pierre Bourdieu's independence of Parisian intellectual modes was never absolute, as shown, for example, by the article "Hommes poltiques, médias, citoyens—La vertu civile," in which he congratulates the French prime minister Michel Rocard for his "exemplary" action in New Caledonia (*Le Monde*, September 16, 1988). But it is true that such susceptibility to the collective enthusiasms of the elites can be found of late in Noam Chomsky, too, for example when he supported Senator John Kerry, the Democratic Party candidate for president in 2004, rather than urging his readers to vote for the alternative candidate, Ralph Nader (*The Guardian*, March 20, 2004).

4. "Appel pour une paix juste et durable dans les Balkans." Cf. Bourdieu (2002) and http://www.humanite.fr/1999-06-11/1999-06-11–291210.

5. http://helicop.chat.ru/rus90.htm.

6. http://www.haaretzdaily.com/hasen/pages/ShArt.jhtml?itemNo=135095&contrassID=2&subContrassID=1&sbSubContrassID=0&listSrc=Y.

BIBLIOGRAPHY

Barkow, J. H., L. Cosmides, and J. Tooby, ed. 1992. *The Adapted Mind: Evolutionary Psychology and the Generation of Culture.* Oxford: Oxford University Press.

Berger, P., and T. Luckmann. 1996. *La Construction sociale de la Réalité.* Paris: Armand Colin, 1996. Original edition: *The Social Construction of Reality.* New York: Doubleday, 1966.

Bourdieu, P. 1994. *Raisons practiques. Sur la théorie de l'action.* Paris: Seuil.

——. 1997. *Méditations pascaliennes.* Paris: Seuil. English edition: *Pascalian Meditations.* Trans. Richard Nice. Cambridge, England: Polity Press, 2000.

——. 2001a. *Langage et pouvoir symbolique.* Paris: Seuil. Revised and augmented edition of *Ce que parler veut dire.* Paris: Fayard, 1982. English edition: *Language and Symbolic Power.* Ed. J. B. Thompson, trans. G. Raymond and M. Adamson. Cambridge, England: Polity Press, 1991.

——. 2001b. *Science de la science et réflexivité.* Paris: Raisons d'agir.

——. 2002. *Interventions 1961–2000. Science sociale et action politique.* Marseille: Agone.

——. 2004. *Esquisse pour une auto-analyse.* Paris: Raisons d'agir.

Chomsky, N. 1996. *Le langage et la pensée.* Paris: Payot. Original edition: *Language and Mind.* New York: Harcourt, Brace, and World, 1969.

——. 2001a. *De l'espoir en l'avenir. Propos sur l'anarchisme et le socialisme.* Marseille: Agone.

——. 2001b. *De la guerre comme politique étrangère des États-Unis.* Marseille: Agone.

Chomsky, N., and J. Peck. 1987. *The Chomsky Reader.* New York: Pantheon.

Herman, E. S., and N. Chomsky. 2003. *La Fabrique de l'opinion publique. La politique économique des médias américains.* Paris: Le Serpent à plumes. Original edition: *Manufacturing Consent: The Political Economy of the Mass Media.* New York: Pantheon, 1988, 2002.

Searle, J. R. 1998. *La Construction de la réalité sociale.* Paris: Gallimard. Original edition: *The Construction of Social Reality.* New York: Free Press, 1995.

Shusterman, R., ed. 1999. *Bourdieu: A Critical Reader.* Oxford: Blackwell.

PART VI

POLITICS: THEORY AND PRACTICE

16

CHOMSKY IN FRANCE

THE RESISTANCE TO PRAGMATIC ANTI-AUTHORITARIANISM

LARRY PORTIS

The political writings of Noam Chomsky have gained a new popularity in France over the past decade, which, given their general rejection before this time, is surprising. Indeed, until recently, mention of Chomsky in leftist circles in France was likely to immediately evoke a mixture of fear and revulsion, so completely had his name come to be associated with a denial of the existence of the Nazi death camps during World War II. This surprising and quite silly situation was the product of a media campaign led, toward the end of the 1970s, by some of the more socially ambitious and politically opportunistic members of the French intelligentsia, the so-called New Philosophers. These people, such as Bernard-Henri Lévy, Jacques Attali, and André Glucksman, realized that personal advantage could no longer be culled from radical social criticism, and they used attacks upon Chomsky to justify their turn to the political right.[1]

Discrediting radical leftist ideas proved to be highly successful for the "New Philosophers" and their followers. Apostate leftists have now dominated the French media—press, radio, television, and the publishing industry—for a generation. Yet it is in this context that Chomsky has become increasingly popular. How to explain this phenomenon? The answer is found in the very political changes that led to the outrageous besmirching of Chomsky's name in the first place.

There is a direct relation between Chomsky's political ideas and the decline of the radical and revolutionary left, and even of the left in general, in

France. The new situation created by the victory of the Socialist Party in the presidential and legislative elections in 1981, the dissolution of the Soviet Union, and general weakening of radical political organizations coincided with rising living standards. Concurrently, labor unions lost membership and moderated their rhetoric and actions. By the end of the 1980s, political terminology was changing, as "socialism" came to mean merely a component of electoral politics in capitalist society, and "communism" was interchangeable with the near-defunct Soviet Union. In the new political context, any reference to "revolution" was perceived as either dangerous or ridiculously utopian. In fact, the triumphant conservatives, now called "neoliberals" in France, proclaimed themselves to be "revolutionary" and the socialists and communists to be "conservative," because they were opposed to the free-market policies now being touted as "radically" new solutions to social and economic problems.

By the mid-1990s, the political and ideological changes in France had advanced to the point that much of the political vocabulary lost its clarity and pertinence. The new interest in Chomsky dates from this time and it represents something of a paradox. Chomsky, radically opposed to the capitalist productive system, has become popular because of the relative collapse of any indigenous opposition to this system. In other words, the ideological void created in France has facilitated the acceptance of his ideas, even though the nature of his ideas and his mode of thought remains alien to most of those who have become receptive to them. Chomsky's writings are increasingly translated and published in France, but cultural resistance to his approach to social and political analysis remains strong.

Chomsky is liked for the clarity and practicality of his analysis, but those attracted by it do not necessarily share his political ideas. For example, his critique of Marxism has not discouraged French Marxists from referring positively to his work. Similarly, although Chomsky proclaims his anarchism, many French anarchists consider him to be insufficiently rigorous in the application of anarchist principles.[2] The reason for these "misunderstandings," I am convinced, is that Chomsky's epistemological and political orientation is generally alien to French political culture. His mixture of political idealism and pragmatism is puzzling to many of his French readers.

In addition, Chomsky's demeanor is confusing to many people in France. Chomsky is impressive when debating, but in a way that is "foreign" to most people in France. Chomsky seems to listen to other people.[3] He actually responds to what they say, and he does so without the verbal aggressiveness typical of political debate in France, where the desire to dominate the opponent seems to be more important than real communication. Although Chomsky presents his arguments with logic and force, he does not use sarcasm or simple polemics (which does not exclude irony) or systematically interrupt when others are speaking. In France, on the contrary, the tendency is first to establish an adversarial relationship and then to attempt to humiliate the opponent. This is a cultural trait related to hierarchical social relations and to the overweening role of the state in French life.

The State Inside the Head

Chomsky is not generally understood in France because of deep cultural differences. Most particularly, these differences are manifest in, and transmitted by, the authoritarian national education system, which is perhaps the most important factor in the formation of "attitudes" in France. A significant example of this system is the importance of the "lesson" (*leçon*) in a preestablished study program (*programme d'étude*).[4] The lesson is learned by heart and becomes a model for subsequent thinking. In this way, the foundations of "rational" thinking are established. The "master" (*maître*) transmits the lesson as principles to be assimilated unquestionably, not as bases for reflection and future discussion. What is essential in all this, its "hidden curriculum," is submission to authority. Here is a "principle" of sound instruction accepted generally throughout all levels of education in France.

The long-lasting consequence of this instruction on the French mentality is the creation of a deep-seated desire to please the *maître* by reproducing the *leçon*. It is virtually inconceivable to call into question its conceptual foundations. One must emulate the *maître* in order to succeed scholastically. The intellectual "brilliance" of a student depends on his or her capacity to reproduce a rhetorical model echoing an established argument or postulate rather than opening up new thinking on the subject at hand.

This pedagogical model exists on all levels of the French educational system. From dictation in primary school to the rigidly constructed "dissertation" in secondary (high) school, and then to the textual commentary (*commentaire de texte*) at the university, the form of the exercise predominates in intellectual "formation." The same methods are applied in the exclusive, elite schools (*grandes écoles*) where the students are preselected; there the students are groomed to constitute a future elite. They acquire a great sense of their own worth and learn the etiquette of leadership.

In this system, nationally competitive examinations (*concours*) have a central role, and training for the exams does much to condition modes of thought. In most cases, these rigorous examinations influence intellectual formation more than do research activities. For teachers, the CAPES (national teaching certificate) is a necessary hurdle qualifying candidates for teaching positions in junior high schools (*colleges*) and high schools (*lycées*). The *agrégation* is on a higher level and is considered the necessary passage to a university post (regardless of whether the candidate has already earned a doctorate). As in earlier instruction, but here far more rigidly, candidates follow a line of thought or study the designated subjects with one objective in mind: to develop the ability to rhetorically dominate the questions asked, first in writing, then verbally in front of a "jury." In this system, success in the *concours* is the key to obtaining job security and a position of power within an academic discipline.

The problem with the *concours* is that creative imagination or nonorthodox thinking is a handicap that eliminates many interesting and otherwise capable students. It is possible that the most creative thinkers are weeded out in this process. As Pierre Hadot, the renowned classicist, remarked in retrospect: "This *concours* system, especially the famous *agrégation*, doesn't it weaken the scientific and humanistic education of students? Doesn't it far too often favor rhetorical talents, the ability to talk about a subject even if one knows little about it, or the art of speaking elegantly and obscurely?"[5]

At the university level, even before the *concours*, the structure of degree programs contributes to narrow thinking. The basic premise of university education in France is that each academic discipline is structured as a program of obligatory courses that leaves little room for explorations outside the discipline itself. From the first year, students are put on a degree track with one principal objective: to obtain the diploma after having completed diverse "modules" composed of different, interrelated courses. Often the grade received for one course is averaged with that of another course. Sometimes, if a student fails one course of the courses linked with another in the module, he or she will be obliged to retake both of them.

Such a curriculum, by its very nature, is oriented toward technical training at the expense of general education based on diversity, reflection, and open-mindedness. It is a kind of education that favors conformism. Following a model established by a *maître* comes down to submitting to a higher authority. Once successful, when a student becomes a *maître*, the multiplicity of professional ranks and grades reinforces the internalization of hierarchical relationships and, consequently, the authoritarian mindset. The transfer of the ego toward authority figures becomes a conditioned reflex.

This question of the cultural specificity of attitudes toward authority has been well elucidated by Edward T. Hall, who invokes the importance of the state when speaking of the difference between United-Statesians and the French.[6] The French accept the presence of authoritarian structures, such as the state and state religion, whereas United-Statesians resist them.[7] This in no way is in itself proof that French people are more conformist, but it can be concluded that the different populations express themselves differently within their respective national contexts. Can it be said, for example, that the French are more "individualistic" than United-Statesians? Certainly not. Individualism is, however, expressed differently in the two national cultures. The haughtiness and arrogance of the French are legendary, as is "Yankee ingenuity," the innovative spirit and the practical sense that are said to be characteristic of United-Statesians.

How do different national populations acquire their traits constituting a distinctive "national culture"? Political and juridical institutions certainly are a major influence in the "cultivation" of particular social relationships: interpersonal expectations, socially accepted behaviors, customs, and traditions. In the United States, the "Founding Fathers" were preoccupied with the task of fashioning both institutions and a corresponding national culture. In France, the process of state building was of far longer duration, and the presence of the state consequently far more anchored in French minds. The centralizing authority has filtered into the lives of people and structured mentality more completely.

Whereas in the United States a relatively limited state authority was created in opposition to what was taken to be tyrannical (British) state authority, in France political change since the French Revolution has tended to extend state administrative authority. Beginning with the social revolution of 1789, successive political revolutions have increased state power in relation to that of individuals and localities. Alexis de Tocqueville's observations concerning the strengthening of state authority in France perceptively described the transition from the Old Regime to postrevolutionary France, and they remain accurate today.[8] The state may be a political problem in France, but its refinement is generally seen as the only solution.

Even the rebelliousness and effervescence of the French can be understood as a result of the omnipresence of the state in daily life. The volatility of French society—its tendency toward periodic explosions of spontaneous rebellion, popular uprisings combining frustration and anger—is a cultural specificity produced by the overweening presence of state authority. State centralization produces popular tensions and then canalizes them. The contradictions and logic of this structure of social and political relationships at times results in unusual state practices, such as when the state-controlled railway authority (the SNCF—Société de chemins de fer) offers low-cost train tickets and increases the number of trains so as to help transport demonstrators to Paris. In this case, the administrative and infrastructural centralization of France combines with political centralization in the management of both protest and social control. The fact that Paris, the seat of government—the State—looms so large in the collective imagination of the population is another element in the psychic structuring of French political perceptions and ideas.

The expression of radical and/or "revolutionary" ideas is not free from such cultural influences. Although France harbors all varieties of socially critical philosophies, until the 1980s it was Marxism-Leninism, among other types of "vanguardism," that dominated the French radical Left. The predominance of these rather "authoritarian" orientations in leftist thinking has been congruent with the hierarchically structured institutions and mentalities, the existence of which are more or less taken for granted. Even the "antiauthoritarian" (*libertaire*) organizations and movements in France have found it difficult to avoid the sectarianism founded upon the theoretical authority of a favored thinker (such as Michael Bakunin or Peter Kropotkin). In such a context, the positions and orientation of Noam Chomsky as an anticapitalist thinker and activist have been generally misunderstood.

To Be an Anarchist in the United States

Chomsky forthrightly says he is an "anarchist" but does not go out of his way to say it at the risk of distracting attention from his social criticism. As a linguist and political activist, he seems to take into consideration how words can be manipulated and how concepts can be reified so as to reduce communication with others. Not only is the term "libertarian" associated with the political far Right in the United States, but the word "anarchy" is generally synonymous almost everywhere with "chaos." It seems clear that his objective is to inform and convince, not to be forced to define terms whose meanings are falsified or co-opted by the powers he criticizes. To begin a discussion of outstanding practical problems with an academic discussion of the meanings of such words could best serve the interests of the political and journalistic institutions responsible for the existing confusion. The risk of such confusion is particularly important in the United States, where the general political culture leaves little space for ideological or philosophical debate about social perceptions and political ideas.

In Europe, where anarchism as a recognized social philosophy emerged directly out of the struggles for a more egalitarian society, the notion of a right-wing anarchism is almost inconceivable. William Godwin and Pierre-Joseph Proudhon, the first anarchist theorists, responded critically to intellectual obscurantism and to the political domination and social exploitation endemic to both feudal and capitalist societies. The anarchist movement was solidly anchored to the Left in that it was founded upon the need to eliminate the foundations of oppression and exploitation exercised by the dominant social classes and the state. This anarchist orientation is part of the historical evolution of industrial capitalist countries in Western Europe. Its most fundamental element is resistance to state power; the nature of this resistance depends upon social, political, and cultural conditions and contexts.

In the United States, largely bereft of these social aspects, anarchism tends to be reduced to a simple opposition to the state. This is so in great part because political institutions are infused with respect for individual liberties. The juridical and political system created by the colonial elites at the end of the eighteenth century was founded on the need to conciliate resistance to arbitrary authority and the protection of property interests while respecting a communitarian Puritan tradition of local autonomy. Far from being the product of historical accretion, as in England, the new laws were conceived in relation to reflected-upon principles and immediate practical necessities. Civil equality was the most fundamental principle and universal (male) suffrage the most far-reaching measure. The result was the creation of the most democratic state in the world at that time. The political culture (mentalities and behavior) that resulted still influence the expression of dissidence in the United States, one characteristic of which is a constitution that is rarely called into question. Political dysfunctions are generally attributed to the faulty application of the law under the U.S. Constitution and not to the institutions themselves.

It is difficult to understand Chomsky's political ideas without placing them within this national context. He readily associates himself with a revolutionary political tradition, that of anarchism, but without any formal affiliation with organizations. He says he is an "anarchist" but avoids theoretical or doctrinal debate.

Chomsky may have said the essential in his preface to the English translation of Daniel Guérin's book *L'anarchisme. De la doctrine à l'action*,[9] published in 1970. For Chomsky, anarchism is a "libertarian socialism" calling for the extension of egalitarian social relations to every area of human activity, especially political decision making, the organization and execution of work, and the control of natural resources. In order to understand the foundations of real democracy, he says, one should study the Spanish Revolution or the *Kibboutzim* in Palestine before the creation of the Israeli state. Chomsky shares Guérin's ecumenical vision, which places anarchism at the heart of an overall movement of which it is one tendency and Marxism-Leninism is another.

Unlike Guérin, however, Chomsky has never accepted Marxian epistemological premises (historical materialism), although his analyses of historical phenomena do not reject the practical thrust of such premises. Despite their divergent philosophical positions, Chomsky did not let the past engagements of Guérin exclude his participation in the English publication of a book presenting the anarchist movement with force and conviction. The fact that Marx and Bakunin had political differences almost a century earlier in the struggle for influence over the revolutionary movement apparently for Chomsky should not be a brake on the movement today. Philosophically and analytically, the anticapitalist writings of Marx and Bakunin remain valuable.

At the same time, and regardless of references to Bakunin, Kropotkin, and other thinkers in his writings and interviews, Chomsky appears to follow no intellectual or political model. In his effort to be factually concrete, he avoids abstractions and doctrinal conflicts because they so easily lead to confusion. To be otherwise, to follow the lead of some designated intellectual guru, is the very contrary of informed, independent thinking:

> In fact, as a rule of thumb, any concept with a person's name on it belongs to religion, not rational discourse. There aren't any physicists who call themselves Einsteinians. And the same would be true of anybody crazy enough to call themselves Chomskian. In the real world you have individuals who were in the right place at the right time, or maybe they got a good brain wave or something, and they did something interesting. But I never heard of anyone who didn't make mistakes and whose work wasn't quickly improved on by others. That means if you identify yourself as a Marxist or Freudian or anything else, you're worshipping at someone's shrine.[10]

In France, where attachment to one or more *maîtres de penser* is an established convention, such apparent humility is rare for an intellectual. Even in the United States there has been a recent trend among university scholars to follow the lead or "model" of a thinker considered to "seminal" in some way in the construction of a new orthodoxy (or "canon"). Ironically, "French philosophers" have been most prominent in this regard. One aspect of the practice is the concomitant exclusion of rival perspectives or modes of analysis, as intellectual criticism is so easily perceived as personal attack. Chomsky does not seem to suffer from either excessive or low self-esteem and seems therefore to have remained unconcerned about his relative rejection by the intellectual establishment.

In any case, Chomsky appears to have overcome the sense of self-importance and the professional ambitions that typify even politically committed university intellectuals. Rejecting mystifying abstractions and free of doctrinal attachments whose primary function is to provide ontological security, Chomsky's ideas and political positions are expressed clearly and directly. Maintaining a discursive level accessible to everyone can, however, appear suspect, and it is certainly taken for demagogy by some and seen to be simplistic by others, especially in France, where verbal elegance and artful verbal constructions often take precedence over the need to communicate clearly.

From this perspective, Chomsky's anarchism is predicated upon what he considers practical imperatives stemming from a realistic assessment of social conditions. He observes, for example, that "Guérin describes the anarchism of the nineteenth century as essentially doctrinal, while the twentieth century, for the anarchists, has been a time of 'revolutionary practice,' " and "it is not very difficult to rephrase these remarks so that they become appropriate to the imperial systems of 1970." Well into the twenty-first century, this assessment has not changed. "The problem of 'freeing man from the curse of economic exploitation and political and social enslavement' remains the problem of our time. As long as this is so, the doctrines and the revolutionary practice of libertarian socialism will serve as an inspiration and a guide."[11]

Being "Pragmatic"

I believe Chomsky's preference for the concrete over the abstract is conditioned by a specific United-Statesian culture. To be problem oriented, to focus on particular problems rather than on general questions, is a national characteristic related to the conditions of "nation building" in the United States and the self-conceptions and social expectations they generated. In this cultural context, once a problem is identified, the most logical course of action is to examine the technical aspects of the question. The need for a solution takes precedence over conceptual fine points, considered "academic," or ideology, seen as producing sectarian strife and thus an impediment to efficiency.

Chomsky's "pragmatism," like that of others, is a kind of "antiphilosophy" from a formal philosophical perspective. As the sociologist Emile Durkheim observed, it is closer to a "method" than a philosophy: "Pragmatism is essentially an attitude, a general orientation that intelligence adopts in the presence of problems to be solved. It is an attitude giving priority to 'results, consequences, and facts.'"[12] Although pragmatism has been taught as a philosophy, first by Charles Peirce and William James and since by others, a "philosophy" predicated on the need to induce action implies a nontheoretical bias.

We could say that pragmatism is a philosophy of action based upon a particular perspective on reality, one that posits the "fact-value dichotomy" in favor of the former. Consequently, it can be seen as a kind of idealism founded upon empiricist values. According to Durkheim, the thinking of William James "is based on a simple principle: the valid idea is that which corresponds to factual reality; it is the image, a faithful reflection of things; it is the mental reproduction of something. The idea is true when this reproduction conforms to the thing perceived."[13]

Chomsky's approach, to my mind, bears an affinity with the pragmatists in that, like Peirce, James, and John Dewey, there seems to be an implicit attempt to reconcile empiricism and idealism (both philosophical and political). James, in fact, promoted "radical empiricism" so as to reinforce skepticism. Moreover, the pragmatists rejected "monism" in favor of a causal pluralism considered as more likely to circumscribe phenomenological complexity. What is perhaps most fundamental in this philosophy is that it favors lived experience over the explanation or justification of this experience. It is in this way that pragmatism can appear to be "anti-intellectual."

However, there is "anti-intellectual" and "anti-intellectual." What can be more "intellectual" than attempting to deepen our understanding by showing that philosophical debates are often based on perceptual differences and ideological confusion? Pragmatist philosophy is not anti-intellectual, but it is certainly "anti-intellectualistic." The problem is not in intellectual activity in itself, but rather in "intellectualisms"—the construction of ideas limiting our apprehension of reality and distorting our perceptions. Intellectualisms are mentally constructed so as to protect us from disturbing realities. They exist as a screen or filter shielding us from uncomfortable facts. Such is the role of ideology, and it is why its maintenance requires, at times, hypocrisy (or, rather, "bad faith"). The whole question is how to escape intellectualisms. A major preoccupation for Noam Chomsky seems to be: how to reduce to a minimum the abstractions that, by their very nature, exclude (or filter out) important parts of reality?

In answering the question, it must be accepted that philosophy should be considered a critical examination of abstractions and not the production of confusing abstractions. From this perspective, abstractions must be understood as components of a system of factual representations. Philosophical inquiry identifies abstractions and subjects them to analysis. The overall objective is to reconstitute the reality of human experience in all its social, political, and historical complexity.

In Chomsky's work, pragmatism seems to take two distinct forms. In his analysis of United States foreign policy, he tends to adopt a strict empiricism. In his discussion of the facts and issues, it exposes the falsehoods and the mystifications spun by decision makers and their servants. The "manufacturing of consent" means the production of lies that, for us, must be "de-constructed" and revealed for what they are. His goal is to establish the facts in order to unveil a whole system of "misinformation" that has been designed to perpetuate elite domination over the population.

Of course, Noam Chomsky is, nevertheless, a theorist himself. In his work in linguistics, there is evidence of idealist abstraction that can be considered surprising. For example, to maintain that language is innate to the human biological organism raises philosophical and ontological questions that that cannot be avoided. Relatedly, the well-known exchange of views on the subject of "human nature" between Chomsky and Michel Foucault went to the heart of the matter when, pushed by Foucault, Chomsky affirmed that there exists "some sort of an absolute basis—if you press me too hard I'll be in trouble, because I can't sketch it out—ultimately residing in fundamental human qualities, in terms of which a 'real' notion of justice is grounded." These qualities "embody a kind of groping towards the true humanity, valuable concepts of justice and decency and love and kindness and sympathy, which I think are real."[14]

In both linguistic research and politics, Chomsky appears to anchor his thinking to a presumption concerning the capacity of human beings to achieve collective liberation from oppression through the application of reason; it is a premise that implies an ethical idealism both moral and philosophical. It is here that we see the link between his empiricism and ethical commitment, both of which infuse his thought and his activism. It is possible that this "attitude," or orientation, is equally related to the serenity and courtesy with which he communicates his knowledge and ideas. In these respects and others, the logic and vision of Noam Chomsky differ greatly from the mechanical use of slogans or conceptual gimmicks typical of the supposedly "rational" mode of French discourse.

The attention given Chomsky and his political work in France in recent years can be understood, at least in part, in terms of the same factors that previously produced resistance to it. This resistance was rooted in an ideological reaction produced by a specific historical conjunction and a culture and a mentality formed by overbearing state institutions. Just as many in France have difficulty accepting the principle of freedom of speech when it concerns antidemocratic and antiegalitarian utterances, so do they find Chomsky's frame of political reference largely impenetrable. And yet, despite these cultural impediments, he has recently come to exert increasing influence in France.

It is the political situation in France, where the dominant orthodoxy on the Left has been in disarray for almost thirty years, which has contributed to Chomsky's popularity and, to a certain extent, discredited his critics. The prominence of the state in French life has so poisoned intellectual life and limited effective communication that, intellectually and politically, Chomsky's ideas and manner have been a breath of fresh air. In this country, Chomsky's work offers a fascinating alternative to rigid responses on the Left at a moment when the global consolidation of capitalism calls for independent and creative thinking.

NOTES

1. I have discussed this in my article, "Noam Chomsky, l'État et l'intelligentsia française," *L'Homme et la Société* 123/124 (1997): 166–171.

2. See, for example, Claude Guillon, "L'Effet Chomsky ou l'anarchisme d'État," *Oiseau-tempête* 9 (Summer 2002): 28–31.

3. My apologies are offered to Noam Chomsky for any inaccurate characterizations of his ideas, opinions, or attitudes.

4. These terms tend to highly reify in the French context, taking on a hallowed status that lends them a kind of sacrosanct authority.

5. Pierre Hadot, *La Philosophie comme manière de vivre (Entretiens avec Jeanne Carlier et Arnold I. Davidson)* (Paris: Albin Michel, 2001), 81–82.

6. Edward T. Hall, *Beyond Culture* (New York: Doubleday, 1976): 108–109.

7. Although it is not yet widespread, I use the term "United-Statesians" to refer to the citizens of the United States of America, just as "Canadians" are the citizens of Canada. For me, the word "Americans" refers to all and any inhabitants of the Americas. The longstanding appropriation of the term "American" contributes to attitudes reinforcing invidious geopolitical power relationships.

8. See Alexis de Tocqueville, *The Old Regime and the Revolution* (1856).

9. Daniel Guérin, *L'anarchisme. De la doctrine à l'action* (Paris: Éditions Gallimard, 1965).

10. Cited in Charles M. Young, "Noam Chomsky: Anarchy in the USA," *Rolling Stone* (May 28, 1992): 47; and quoted in Peter Wintonick and Mark Achbar, eds., *Manufacturing Consent: Noam Chomsky and the Media* (Montréal: Black Rose Books, 1994), 219.

11. Noam Chomsky, "Introduction" to Daniel Guérin, *Anarchism* (New York: Monthly Review Press, 1970), xviii–xix.

12. Emile Durkheim, *Pragmatisme et sociologie: Cours inédit donné à la Sorbonne en 1913–1914 (d'après des notes d'étudiants)* (Paris: Vrin, 1955), 44.

13. Durkheim, *Pragmatisme et sociologie*, 45.

14. Wintonick and Achbar, eds., *Manufacturing Consent*, 33. This debate occurred previously between Chomsky and the child psychologist Jean Piaget. See *Théories du langage. Théories de l'apprentissage. Le débat entre Jean Piaget et Noam Chomsky* (Paris: Editions du Seuil, 1979), 53–87. On the concept of "nature" in general, see Noam Chomsky, *On Nature and Language* (Cambridge: Cambridge University Press, 2002).

17

TESTIMONY

SUSAN GEORGE

Others will certainly contribute scholarly analyses to this volume in honor of Noam Chomsky, dealing with his work as a major theorist in linguistics or as a political analyst who has accompanied progressive struggles of every kind for decades. For my part, I have a more modest aim, which is to evoke the humanity and the openness of this great mind, from my personal experience. I am thinking of a gesture Noam has certainly forgotten but that was of great, indeed crucial, importance to me.

In 1967, I was well settled in France, where I had just completed a graduate degree (*licence*) in philosophy. I was increasingly outraged by the war in Vietnam, which was breaking my American heart. This war seemed to me abominable, of course, but also aberrant: it could only be a monstrous mistake; the country of my birth was incapable by definition of doing "that." As one can see, I was still full of illusions. In any event, I wanted, as they say, "to do something," but I had no clear idea of what, or how. A married *bourgeoise* with three young children, I had no affinity with the Maoists, the young Communists, or the infinite variety of Trotskyites who proliferated in the Sorbonne, but I had no idea either of how to express my opposition without joining one of these noisy groups, whose style of political thought was totally impenetrable to someone like me.

What to do in the circumstances? I had read, I no longer remember exactly where, Chomsky's "The Responsibility of Intellectuals."[1] I believe it must have been in the *New York Review of Books*, to which I had subscribed from the start. "Intellectual" was a status to which I could not really lay claim, but at least I had two advanced degrees, one American and one French: perhaps it would not be completely pretentious to ask the author of this brilliant piece for some advice.

Gathering my courage, I wrote to Noam Chomsky. And he wrote back, practically by return post, to let me know that there was a group of Americans in Paris calling themselves PACS, the Paris American Committee to Stop War. So it took a professor at MIT in Cambridge, Massachusetts, to explain to an American Parisian what was going on among the antiwar Americans in Paris. I made contact with PACS immediately and threw myself into the opposition with these new comrades.

A year later the French government, in its eagerness to please the Americans, outlawed PACS. Few American members of PACS felt they were free, as foreigners, to continue militant opposition to the war in France without risking their job or their residency permit. I had not yet acquired citizenship, but as the wife and mother of tried and true French citizens, I felt more secure than most of my comrades, and with some other ex-members of PACS, I founded a new group sheltered by the Movement for Disarmament, Peace, and Liberty, led by Claude Bourdet, a great figure of the Resistance. Thus I was able to carry on, doing mostly low-profile but useful tasks and often acting as organizer and translator at meetings between antiwar Americans who were visiting Paris and the Vietnamese. Subsequently, I took part in the Front Solidarité Indochine, mostly run by the energetic members of the Fourth International, which later became the Ligue communiste revolutionnaire (the Communist Revolutionary League), giving me the opportunity to discover that at least some Trotskyites were actually human beings with whom one could have cordial relations . . .

After years of fighting against the war in Vietnam, taking part in welcoming the Chileans exiled by the Pinochet-U.S. coup d'état, or participating in a working group on the real reasons for world hunger, then going on from there to publish my first book, followed by a dozen others—all that seemed perfectly natural, and it was all due, at least in part, to the fact that Noam Chomsky had taken the trouble to answer a letter from a perfect stranger.

Since then I haven't seen him often, three or four times perhaps (although we sometimes correspond), and this despite the great kindness that has led him to write a few warm words of recommendation for a number of my books. More than once, for example during an Amnesty lecture series at Oxford in which we both took part, we missed each other by a few hours. At the World Social Forum in Porto Alegre, I came early to get a good seat in the immense hall where Noam Chomsky was supposed to speak. Then the organizers announced that the audience struggling to get in was so huge that his lecture would be held in an even larger hall. By the time I got there, I could only find a seat a few miles from the speaker and couldn't get near him.

Nor is France exactly the best place to meet Noam—as his readers will know, he has good reason not to come here. Many French people have never understood how one of his texts could possibly have been used as a preface, even without his knowledge, to a book by the Holocaust denier Robert Faurisson.[2] Their reaction is perhaps understandable from an exclusively French point of view. But to grasp Chomsky's attitude, you have to realize that he is an American progressive; you have to know the Constitution of the United States and the whole Enlightenment tradition of freedom of opinion and expression. In France, there are laws prohibiting racist, anti-Semitic, or denialist statements. In the United States, freedom of expression permits the utterance of the most odious opinions, and you can see genuine Nazis parading in public with their slogans and banners. Chomsky, himself Jewish, obviously has no sympathy for Faurisson's position. But as a libertarian and Voltairean, he will defend to his last breath Faurisson's right to express himself.

It seems superfluous to recall here how much Chomsky has helped all of us to polish and shape our arguments concerning any number of abominable events in the world, especially those in which the American government has been involved. One has to wonder, too, how he always manages to have files stuffed with evidence that the Masters of the World would much prefer to keep hidden away in the shadows. They are full of facts, facts, and more facts, each more damning than the last and all arrayed with implacable logic.

If this remarkable man has a fault, and in my opinion he doesn't have many, it is that he seems to be convinced—or so at least his method implies—that people are rational, that they genuinely want to know the truth, and that they will change their opinions and their conduct once he, Chomsky, has submitted his irrefutable evidence and reasoning to them. If only it were true! Politics, militancy, and activism would be a whole lot easier. One of our greatest problems remains the desire of the majority to bask in illusions. Otherwise the American press, the willing servant of the powers that be, would be facing less rosy prospects.

But when it comes to intellectual heroes, you can't do better than Chomsky. I will never be able to thank him enough, but at least I can do so here, in print, without his hushing me out of modesty. It was he who got me started in the struggle against poverty, inequality, and injustice. Nor will I ever do as well as he has in researching and revealing the misdeeds of power, but thanks to his letter to a stranger forty years ago, I have been able to play my part.

So thank you, Noam.

NOTES

1. *New York Review of Books* (February 23, 1967), http://www.chomsky.info/articles/19670223.htm. Published in French in Noam Chomsky, *L'Amérique et ses nouveaux mandarins* (Paris: Seuil, 1969).

2. See chapter 14, "Chomsky, Faurisson, and Vidal-Naquet," which discusses Chomsky's views on freedom of expression.

18
TRUTH, BALANCE, AND FREEDOM

AKEEL BILGRAMI

I

Though there is much radical—and often unpleasant—disagreement on the fundamental questions around academic freedom, these disagreements tend to be between people who seldom find themselves speaking to each other or even, in general, speaking to the same audience. On this subject, as in so much else in the political arena these days, one finds oneself speaking only to those with whom one is measurably agreed, at least on the *fundamental* issues. As proponents of academic freedom, we all recognize who the opponents of academic freedom are, but we seldom find ourselves conversing with them. We only tend to speak to them—or *at* them—in heated political debates when a controversy arises, as for instance at Columbia University over the promotion of faculty in Middle Eastern studies or in those states where the very idea of a curricular commitment to modern evolutionary biology is viewed with hostility. I will not be considering such controversial cases of overt political influence on the academy. *This is not because they are not important*. The threats they pose are very real, when they occur, and the need for resistance to these threats is as urgent as anything in the academy. But they raise no interesting intellectual issues at a fundamental level over which anyone committed to academic freedom is likely to be in disagreement. If there is disagreement among the proponents of academic freedom, it is likely to be on relatively *marginal* questions such as, for instance, whether academic freedom is a special case of the more basic constitutional right to free speech or whether instead it is a special form of freedom tied to the specific mission of universities.

What might a philosopher—I am announcing my disciplinary location—contribute to these more marginal questions? In this brief essay, I would like to make a fuss about standard arguments for a conception of academic freedom that we all seem to subscribe to when it is coarsely described but that, when we describe it more finely and look at the arguments more closely, is quite implausible and leads directly to thoroughly confused ideas about displaying "balance" in our classrooms and our pedagogy quite generally. I will then use some of the points and distinctions I make in this critique to explore whether there is scope for locating more subtle and interesting kinds of threats to academic freedom than the obviously controversial ones that I mentioned above, which raise no interesting issues for proponents of academic freedom, even if they understandably ring urgent alarms. At the very end, I will venture to advocate imbalance of a very specific kind in the "extramural" domain, when it is neither inquiry nor classroom curriculum that is at stake but rather the effort to engage the intellectual and political culture at large.

Much of this essay, especially the second half of it, has been inspired in detail by Noam Chomsky's passionate and penetrating writing on freedom in modern Western societies. I fear, however, that my epistemological critique of a certain classical liberal argument for academic freedom may not sit perfectly at ease with his own epistemological commitments, and I would be curious to know whether or not that is so.

2

No matter which stand is taken on the marginal question as to whether academic freedom is a special case of the constitutional right to free speech or something special and apart, there is a great and recurring tendency in the literature on the subject to appeal to the same arguments, metaphors, and intuitions to describe the justifications for academic freedom, along roughly the following lines: First, there is a statement of purpose or *goal*: academic institutions are sites for intellectual inquiry and research, and therefore one of their chief goals is the pursuit of truth and the pedagogical project of conveying the truth, as one discovers it and conceives it in one's research, to students, and to set students on the path of discovering further truths in the future on their own. And second, there is a statement of the *conditions for the possibility of the pursuit of that goal*: this pursuit of truth is best carried out, it is said, under conditions where a variety of opinions are allowed to be expressed on any subject, even if one finds some of them quite false, since it is possible that they might be true and one's own view might turn out to be false.

Often, the metaphor used to capture this ethos and its efficacies in the matter of truth is that truth surfaces in a *"market place of ideas."* When Justice Holmes first put that idea into the air, he was not particularly thinking about the academy but quite generally about the shape of a free society.[1] In fact, as two Columbia historians (Richard Hofstader and Walter Metzger)[2] pointed out, Holmes was really expressing in more intuitive and metaphorical terms the justification for tolerance in speech quite generally, for which John Stuart Mill had earlier in *On Liberty* given a more structured argument, with premises and a conclusion.[3] So even if one thought that academic freedom was set apart from the articulations of the First Amendment, the structure of the *underlying* philosophical argument is the same as to be found in Mill's more general argument for liberty of speech as a fundamental principle. I want to spend some time on this underlying argument, but before I do, it is worth emphasizing that it is not just given by professional and lay philosophers; it is also found in the case law in which universities have figured, repeatedly. Thus for instance in *Keyishan v. Board of Regents of the State University of New York* (1967), the language of the Supreme Court explicitly cites the phrase "marketplace of ideas" and talks of the "robust exchange of ideas which discovers truth out of a multitude of tongues." That is just one example. There are also literally scores of cases in the lower courts as well that appeal to Millian considerations, and they too begin by defining the goal of universities as being one of seeking the truth in intellectual inquiry.

What is Mill's argument and why does it have such a strong appeal for law, philosophy, and even our everyday understanding of the justifications for academic freedom? Its appeal is the appeal of a certain fallibilist epistemology that widely underlies the classical and orthodox liberal mentality. Curiously, this form of fallibilism clashed starkly with the pragmatist epistemology of American thinkers such as Peirce and also with the heterodox form of liberalism that one finds in American thinkers such as Dewey. Yet the American courts and American quotidian opinion cite Holmes and Mill like a mantra.

Mill's argument has two premises and a conclusion. The premises are: (1) Many of our past opinions, which we had held with great conviction, have turned out to be false. (2) So some of our current opinions that we hold with great conviction may also be false. From these premises, he drew a conclusion about tolerance and free speech: (3) Therefore, let us tolerate dissenting opinions just in case our current opinions are wrong and these dissenting opinions are right.

The marketplace of ideas keeps us honest. Since we can never be sure that we are right, a marketplace of opinions, many of which may oppose our own opinions, may well throw up the truth, displacing our own convictions about it. Metzger and Hofstader make this connection between Holmes and Mill explicit, and there is no doubt that something like this justification, if true, would hold for free speech in the academy with particular force, even if we saw the academy as standing apart from constitutional contexts for free speech, because the academy is specially geared to pursue the truth in its various disciplinary pursuits.

Let's stare at the argument for a while.

Mill's argument is based on an induction. It is often called Mill's "metainductive argument." The induction is found in the transition from the first premise to the second. It is called a *meta*induction presumably because whereas most inductions go from observations about the *world* in the past to conclusions about the future, his induction goes from an observation about our past *beliefs* about the world to a conclusion about our future *beliefs* (viz., that they may be false).

There is extraordinary ambition in this argument. It hopes to persuade us of a value, the value of free speech, as something for a polity to embrace on the basis of something that is pure rational argument. By this I mean that it does not aim to convince us to adopt a value (the value of free speech) *on the basis of any other moral or political values*. It hopes to convince us on grounds that are, in that sense, value free. It does not matter what moral or political values we have. So long as we are capable of induction, we are supposed to see the force of the argument. And since inductive capacities, like deductive capacities, are part of general rational capacities possessed by all human beings, if the argument is right, everyone should see the value in free speech just in virtue of their rationality. To fail to do so, therefore, is nothing less than irrational. Mill gives quite other arguments for free speech in that careless masterpiece, such as, for instance, that free speech is a value to live by because it encourages diversity and creativity in society and that a willingness to submit to the clash of ideas is essential to the moral courage of human beings and prevents their mental pacification. But such an argument is inherently less ambitious. Its appeal is confined to those who value individual creativity, or variety, or what Blake called "mental fight." There is a risk in any argument that comes to an evaluative conclusion by appealing to another value. Values are things that have variable appeal. And so those who do not subscribe to the other value will not be convinced by it. The metainductive argument, by contrast, if successful, is supposed to knock us down with a much more general logical force.

But is it successful? The incessant sloganeering about the marketplace of ideas depends centrally on its success.

Deep though it goes in liberal culture and sensibility, I think Mill's argument is a numbing fallacy. To begin with, even at a cursory glance, you will notice that the judgment in the first premise is made from the point of view of one's current opinions and convictions. It is from our present point of view, from what we *currently take to be true*, that we are able to say that our past opinions are false. But the judgment in the second premise is telling us that our current point of view may contain false views and therefore to be unsure and diffident about them. Now, if we are unsure about our current beliefs, and our judgment in the first premise is made on the basis of our current beliefs, then to that extent we must be unsure of our first and basic premise. Any conclusion based on it therefore is bound to be, to that extent, itself shaky and uncertain.

There is another more fundamental internal problem with the argument. In characterizing it, I have said that it comes to a value conclusion on the basis of premises that appeal merely to an induction and not on the basis of any other political or moral value. Is this not to derive an *ought* from an *is*? Is it not a form of naturalistic fallacy? Now, I myself do doubt that one can get norms out of a normative void. That would be to think that values are something that we can dig deeper than and ground on some foundation that is not evaluative at all. If you believe in the irreducibility of values, as I do, then values can only be justified by other values, and there is no escaping or getting behind or underneath values to see their point and rationale. But this sort of abstract objection to Mill's argument is less than satisfactory, and I don't want to rest my case on it. I think there are more *internal* flaws in it.

For one thing, it is not obvious that there is *no* value at all hidden in what the argument appeals to in its justification for free speech. After all, since it says that one should adopt free speech because it creates a marketplace of ideas from which the truth, even if it goes against one's convictions, will emerge, one is assuming at least that there *is value* in pursuing the truth. Admittedly, the value of truth is a *cognitive* value, and the value of free speech is a moral and political value. But even so, the fact is that we *are* appealing to another value to justify the value of free speech. It is only because we value truth and have it as a goal that we will be moved by the idea that a marketplace of ideas engendered by freedom of speech is something that we should adopt. So let us grant that the qualm about coming to norms from within a normative void does not apply to this argument.

Even if we did grant this, there is something internally peculiar about an argument that appeals to the value of truth and the goal of pursuing the truth, as it does, while also asserting, as the second premise does, that we can never know that we have achieved the truth. How can we claim to have a goal that we can never know we have achieved, even when we have achieved it? What sort of goal is that? It is not perhaps as peculiar as having a goal that we know that we can never achieve. That is outright incoherent. You cannot coherently strive to achieve what you know to be impossible. But to allow that we can achieve a goal and yet insist that we can never *know* we have achieved it when we have, though not perhaps outright incoherent, is a very peculiar understanding of what goals are.

To put it explicitly, the internal tension is this: The argument's second premise says that beliefs about whose truth we are utterly convinced may turn out to be false. This strictly implies that we can never be sure that we have achieved the goal of truth, not even when we are quite convinced we have. And yet the argument presupposes that the pursuit of truth is a value and that we have it as a goal to pursue. If the goal of inquiry into the truth that all academic institutions embrace is really to pursue in this way something that we never can be sure we have achieved, then we must be assuming that what we do, in pursuing it, is a bit like sending a message in a bottle out to sea. We never know what comes of it; we never know that it has arrived. What sort of epistemological project is that? It is a conception of inquiry in which we have no control over its success. If inquiry is successful, that success is, from our hapless point of view as inquirers, necessarily some sort of bonus or fluke.

The argument demands that our point of view of inquiry have a built-in diffidence: we are supposed to be diffident even about our most well-established claims. But such diffidence yields no instruction. The doubt expressed by the thought "for all one knows even our strongest convictions as to what is true might be false" is an idle form of doubt. Consider the paradox of the preface, in which the author says coyly, "Something or other that I say in the next four hundred pages is bound to be erroneous or false" (and then typically adds that "for those errors I alone am to blame and not all those nice people I have just acknowledged as having aided my thought and argument"). The author's declaration of impending falsity in the pages to come is idle, because it gives no instruction about what to do to remedy things. It is not as if he knows what it is that is bound to be false, and why. Like Mill and Holmes, he just thinks that that is the tentativeness and diffidence with which we must hold the views we have written down. But a doubt that gives no instruction in his practice of writing is a doubt that does not make any epistemic difference. And as pragmatists say, something that makes no difference to practice (not even to cognitive practice, as in this case) makes no difference to inquiry and epistemology at all.[4] Any argument that arrives at a commitment to free speech on the basis of a conception of inquiry that has such precarious coherence hardly deserves the centrality that it has been given in the liberal tradition of political thought.

In the immediate context of the political controversies we find ourselves in, in university life, the conception of academic freedom based on such a classical liberal form of argument leads *directly* to the advice we often get, sometimes even from university presidents, about how we should be *balanced* in what we say in our classrooms, showing consideration to all points of view, even those that from our point of view we confidently know to be wrong. This directive wholly fails to understand what sort of role the ideal of "balance" ought to play in the academy. It is a worthy ideal, but we have to understand the right place and context for it in the academy.

Let's go along, as we have been doing, with the assumption that a primary aim of universities is to pursue the truth in our various disciplinary inquiries, and that the point of pedagogy is to try to present the truth we have found by presenting evidence and argument for it. Now, if "balance" has any role to play in all this, its role is entirely *nested within* this primary goal, *not* something *independent* of this goal. And within this primary goal, the only thing that "balance" *could* mean is that one must look at *all* the evidence that is available to one in our inquiries. (This is the cognitive counterpart to what decision theorists call "the total evidence requirement.") What "balance" cannot possibly mean is the nonsensical thing that the directive we are considering tells us, viz., the equal presentation in the classroom of two contradictory views. No educator with any minimal rationality would do that on the elementary grounds that if there are two contradictory views, only one can be right. Of course, if she cannot make up her mind on the evidence as to which one is right, she might present the case for both views even-handedly. But presumably such undecidedness is an *occasional* phenomenon. If so, balance cannot be put down as a *requirement* for pedagogy in the classroom. Hence, the constant demand that we always present both sides of a disagreement presupposes a conception of education as a sort of chronic dithering. It is far more sensible to say that "balance" allows that an educator presents her judgment with complete conviction because "balance" in the academy is nothing other than a synonym for the idea that we must look at *all* the evidence before coming to our convictions. It has no other role or meaning . Attempts to give it another meaning (as in the directive with which I am finding fault) are drawn from a fault line that has its beginnings in the canonical Millian form of liberal argument for free speech.[5]

3

I have been inveighing against a very standard liberal argument and a metaphor that it yields about truth emerging from a marketplace of ideas, which goes deep in the sensibility of our self-understanding in the academy—and in the courts that have pronounced judgment in controversial cases that the academy has thrown up. This may have given the impression that I am recommending more dogmatism regarding our own convictions than a commitment to academic freedom can allow. That impression would be wrong.

The criticisms I have just made of Mill's argument are quite compatible with the view (which is my own view) that there is far too much dogmatism in the academy, especially in the social sciences and even in the humanities. (And if it is less so in the natural sciences, still, as Kuhn pointed out almost five decades ago, there is some there too.) As a matter of fact, my view is that if we could characterize more or less exactly what this dogmatism is, we would have identified the most pervasive, insidious, and interesting threat to academic freedom.

As I said at the outset, this paper was going to raise a typical philosopher's fuss about how to rigorously characterize the arguments by which we justify academic freedom, and I have said that I find Holmes's metaphor and Mill's argument less than exact and plausible, and this implies that theirs is not the way to understand the dogmatism that thwarts academic freedom. To be fussy is to demand that one gets certain distinctions carefully right. And I am claiming that to diagnose and combat the overly high levels of dogmatism in the academy, we do not have to assume a fallibilist notion of diffidence and doubt. It is one thing to be undogmatic in the way that academic freedom demands and quite another to have the sort of notion of inquiry suggested by Millian and classical liberal arguments for academic freedom.

Let me convey what I have in mind by the dogmatism that constitutes a threat to academic freedom by returning to the paradox of the preface. The paradox offers us a site for locating a useful taxonomy via which we can identify what sort of dogmatism amounts to such a threat.

I had said about the paradox that the *generalized*, that is the *unspecific* form of doubt that is stated in the preface ("something or other in what follows in these pages may not be true," echoing Mill's argument that our strongest convictions may turn out to be false), gives the author no instruction as to what to do about it. He cannot possibly be moved to do anything about his text by a doubt such as this. What the author will be moved and instructed by is not this sort of doubt but rather—if he or she is not dishonest and not obtuse—by some *specific* evidence or argument that is provided against one or other of his *specific* conclusions or claims. Now, both these qualifications, "if he or she is not dishonest and not obtuse," are revealing.

They show that there is no direct relevance of this issue I have just raised (about ignoring *specific* counterevidence and counterargument presented to one) to the question of academic freedom. Suppose someone failed to recognize counterevidence that was presented to him. That would be a sign of his obtuseness. Suppose again that someone did recognize that counterevidence had been presented to him by some colleague and he simply ignored it. That would be a sign of his intellectual dishonesty. But both these things are quite *separate* kinds of wrong from thwarting academic freedom. Now, it is true that sometimes those who are dishonest in this way are caused by this dishonesty to suppress or hound out someone who presented that evidence and that would, of course, be threatening to academic freedom, but suppressing or hounding someone out is a separate matter from what we are concerned with, which is the ignoring of evidence that is provided against what one takes to be the truth.

If this is right, we have identified so far three different phenomena. *First*, there is academic *dishonesty*—to recognize evidence or argument that goes against one's conclusions but ignore it. This in itself is *not* academic unfreedom. *Second,* there is the inability to even recognize the force of counterevidence and counterargument. Let's call this academic or intellectual *obtuseness.* And, even more obviously, that is not a case of academic unfreedom either. *Third,* there is the suppression of those who present counterevidence and counterargument that one has recognized to be so and dishonestly evaded. This, I have said, *is* a case of academic unfreedom. But, as I said at the beginning of the essay, it is a very obvious case and not a very interesting one, so I will simply put such cases aside since they raise no difficult questions. It is not even clearly characterizable as a case of dogmatism, though it bears some relations to dogmatism.

We, then, still do not have the kind of academic unfreedom that is genuinely and clearly also a case of dogmatism. So now, finally *fourth* in our taxonomy, I want to present that kind of dogmatism and show why it is a far more interesting, unobvious, and more pervasive threat to academic freedom than is identified in the third category, and in presenting it, it will become clear what its relation is to the first and second phenomena in the taxonomy, from which it is also important to distinguish it—especially the first phenomenon, with which it is too often conflated.

The dogmatism that interests me is found in submerged forms of academic *exclusion*: when we circle the wagons around our own frameworks for discussion so that *alternative frameworks* for pursuing the truth simply will not even become visible on the horizon of our research agenda. This form of dogmatism is distinguishable from the first of our four phenomena, academic dishonesty of the kind that refuses to accept counterevidence and argument presented in refutation of some specific conclusion of our inquiry. Why? Because alternative frameworks *do not refute our conclusions directly* with counterevidence or arguments so much as they point to other, possibly deeper and more interesting ways of looking at what we are studying. And here is the crucial point. *If* they *do* contain counterarguments and counterevidence to our own claims and convictions, those will only surface *further downstream,* well *after* they are recognized by us upstream as possibly fruitful forms of investigation. But it is this recognition *upstream* that the dogmatist in us finds so hard to confer, and it is in this failure that academic unfreedom (rather than intellectual dishonesty) is located.

These are cases in which a discipline discourages the development of frameworks outside of a set of assumptions on which there is mainstream consensus—and the political influence on the formation and maintenance of these exclusive assumptions, where it exists, is very indirect indeed, so indirect that it would need a fair amount of diagnostic work to reveal it, since the *practitioners themselves are often quite innocent of the influence.* (On the other hand, it is not as if this is a rare or unusual phenomenon. It is widespread and is quite well known, and many of you know it closely, since what has made The New School, where we are gathered, one of the most valuable institutions of higher learning in this country in that it has valiantly housed—indeed it has been something of a hospice for—those suffering from an exclusion of unorthodox frameworks for thinking about a range of themes in a range of different disciplines.)

Dogmatism of this kind is also distinguishable from the third sort of flaw, obtuseness. To be dogmatic in this way is not at all to be lacking in the acuity that would recognize the force of counterevidence and counterargument. If one has failed to recognize any counterevidence (downstream, in my metaphor), that is because one has (further upstream) not even so much as recognized the possibility of the framework from which it flows. It is not as if the counterevidence is there for us to see downstream and we are not perceptive enough to see it. Rather it is *not there for us to see* downstream because we have not recognized the *framework* upstream from *within which it is visible*. And this last failure is a kind of dogmatism, not stupidity.

Among disciplines, economics provides the most gorgeous examples of this. It is perhaps the worst offender in inuring itself against alternative frameworks of thought and analysis. In fact, I will venture to say that I have never come across a discipline that combines as much extraordinary sophistication and high-powered intellect and intelligence with as much demonstrable falsehood. So, for instance, some of the most brilliant intellectuals I have known to this day make claims about the trickle down of wealth in capitalist economies and present them with the most sophisticated quantitative methods, despite the plain fact that wealth has not trickled down (at least not to the places where it needs to trickle down) *anywhere in the world* in the *entire history of capitalist political economy*. If a physicist were to make some of the claims that economists have made, which have been falsified as repeatedly as they have, they would not only have their careers terminated, they would properly be the laughingstock of the profession. Now, there is no direct political influence that forces this sort of refusal to question, let alone give up, one's assumptions in a discipline such as economics. The regulation is wholly *within* the discipline's profession, and even there, there may be very little browbeating or intellectual bullying, that is to say, very little *explicit* regulation. It is largely unconscious self-censorship—often done with career advancement in mind—that threatens academic freedom in such disciplines.

On the very evening after I wrote these words in a draft of this essay, I was at a dinner at my economist colleague Joe Stiglitz's apartment, and I impertinently told him that I was going to raise this point in an essay I was writing. His response was memorable. "Akeel, I agree with you about economists but I don't understand why you are so puzzled. One would only be puzzled if one were making the wrong assumption about economics. What you should be assuming is that—as it is done by most economists—economics is really a religion. And so why should you be puzzled by the fact that they cling to and never give up their views despite their frequent falsification?" So I will rephrase my point: one apparently makes one's way up in a church hierarchy by clinging fast to the orthodox faith.

But there is the following difference. The church has had a history of explicit and rigid regulation of what may or may not be said and pursued in its fold. But, as I said, if there is political (or corporate) influence in play in the sort of dogmatism I have described in economics, it is not obviously visible and direct, and the protagonists in economic inquiry in universities would be quite genuinely clueless about it and, with no dishonesty, deny its influence. Sometimes, as in my own subject of analytic philosophy, where there is a great deal of exclusion of alternative frameworks for discussion, there is no political influence, *however indirect*, in play. If there is a question of power and politics involved, it is entirely internal to the discipline, the power that is felt and enjoyed simply in keeping certain ways of thinking out of the orbit of discussion, forming small coteries of people referring to one another's work with no concern that the issues they discuss are issues that have no bearing on anything of fundamental concern to any of a number of disciplines with which philosophy had always been concerned before, say, even fifty years ago, and from which it has now managed to isolate itself for the most part. Richard Rorty had tried to raise Kuhnian questions for the discipline of analytic philosophy[6] and spoke with eloquence about its insularity, and he was certainly right to notice just how exclusionary the subject had become the more it had become a *profession in universities*.

Moving away from specific disciplinary examples, the general point that emerges from these examples can be made if one recalls that De Tocqueville famously said, "I know of no country where there is so little independence of mind and real freedom of discussion as in America." And here is a wildly curious thing. At the same time, it is America that has more free institutions (including academic institutions) than anywhere else in the world. How can this extraordinarily paradoxical duality coexist? What explains this paradox? I can't possibly try to provide an explanation here,[7] but whatever it is that explains it will provide a very good sense of the deep, that is to say submerged, forms of academic unfreedom that exist in this country. When the freest academic institutions coexist with some of the highest levels of academic unfreedom in the democratic world, the sources of unfreedom are bound to be far subtler than is captured by the standard vocabulary of "suppression," "brainwashing," "political pressure," "manipulation," and so on. That is why this fourth phenomenon, this pervasive sort of dogmatism, is a far more interesting case of academic unfreedom than the third phenomenon in our taxonomy. When a person working with unorthodox frameworks of research is looked upon with *perfect sincerity* by professionals, as someone unfortunate and alienated and to be pitied as irrelevant rather than bullied and hounded, we know not only that the political influences on these professionals are not even easily identified—let alone easily confronted—but that this kind of thwarting of academic freedom needs a quite different descriptive vocabulary than I used in describing the third phenomenon.

Equally, I would insist that it is different from the first phenomenon of academic dishonesty as well, because to accuse these professionals of *dishonesty* (rather than in their dogmatism unconsciously perpetrating academic *unfreedom*) would be to be glibly moralistic, since (if I am right in making the upstream/downstream metaphor) it is not *honesty* that requires that people should be willing to allow frameworks of investigations other than their own. At any rate, it is not honesty in the sense that is required to admit that one's views have been refuted, when one has been shown evidence against them and one is not too obtuse to recognize the evidence. To insist that they are both a case of dishonesty would be to perpetrate a (not very good) pun.

The interest and subtlety of this exclusionary phenomenon, then, lies in its ability to be distinguished from the other three in our quartet: academic dishonesty, academic stupidity, and straightforward and obvious forms of academic suppression. Despite its subtlety, it *is* a recognizable assault on academic freedom, and it is the more important to analyze in detail precisely because there is nothing as obvious (or infrequent) about it as there is about the efforts at external influence of Christian groups on science curricula or of Zionist groups on Middle East studies departments. Being much more subtle, it is also much more pervasive than these more obvious phenomena—and much harder to resist. Different people feel it differently at different times. Frameworks for serious research in race and gender felt it constantly for decades, as late as the 1970s. Quite possibly more old-fashioned forms of humanistic scholarship in a range of literary disciplines began to feel it since the late 1980s and 1990s. And I daresay, research programs that pursue seriously socialist forms of analysis feel it more than ever today in economics departments.

I have tried in this essay to shift attention away from the fallibilist epistemological presuppositions of metaphors such as "the marketplace of ideas" and its classical Millian arguments for academic freedom, and I have tried to focus instead on the need to diagnose the sorts of unconscious attitudes that make for unwitting disciplinary mandarins and gatekeepers, "normal scientists," as Kuhn called them. One doesn't need to be diffident in the conviction with which one holds one's views in order to resist such attitudes. We should allow alternative frameworks not because we have some generalized doubt that we ourselves might be holding false views. We should allow alternative frameworks for quite different kind of reasons, also found in Mill's writing on liberty, as I said, having to do with the fact that if we allow for frameworks of investigation other than our own, we make for an attractively diverse intellectual ethos and in doing so allow the creativity of different sorts of people and minds to flower. These sorts of consideration in favor of academic freedom, unlike Mill's argument considered in section 2, which appeals to all those capable of inductive reasoning and in pursuit of the truth, gives rise to a picture of academic freedom that appeals *only* to those who think that there is value in creativity and diversity in the academy. The appeal therefore is frankly disadvantaged by its less than universal reach. But, on the other hand, the picture can claim the advantage of not being landed with a bizarre conception of inquiry presupposed by Mill's more ambitious argument and the metaphorical cliché that it has yielded about the "marketplace of ideas."

The point I want to repeat in marking this difference from Holmes's metaphor and from Mill's metainductive argument on which that metaphor is based is that if considerations about truth and falsity enter this picture, it is only, as I said, *farther downstream*, when something that other frameworks deliver might claim to be a truth that clashes with ours. But since those considerations do not surface upstream where we are ourselves currently investigating, the pursuit of truth in our investigations is not to be conceived of as a goal whose success is necessarily opaque to its seekers, as in Mill's argument for freedom. In our own pursuits, we may be as confident in the truth of the deliverances of our investigations as is merited by the evidence in our possession, and we need feel no unnecessary qualm about displaying balance in the classroom, if we have shown balance and scruple in our survey of the evidence on which our convictions are based, the only place where balance is relevant in the first place.

4

Having said that, I should like to conclude with a point that rotates the angle a bit on the question of balance.

One of the questions that has most exercised scholars of academic freedom is the extent to which the concept and the policy applies to the utterances of a scholar not within the university but in what is called the *extramural* context. Is a professor free to say things outside the university in public forums that would be unsuitable for one reason or another in the classroom or at official university events? There is a lot of interesting writing on this subject, some of the most interesting by scholars of the law. But I want to say something here that is a bit off that beaten track.

When it is not classroom curriculum and intellectual inquiry in the university but political debate in general outside the study and the classroom that is in question, there are good reasons why the views one expresses can and often should be substantially *im*balanced. And by imbalance here, I don't just mean that they should speak with conviction for one side of a disagreement if that side has the preponderance of evidence on its side. That form of imbalance is what my critique of Mill and Holmes has tried to establish as perfectly appropriate in the classroom. But for extracurricular and extramural public speech by academics, I have in mind the moral appropriateness of a *further* and more *willful* kind of imbalance. To conclude in one's thought what the evidence in one's investigations dictate is not really a matter of choice or will.[8] The evidence *compels* us, as it were. But to be imbalanced in the *further* way I am about to mention *is* a matter of will and moral decision. Let me explain.

I find it not only understandable but honorable if someone speaking and writing in America finds it important to *stress much more* the wrongs of the American government and its allies and clients, like Israel, Saudi Arabia, Egypt, Pakistan, Indonesia under Suharto, Chile under Pinochet, and so on, rather than speak obsessively, as is so often done, about the wrongs done by Muslim terrorists or Islamic theocratic regimes or, for that matter, Cuba and North Korea and so on. But if the same person was speaking or writing, say, in the Palestinian territories or in Arab newspapers, it would be far more admirable if he were to criticize Hamas or Islamic regimes like Iran's.

It is said that whenever Sakharov criticized the Soviet Union's treatment of dissidents in the 1950s, he was chastised by his government for showing an imbalance and not speaking out against the treatment of blacks in the American South. That is precisely the kind of imbalance that courageous academics are going to be accused of by the enemies of academic freedom in this country, and I hope that all of us will have the courage to continue being imbalanced in just this way.

NOTES

This essay appeared first as a lecture in a conference at the New School of Social Research in 2008 and will appear in a volume of its proceedings in the journal *Social Research* edited by Arien Mack. I am grateful to Jonathan Cole, Isaac Levi, and Carol Rovane for very helpful comments on a draft of this paper. I have also benefited from the discussion of a fragment of this paper presented to a workshop on pragmatism at the Institute of Public Knowledge at New York University, in particular the verbal commentary made during the discussion on that occasion by Craig Calhoun, Benjamin Lee, and Richard Sennett.

1. Actually, it was the *idea* and not the *expression* that Oliver Wendell Holmes put into the air in his dissenting opinion in *Abrams v. the United States*. His own expression was "free trade in ideas." The expression "marketplace of ideas" was first used in the language of the Supreme Court in *Keyishan v. Board of Regents of the State University of New York* (1967).

2. R. Hofstadter and W. P. Metzger, *The Development of Academic Freedom in the United States* (New York: Columbia University Press, 1955), 527.

3. John Stuart Mill, *On Liberty* (London: Longman, Roberts and Green, 1869), chap. 2.

4. To be more precise and detailed, the pragmatist—Peirce, for instance, in his remarkably profound and original paper "The Fixation of Belief," in his *Collected Papers* (Cambridge, Mass.: Harvard University Press), vol. 5—makes a distinction within inquiry between our settled beliefs and our hypotheses. See also Isaac Levi, *The Enterprise of Knowledge* (Cambridge, Mass.: The MIT Press, 1986), for an elaborate and interesting deployment of this distinction to construct a theory of the dynamics of knowledge or what is sometimes called a theory of "belief revision." Hypotheses don't command our confidence in the same way that our beliefs do. Mill might be right to ask us, as inquirers, to have some diffidence in the way we hold our hypotheses to be true since, unlike a settled belief, a hypothesis, even by our own lights, is still up for grabs; it is

not something we have decisively counted as having the full prestige of "truth." But making this distinction and granting that he has a point about diffidence for one half of the distinction (hypotheses) does not help with Mill's metainductive argument *for liberty*. The distinction merely says that unlike a settled belief (such as, say, her belief that the Earth is not flat), a scientist today might make a hypothesis that she is hoping to have confirmed by the evidence. She will hold the latter with diffidence but not the former. But, if the former is held without diffidence, then Mill's argument for liberty does not hold for such settled beliefs. That is absurd, from Mill's point of view. He would not have wanted "flat-Earthers" to be tyrannized and suppressed, so he would not have been prepared to restrict his argument for liberty to just hypotheses. He would have wanted freedom of speech and discussion to apply to the expression and discussion of *all* beliefs, settled and hypothetical. The trouble with his argument is that he extrapolates fallaciously from the diffidence that is properly advocated for hypotheses to *all* beliefs, even settled ones. And he *must* do so, since liberty presumably will apply to the expression of all beliefs. That is why I am suggesting that one should simply abandon this argument, with its fallibilist appeal to diffidence, *altogether* as providing the basis for free speech. It is the wrong basis for liberty. We should be looking for entirely different grounds and arguments for liberty, some of which Mill himself provides elsewhere in that work.

5. It might be thought that there is no very direct link between the broad liberal mentality toward freedom of speech and academic freedom that I am situating in Mill's argument and this talk of "balance" in the university. After all, there are much more straightforward political motives for those interested in protecting Israel from the harsh (though deserved) criticisms of its actions toward Palestinians than to insist on "balance" in the way contemporary Middle Eastern politics is taught. If both sides are constantly being presented equally, as is demanded by "balance," then the force of such decisive criticism can be softened. I don't deny that there are these political motives for demanding balance in cases of this kind as well as other cases. But we can't forget that many political motives of this kind are cloaked in high-sounding intellectual arguments so that their nakedness, qua political motives, is hidden. Just think of the way slaves were said to be not quite "persons" by ideologues rationalizing slave ownership or the way natives were said to be lacking "rationality." These philosophical arguments are a constant factor in the pursuit of political motives. Millian forms of liberalism similarly often underlie (as rationalizations) the political motives for demands for balance in pedagogy.

6. Richard Rorty, *Philosophy and the Mirror of Nature* (Princeton, N.J.: Princeton University Press, 1981).

7. Noam Chomsky and Edward Hermann in their rightly celebrated *Manufacturing Consent* have addressed this subject with the focus primarily on the media in this country, and I am hoping that Jonathan Cole in his forthcoming magnum opus on the research university in America will—among the many other things about the American academy that book is ambitiously intended to address—speak to this issue, with the focus on this country's universities.

8. I realize that it *is* a matter of will whether one *presents* in a classroom (or indeed in one's research) what one has evidence for. That is why failing to do so is to be described as "dishonesty." I am only saying here that when the evidence compels us to draw a conclusion, the will is not in play—it is not a matter of choice—even though coming to believe something on the basis of evidence is in the realm of the intentional.

CONTRIBUTORS

NORMAND BAILLARGEON is a philosopher of education and professor in the département d'éducation et pédagogie of the University of Quebec at Montreal. His books include *De l'Ordre moins le pouvoir* (Marseille: Agone, 2001), *Éducation et liberté*, and *Petit Cours d'autodéfense intellectuelle* (Montréal: Lux). The latter has been translated as *A Short Course in Intellectual Self-Defense* (New York: Seven Stories).

AKEEL BILGRAMI is Johnsonian Professor of Philosophy at Columbia University, a fellow at the Center for Scholars and Writers at the New York Public Library, and the author of the books *Belief and Meaning* (New York: Blackwell, 1992), *Self-Knowledge and Resentment* (Cambridge, Mass.: Harvard University Press, 2006), and a forthcoming volume of essays on politics and political philosophy entitled *Politics and the Moral Psychology of Identity* (Cambridge, Mass.: Harvard University Press).

CEDRIC BOECKX is an associate professor of linguistics at Harvard University. He is a member of the Mind-Brain-Behavior Interfaculty Initiative. His research fields include theoretical syntax, comparative grammar, and the architecture, origins, development, and neurobiological basis of language.

JEAN BRICMONT is a professor in the département de physique théorique et mathématique of the Université Catholique de Louvain. In addition to a number of articles on Chomsky, he is the author, with Alan Sokal, of *Impostures intellectuelles* (Paris: Odile Jacob, 1997), translated as *Fashionable Nonsense* (New York: Picador Press, 1998); "Folies et raisons d'un processus de dénigrement. Lire Noam Chomsky en France," the afterword to Noam Chomsky, *De la guerre comme politique étrangère des États-Unis* (Marseille: Agone, 2004); with Régis Debray, *À l'ombre des lumières* (Paris: Odile Jacob, 2003); and *Impérialisme humanitaire* (Aden, 2005), translated as *Humanitarian Imperialism* (New York: Monthly Review Press, 2007).

GENNARO CHIERCHIA is the Haas Foundations Professor of Linguistics at Harvard University. He has taught linguistics in various universities in the United States (Brown, Cornell) and in Europe (Milan-Bicocca, École Normale Supérieure). His research fields include semantics and its interfaces with syntax, the philosophy of language, and cognitive psychology.

FRÉDÉRIC DELORCA is a graduate of the Institut d'études politiques, Paris; he holds a master's degree in philosophy and a doctorate in sociology. He coordinated the collective publication *Atlas alternatif* (Pantin: Le Temps des Cerises, 2006).

JULIE FRANCK holds a doctorate in psychology and is maître d'enseignement et de recherche at the Laboratoire de psycholinguistique experimentale, Université de Genève. Her research fields include syntactic processing in adults and the normal and pathological development of syntax.

CHARLES R. GALLISTEL is a professor of psychology and co-director of the Rutgers University Center for Cognitive Science. His research fields include genetic approaches to the neurobiological mechanism of memory and spatial, temporal, and numerical learning and reasoning aspects of animal cognition.

SUSAN GEORGE was born in the United States and has lived in France for many years. She is the author or co-author of over a dozen books and is president of the board of the Transnational Institute in Amsterdam. She is also honorary president of ATTAC-France (Association pour une Taxation des Transactions financières pour l'Aide aux Citoyens; Association for Taxation of Financial Transaction to Aid Citizens), where she also served as vice president between 1999 and 2006 and remains a member of the scientific council. Her most recent book is *Hijacking America: How the Religious and Secular Right Changed What Americans Think* (New York: Polity Press, 2008).

YOSEF GRODZINSKY is a professor in the department of psychology, Tel Aviv University, and Canada Research Chair in the department of linguistics, Mc-

Gill University, Montreal. His specializations include neurolinguistics, syntax, comparative aphasiology, and functional neuroimaging.

PIERRE GUERLAIN is a professor of American civilization at the University of Paris-X (Nanterre); he also teaches at the Institut d'études politiques, Paris. He specializes in the relations between France and the United States and social questions in the United States, and his research focus is on American foreign policy.

SERGE HALIMI has been the editorial director of *Le Monde Diplomatique* since March 2008; he was a journalist with the paper prior to that. His books include *Nouveau chiens de garde* (Raisons d'Agir, 1997); *L'opinion, ça se travaille*, with Dominique Vidal (Marseille: Agone, repr. 2002); and *Le grand bond en arrière* (Paris: Fayard, 2004).

NORBERT HORNSTEIN is a professor in the department of linguistics at the University of Maryland. He works on formal syntax.

PIERRE JACOB is a professor of philosophy and the director of the Institut Jean-Nicod, Paris. His work on the philosophy of mind integrates the latest developments in cognitive neuroscience. He is also president of the European Society for Philosophy and Psychology.

LARRY PORTIS is a professor of American studies at Université-Paul-Valéry, Montpellier-III. He is author or co-author of numerous works, including *La main de fer en Palestine. Histoire et actualité dans les territoires occupés* (Paris: Monde libertaire, 1992), *La politique étrangère des Etats-Unis depuis 1945* (Ellipses, 2000), and *French Frenzies: A Social History of Popular Music in France* (College Station, Md.: Virtualbookworm, 2004).

ARNAUD RINDEL is a member of Acrimed (Action Critique Média) and collaborator with *Plan B*, a critical journal focusing on French media and social research. He is the author of numerous texts on Noam Chomsky, including "Noam Chomsky et les médias français."

ELIZABETH SPELKE is a professor in the department of psychology at Harvard University and is director of the Laboratory for Developmental Studies. Her research fields include perceptual and cognitive development in infants. She is D. Phil. *honoris causa* of the École Pratique des Hautes Études, Paris.